Albumin in Medicine

Masaki Otagiri • Victor Tuan Giam Chuang
Editors

Albumin in Medicine

Pathological and Clinical Applications

Springer

Editors
Masaki Otagiri
Faculty of Pharmaceutical Sciences
Sojo University
Kumamoto, Japan

Victor Tuan Giam Chuang
School of Pharmacy,
Faculty of Health Sciences
Curtin University
Perth, WA, Australia

ISBN 978-981-10-2115-2 ISBN 978-981-10-2116-9 (eBook)
DOI 10.1007/978-981-10-2116-9

Library of Congress Control Number: 2016953185

Printed on acid-free paper

This Springer imprint is published by Springer Nature
The registered company is Springer Science+Business Media Singapore Pte Ltd.

Preface

Human serum albumin (HSA) is the most extensively researched plasma protein to date. Technological advancements in genetic engineering and molecular and structural biology have progressed in tandem with albumin research, especially in the area of applications, where rapid progress and development have resulted in massive favorable outputs for which albumin has clearly been proven to be a robust biomaterial.

Owing to its relatively long in vivo half-life of approximately 19 days, albumin is an attractive recombinant genetic fusion partner for extending the half-life of peptides and small proteins. The genetic modification of albumin also allows it to be applied to a wide variety of in vivo purposes including the targeting of specific types of cells or organs for the delivery of albumin-bound drugs. A recent landmark finding in the metabolism of HSA is the discovery of its pH-dependent interaction with the intracellular neonatal Fc receptor. The Fc receptor–albumin interaction can be intervened in a therapeutically useful manner to manipulate the half-life of albumin-bound drugs and albumin fusion proteins. The enormous ligand-binding properties of HSA can be applied in extracorporeal albumin dialysis, a procedure that involves the removal of toxins and drugs that are known to bind to albumin from the body via an external dialyzing solution that contains albumin.

This book summarizes medical and pharmaceutical applications of HSA in which current albumin-based products are presented in a significant number of chapters. The book is intended for use by pharmaceutical and medical scientists including pharmaceutical chemists, pharmacokineticists, toxicologists, and biochemists in both academia and the private sector.

We take this opportunity to thank all of the scientists who contributed to the successful publication of this book, and we hope that this work will provide useful insights that will stimulate further progress in the field of albumin research and development.

Kumamoto, Japan
Perth, Australia
March 2016

Masaki Otagiri
Victor Tuan Giam Chuang

Contents

Contributors

Jan Terje Andersen Centre for Immune Regulation (CIR) and Department of Immunology, Oslo University Hospital Rikshospitalet, Oslo, Norway

Department of Biosciences, University of Oslo, Oslo, Norway

Makoto Anraku Faculty of Pharmaceutical Sciences, Sojo University, Kumamoto-shi, Kumamoto, Japan

Malin Bern Centre for Immune Regulation (CIR) and Department of Immunology, Oslo University Hospital Rikshospitalet, Oslo, Norway

Department of Biosciences, University of Oslo, Oslo, Norway

Jason W. Castle General Electric Global Research, Niskayuna, NY, USA

Victor Tuan Giam Chuang School of Pharmacy, Faculty of Health Sciences, Curtin University, Perth, WA, Australia

Neil Desai Vice President, Strategic Platforms, Abraxis BioScience, A wholly owned subsidiary of Celgene Corporation, Los Angeles, CA, USA

Yu Ishima Department of Pharmacokinetics and Biopharmaceutics, Institute of Health Biosciences, The University of Tokushima, Tokushima, Japan

Yasunori Iwao Department of Pharmaceutical Engineering, School of Pharmaceutical Sciences, University of Shizuoka, Shizuoka, Japan

Alex Jackson GE Healthcare, Life Sciences, The Grove Centre, Amersham, Buckinghamshire, UK

Christina K. Kalli GE Healthcare, Life Sciences, The Grove Centre, Amersham, Buckinghamshire, UK

Teruyuki Komatsu Department of Applied Chemistry, Faculty of Science and Engineering, Chuo University, Tokyo, Japan

Ulrich Kragh-Hansen Department of Biomedicine, University of Aarhus, Aarhus C, Denmark

Peng Lei Centre for Immune Regulation (CIR) and Department of Immunology, Oslo University Hospital Rikshospitalet, Oslo, Norway

Department of Biosciences, University of Oslo, Oslo, Norway

Toru Maruyama Department of Biopharmaceutics, Graduate School of Pharmaceutical Sciences, Kumamoto University, Chuo-ku, Kumamoto, Japan

Center for Clinical Pharmaceutical Sciences, School of Pharmacy, Kumamoto University, Chuo-ku, Kumamoto, Japan

Jeannette Nilsen Centre for Immune Regulation (CIR) and Department of Immunology, Oslo University Hospital Rikshospitalet, Oslo, Norway

Institute of Clinical Medicine, University of Oslo, Oslo, Norway

Masaki Otagiri Faculty of Pharmaceutical Sciences, Sojo University, Kumamoto, Japan

Kine Marita Knudsen Sand Centre for Immune Regulation (CIR) and Department of Immunology, Oslo University Hospital Rikshospitalet, Oslo, Norway

Department of Biosciences, University of Oslo, Oslo, Norway

Inger Sandlie Centre for Immune Regulation (CIR) and Department of Immunology, Oslo University Hospital Rikshospitalet, Oslo, Norway

Department of Biosciences, University of Oslo, Oslo, Norway

Adrian Smith GE Healthcare, Life Sciences, The Grove Centre, Amersham, Buckinghamshire, UK

Hiroshi Watanabe Department of Biopharmaceutics, Graduate School of Pharmaceutical Sciences, Kumamoto University, Chuo-ku, Kumamoto, Japan

Center for Clinical Pharmaceutical Sciences, School of Pharmacy, Kumamoto University, Chuo-ku, Kumamoto, Japan

Keishi Yamasaki Faculty of Pharmaceutical Sciences, Sojo University, Kumamoto-shi, Kumamoto, Japan

Chapter 1
Human Serum Albumin: A Multifunctional Protein

Ulrich Kragh-Hansen

Abstract Human serum albumin is synthesized in the liver and continuously secreted into the bloodstream. Several receptors are strongly involved in the following distribution and metabolism of the protein. The receptor-albumin interactions can be modified by specific mutations, a finding which could be of pharmaceutical and medical interest.

The largest pool of albumin is found in the extravascular spaces although at a lower concentration than in the bloodstream. The higher concentration in the circulation is the main contributor to plasma's colloid osmotic pressure and to the Gibbs-Donnan effect in the capillaries.

Albumin seems to be the quantitatively most important circulating antioxidant, and it has enzymatic properties which are so pronounced that they most probably are of biological importance. The protein's ability to bind ligands and thereby to serve as an important depot and transport protein for numerous endogenous and exogenous compounds is well studied. Recent work has given much new information about the location and structure of binding sites and about potential ligand interactions. Structural information is also useful when designing new drugs whether the aim is to avoid binding or to make use of the protein's depot function. Nonbinding therapeutics can get improved stability and benefit from the long biological half-life of albumin by forming complexes with it. The complex formation can take place by enriching the therapeutic with an organic molecule which can bind reversibly or covalently to the protein. If the therapeutic is a polypeptide or protein, fusion proteins can be produced.

Albumin also shows promises for targeted drug delivery. This process can be passive and based on the enhanced permeability and retention effect. The effect can be increased by using dimers, polymers, or albumin-based nanoparticles. The targeting process can also be active and based on an interaction between albumin carrying a targeting ligand and cellular receptors.

Keywords Albumin-receptor interactions • Ligand binding • Stability • Half-life • Drug targeting

U. Kragh-Hansen (✉)
Department of Biomedicine, University of Aarhus,
Ole Worms Allé 6, Building 1180, DK-8000 Aarhus C, Denmark
e-mail: ukh@biomed.au.dk

M. Otagiri, V.T.G. Chuang (eds.), *Albumin in Medicine*,
DOI 10.1007/978-981-10-2116-9_1

1.1 Introduction

Human serum albumin (HSA) is a multifunctional protein exclusively synthesized by liver hepatocytes and continuously secreted into the circulation. Here, it is the most abundant protein and comprises 60–65 % of total plasma protein. Many observations propose the existence of an important link between the concentration of HSA and health (Peters 1996). Due to a large number of acidic (98 Glu + Asp) and basic residues (83 Lys + Arg), the protein is highly soluble in aqueous media. Thus, its concentration in plasma is ca. 0.6 mM (4 % w/v), but solutions of 20 % can be made for clinical use. Actually, it is possible to make preparations of up to 50 % (Peters 1996). The presence of the many titratable amino acid residues also implies that HSA has an important buffering capacity. The uneven content of acidic and basic residues results in a net charge of ca. −15 at physiological pH, a fact that renders HSA important for the Donnan effect in the capillaries. Finally, under certain circumstances, the protein can serve as a source of amino acids or energy.

1.2 Synthesis and Structure

HSA is a member of the albumin superfamily, which also includes the transport proteins α-fetoprotein, vitamin D-binding protein (Gc-globulin), and afamin (α-albumin) (Kragh-Hansen et al. 2013). In addition, the superfamily includes the α-fetoprotein-related gene, but due to multiple mutations, this gene is an inactive pseudogene in humans. All the genes are single-copy genes, and the four active ones in the human are expressed in a codominant manner, ie, both alleles are translated. The genes lie on chromosome 4, near the centromere for the long arm, at position 4q11–13. The albumin gene is 16,961 nucleotides long from the putative cap site to the first poly(A) addition site. It is split into 15 exons that are symmetrically placed within the three domains thought to have arisen by triplication of a single primordial domain.

The mRNA for HSA encodes for a precursor protein (pre-pro-albumin) of 609 amino acids. The N-terminal pre-peptide of 18 amino acids guides the nascent albumin peptide chain from the ribosome, where it was synthesized, through a receptor on the membrane of the endoplasmic reticulum into the lumen of the reticulum. Afterward, it is rapidly cleaved off. The N-terminal, basic pro-peptide of six amino acids, is cleaved in one of the last steps before secretion of the mature protein into the space of Disse and the hepatic sinusoid. Thus, HSA consists of 585 amino acids; the molecular mass is ca. 66.5 kDa.

Normally, wild-type pro-albumin is not secreted from the liver cells in a detectable amount. However, it can be found in the circulation in certain pathological conditions (Kragh-Hansen et al. 2013).

HSA is produced as a simple, monomeric protein, ie, without prosthetic groups and covalently bound lipid or carbohydrate. The three-dimensional structure of the single polypeptide chain, and of its recombinant version (rHSA), has been determined crystallographically, and the structure is now known to a resolution of 2.3 Å (He and Carter 1992; Sugio et al. 1999; Hein et al. 2010). The polypeptide chain forms a heart-shaped protein with approximate dimensions of 80 × 80 × 80 Å and a thickness of 30 Å. It has about 67 % α-helix but no β-sheet and can be divided into three homologous domains (I–III). Each of these is comprised of two subdomains (A and B). The A and B subdomains have six and four α-helices, respectively, connected by flexible loops. All, but one, Cys34, of the 35 cysteine residues are involved in the formation of 17 stabilizing disulfide bonds. Small-angle X-ray scattering studies of HSA in solution show general agreement with the crystal structure (Olivieri and Craievich 1995). Also, a combined phosphorescence depolarization-hydrodynamic modeling study has proposed that the overall conformation of HSA in neutral solution is very similar to that observed in crystal structures (Ferrer et al. 2001).

In addition to HSA, the crystal structure of albumin from cattle, horse, rabbit, and hare has been determined (Bujacz 2012; Majorek et al. 2012). Although a number of differences were found in the binding pockets, as well as variations in surface structure and charge distribution, structural alignments of the crystal structures with HSA showed strong structural similarities between the albumins. This finding is probably mainly due to a conserved set of disulfide bridges.

At present, 70 mutations of the HSA gene are known which result in a circulating variant of pro-albumin or albumin (alloalbumins) (Kragh-Hansen et al. 2013; The Albumin Website). Because both alleles of the gene are translated, most genetic variants have been detected in heterozygotes, ie, in persons having both a variant and wild-type (normal) HSA. In addition to single-amino acid substitutions, glycosylated variants, N-terminally and C-terminally modified alloalbumins, have been found. Mutations can also compromise the protein synthesis to such an extent that HSA is completely absent or strongly decreased in affected individuals leading to the condition known as analbuminemia. To date, 22 such molecular defects have been reported (Minchiotti et al. 2013; The Albumin Website).

Because alloalbumins do not seem to be associated with disease, they can be used as markers of migration and provide a model for study of neutral molecular evolution. They can also give valuable molecular information about binding sites, antioxidant and enzymatic properties, as well as in vivo and in vitro stability. Mutants with increased affinity for endogenous or exogenous ligands could be therapeutically relevant as antidotes, both for in vivo and extracorporeal treatment. Variants with modified biodistribution could be used for drug targeting. In most cases, the desired function can be further elaborated by producing site-directed, recombinant mutants.

1.3 Distribution and Circulatory Half-Life

HSA is solely synthesized in the liver. By contrast, its sites of degradation are widespread. Most of the protein is hydrolyzed in the muscle and skin, but some leaks into the gut, some is taken up by Kupffer cells of the liver, and a small amount is degraded elsewhere or lost with shed dermis, saliva, sweat, tears, or milk (Fig. 1.1) (Peters 1996).

In healthy adults, ca. 13.8 g is made per day and secreted into the bloodstream. This amount corresponds to ca. 25 % of the protein synthesis activity of the liver. However, under physiological circumstances, only 20–30 % of the hepatocytes produce albumin, and synthesis can therefore be increased on demand by a factor of 200–300 % (Evans 2002). From the liver, HSA is distributed in the bloodstream but also to several extravascular spaces, some of which are poorly accessible; these are mainly found in the skin (Fig. 1.1). The total amount of HSA in the body is ca. 360 g, of which about two-thirds is outside the bloodstream and about one-third is in the bloodstream. However, the concentration of HSA is higher in the bloodstream, and that is why the protein can contribute with ca. 80 % of the colloid osmotic pressure of plasma (ca. 15 mm of Hg).

HSA leaves the intravascular space in different ways. For example, fenestrated capillaries and, especially, sinusoidal capillaries allow the protein to pass. The latter are mainly found in the liver and spleen but also in bone marrow, lymph nodes, and adrenal glands. In other situations, the escape is transcellular and mediated by a receptor.

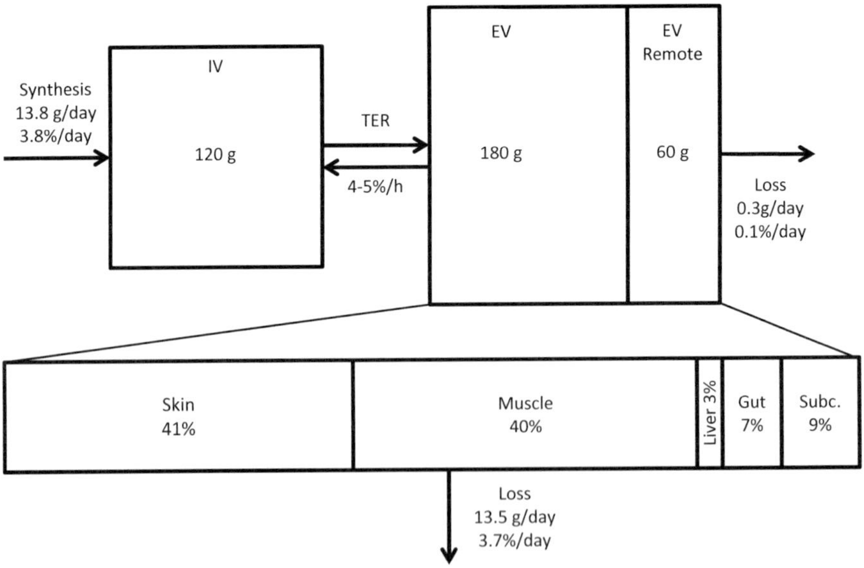

Fig. 1.1 Distribution and dynamics of HSA in a healthy person of 70 kg. *IV* intravascular, *EV* extravascular, *TER* transcapillary escape rate, *subc*, subcutaneous (The illustration is based on information found in Peters (1996))

Thus, the protein can interact with the receptor gp60 (60-kDa glycoprotein), also called albondin, situated in the plasma membrane of continuous endothelium (except for the brain) and alveolar epithelium (Sleep 2014; Merlot et al. 2014). Binding results in internalization of the complex by a caveolin-dependent endocytotic process and ultimately to transcytosis of albumin. It has been proposed that ca. 50 % of albumin leaves the capillary lumen in this way. Despite extensive studies, the gene for and the structure of gp60 are still unknown, and the molecular mechanisms of this transcytosis are still poorly understood.

HSA can most probably also leave the bloodstream by a process involving the intracellular receptor FcRn (neonatal Fc receptor) (Fig. 1.2a). This type of transcytosis is initiated by pinocytosis of HSA at the luminal membrane of the endothelial or epithelial cell (Bern et al. 2015). FcRn is placed in the membrane, but at physiological pH, there is no significant interaction between receptor and protein. After being taken up, HSA enters early endosomes. From there, the protein is transferred to acidified endosomes which have FcRn in their membrane, and at that pH (5–6), HSA binds strongly to FcRn. The endosomes with the protein-receptor complexes can now fuse with the basolateral side. This event results in exocytosis of HSA, because the pH of the cellular surroundings is neutral. Interestingly, IgG can be transcytosed by the same mechanism (Bern et al. 2015).

When albumin is saturated with fatty acids, transcytosis is two to three times higher than that of defatted albumin (Galis et al. 1988). The preferential transcytosis of cargo-carrying albumin is an interesting aspect, because it could help albumin to transport cargo into extravascular compartments.

HSA has an approximate plasma half-life of 19 days (Peters 1996). This half-life is extraordinary long for a circulating protein and is partly due to a return from the extravascular space to the circulation via the lymphatic system. The return amounts to 4–5 % of intravascular albumin per hour (Fig. 1.1), and the protein makes ca. 28 "trips" in and out of the lymphatic system during its lifetime (Peters 1996). Another contributing factor to the long circulatory half-life is that normally HSA is not lost in the urine. One reason for this is that, due to its size and charge, the filtration of HSA in the glomeruli is low. In addition, any protein filtered is reabsorbed in the proximal tubuli and transferred to the bloodstream via endocytosis by a receptor complex formed by cubilin and megalin (Merlot et al. 2014; Bern et al. 2015). Interestingly, results of animal studies have suggested an important role of FcRn for the renal retrieval of albumin (Sand et al. 2015; Bern et al. 2015).

FcRn is expressed in multiple cell types and tissues and is most important for the half-life of HSA, because it protects the protein (and IgG) from degradation in the lysosomes. Actually, FcRn rescues as much albumin as the liver produces (Bern et al. 2015). The initial steps in the protection are similar to those leading to transcytosis (Fig. 1.2a): the protein is taken up by the cell in question by pinocytosis and ends up as FcRn-bound in acidified endosomes. Now, in the present situation, the endosomes migrate to the membrane, where HSA originally was taken up, and release it by exocytosis. Thus, HSA is recycled back to the circulation. Albumin that does not bind to FcRn in the acidified endosomes, because it is conformationally

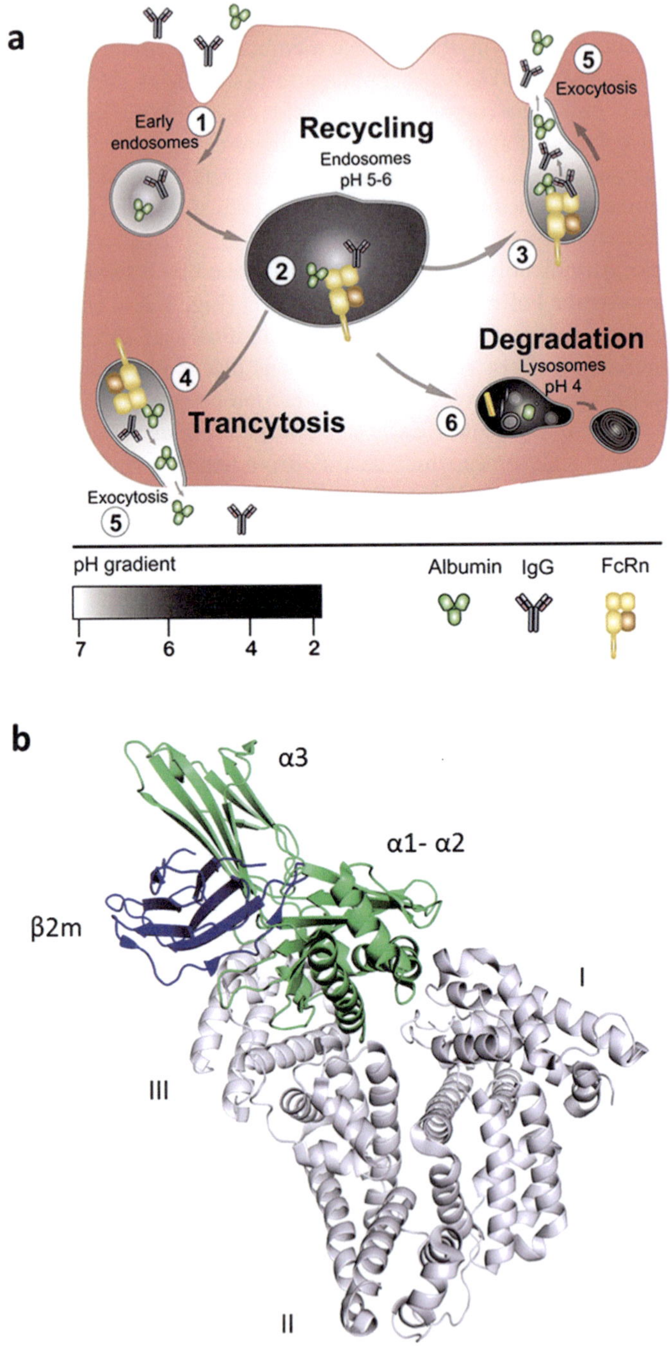

Fig. 1.2 (**a**) pH-dependent, FcRn-mediated cellular transport of HSA (and IgG). Initially, HSA is taken up by pinocytosis from the blood or luminal space of the cell and enters early endosomes (step *1*). From there it is transferred to acidified endosomes having FcRn in the membrane. At this pH of 5–6, the protein binds strongly to the receptor at a 1:1 stoichiometry (step *2*). At this point, three possible fates seem to exist for HSA. First, it can be exocytosed back to the blood or

modified or in surplus, will neither be transcytosed nor recycled but will be destined for lysosomal degradation.

In contrast to the other receptors interacting with albumin, the crystal structure of FcRn, and of its complex with HSA, is known (Fig. 1.2b). This detailed information is of great pharmaceutical and pharmacological interest, because on the basis of this knowledge, it is possible to construct albumin mutants with modified affinities for the receptor (Sand et al. 2015; Bern et al. 2015). Thus, it should be possible, for example, to design albumins with improved recycling and thereby increased circulatory half-life. Such a possibility is of great clinical impact, because in addition to increase the half-life of albumin itself, it should also increase very much the half-life of any bound cargo. An increased affinity for FcRn could perhaps also facilitate transcytosis of HSA with or without bound drugs or therapeutics. These possibilities are increasingly explored, but the influence of other factors has to be addressed. For example, the affinity of albumin for FcRn is strongly species dependent, a fact that has to be taken into account when using laboratory animals in preclinical investigations. In addition, engineered HSA mutants may become immunogenic, and the albumin-FcRn interaction could be influenced by protein-bound drugs or therapeutics. Thus, although very promising, there is still a long way to go before manipulating the interaction with FcRn can be used in the clinic.

1.4 Binding to Other Receptors

Several types of cancer secrete a glycoprotein known as secreted protein acidic and rich in cysteine (SPARC). Among its functions is albumin binding, a function which results in enhanced accumulation of albumin in the tumors, but apparently not in increased uptake of the protein into the tumor cells (Merlot et al. 2014). The albumin-receptor binding could also result in accumulation of albumin-bound drugs and therapeutics in the tumors and thereby result in better therapeutic effects. For example, the response of human head and neck cancers to nab-paclitaxel (nanoparticle albumin-bound paclitaxel) was reported to correlate with SPARC expression (Desai et al. 2009).

Fig. 1.2 (continued) luminal space from where it was taken up. Such an event results in recycling of the protein (steps *3* and *5*). Second, the exocytosis can take place at the basolateral membrane resulting in transcytosis (steps *4* and *5*). Finally, HSA which does not bind to FcRn in the acidified endosomes goes to degradation in the lysosomes (step *6*). Whether the HSA-FcRn complexes go to the luminal or to the basolateral membrane depends on, among other factors, binding of intracellular proteins such as adaptor protein-2 and calmodulin to sorting motifs of the cytoplasmic tail of the α-3 subunit (Sand et al. 2015). Principally the same mechanisms exist for IgG. IgG and HSA bind independently and noncooperatively to FcRn (The illustration is a modification of Fig. 2 in Bern et al. (2015).). (**b**) Crystal structure of the HSA-FcRn complex (Oganesyan et al. 2014). The domains (I–III) of HSA are indicated. FcRn is composed of a long α-chain of 44 kDa (domains α1–α3) and a short β2-microglobulin unit of 12 kDa (β2m). Of these structures, only α3 has a transmembrane fragment and a cytosolic part. Domain III of HSA plays the essential role for binding, but domain I is also necessary. The figure was made with PyMOL on the basis of the atomic coordinates (PDB ID: 4N0F) available at the RCSB Protein Data Bank

The membrane-bound receptors gp18 (18-kDa glycoprotein) and gp30 (30-kDa glycoprotein) are widely distributed scavenger receptors which bind and internalize chemically modified albumin. Thus, the receptors recognize damaged or changed albumin and target it for lysosomal degradation (Merlot et al. 2014).

1.5 Clinical and Pharmaceutical Uses

Huge amounts of HSA, ca. 500 ton per year worldwide, are used for improving clinical conditions such as shock, burns, trauma, surgical blood loss, hypoalbuminemia, decompensated cirrhosis, cardiopulmonary bypass, and acute respiratory distress (Mendez et al. 2005). HSA can also be used as a part of a hemodialysis regimen especially in patients with hepatic failure. Several such systems exist, but the molecular adsorbent recirculating system (MARS) is currently the most effective liver support device and can effectively remove protein-bound and water-soluble substances (Mitzner 2011). The efficiency of the system can most probably be increased by using tailor-made mutants of HSA itself or of one of its domains (Minomo et al. 2013).

HSA is also widely used as a stabilizing agent in pharmaceutical and biological products like vaccines, recombinant therapies, drug formulations, and coatings for medical devices (Chuang et al. 2002). Furthermore, the protein is used as a component in serum-free cell culture media, as a component for imaging agents, and, possibly, for therapeutic apheresis where plasma exchange might be desirable (Chuang et al. 2002).

Traditionally, HSA is obtained by fractionating human plasma. The risk by using this approach is that such preparations can be contaminated with blood-derived pathogens, for which reason the preparations have to be heated at 60 °C for 10 h in the presence of sodium octanoate and N-acetyl-L-tryptophanate. Of these ligands, octanoate has the greatest stabilizing effect against heat, whereas the presence of N-acetyl-L-tryptophanate diminishes oxidation of the protein (Anraku et al. 2004). However, recent studies have revealed that N-acetyl-L-tryptophanate should be replaced by N-acetyl-L-methionine, because the latter is a superior antioxidant and protects HSA against light and thereby photo-irradiation (Kouno et al. 2014). Although this pasteurization procedure is very effective, the potential risk of the presence of pathogens still exists, and the associated screening costs are substantial. In addition, in several countries, supplies of human plasma are limited. Therefore, many and large efforts are being made to produce rHSA as a substitute. rHSA can be highly expressed in various hosts including bacteria, yeast, and transgenic animals and plants (Chen et al. 2013). Of these, yeast (*Pichia pastoris*) and plants (Asian rice) seem to be the most promising for large-scale production. However, challenges like purity and production costs have to be met. For example, HSA expressed in rice has extensive glycation which showed supplier-to-supplier and lot-to-lot variability (Frahm et al. 2014). Therefore, it has not yet been possible to do large-scale production for clinical uses. Nevertheless, rHSA is currently used for different pharmaceutical purposes. This is the case with, for example, Recombumin® from Novozymes Biopharma, Albagen™ from New Century Pharmaceuticals, and recombinant human albumin from Akron Biotech and Sigma-Aldrich.

1.6 Ligand Binding

HSA serves as a depot protein and transport protein for numerous endogenous and exogenous compounds. In this way, the protein has a major impact on the pharmacokinetics and pharmacological effects of drugs (Yamasaki et al. 2013). Albumin binding can also affect the ligands in a more individual manner. Thus, binding can result in an increased solubility in plasma of the compound, in a reduced toxicity, or in protection of the ligand against oxidizing agents.

Although the ligand-binding properties of HSA are very versatile, it mainly binds organic anions and inorganic cations. This unique ability of the protein is due to a favorable combination of hydrophobic pockets and side-chain charges and to a pronounced flexibility, which, to a large degree, is caused by fairly long interdomain and intradomain polypeptide linkers and flexible loops.

Detailed molecular information about the binding sites is very helpful in the assessment of displacement effects. However, it should be born in mind that displacement effects may as well be caused by ligand-induced conformational changes of the protein. Structural information is also useful when designing new drugs whether the aim is to avoid binding or to make use of the protein's depot function.

For comprehensive tabulations of ligands, see the reviews of, for example, Kragh-Hansen (1981), Peters (1996), Kragh-Hansen et al. (2002), and Fanali et al. (2012). In the following, the known binding sites and regions for high-affinity binding will be presented.

1.6.1 N-terminal End

Cu^{2+} and Ni^{2+} are strongly bound in a square-planar ligand arrangement formed by the three N-terminal amino acids Asp1, Ala2 and His3. The metal ions are held tightly in a chelate ring involving the α-amino nitrogen of Asp1, the first two peptide nitrogens, and the N^1 imidazole nitrogen of His3. Co^{2+} binds with a lower affinity to principally the same site in an octahedral environment but with the β-COO^- group of Asp1 and the ε-group of Lys4 axially contributing to the metal ion coordination sphere (Fanali et al. 2012; Bal et al. 2013).

Co^{2+} binding to HSA from patients with cardiac ischemia is diminished. Perhaps it is not the site at the N-terminal end but other sites which are affected (Bal et al. 2013). Anyway, the observation has led to the development of assays for the condition. However, albumin's affinity for Co^{2+} can also be affected by several other conditions (Gaze 2009).

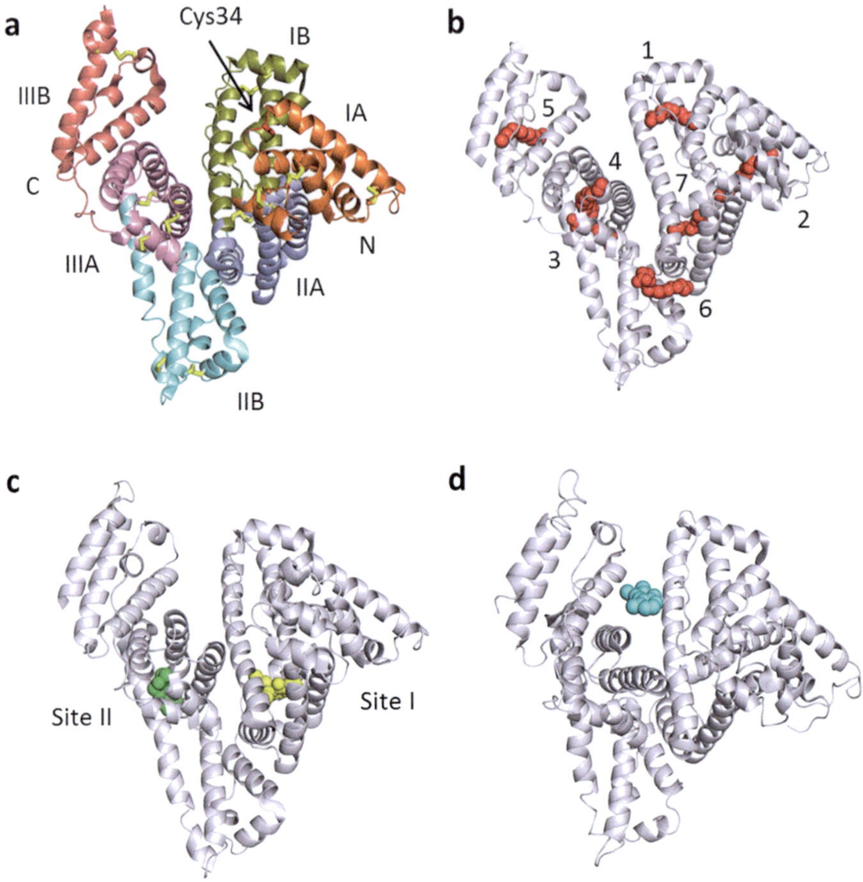

Fig. 1.3 (**a**) Crystal structure of HSA with the position of Cys34 indicated in *red* and the 17 disulfide bridges marked in *yellow*. The subdivision of the protein into domains (I–III) and subdomains (**a**, **b**) is shown. *N* and *C* represent the N-terminal and C-terminal ends, respectively. PDB ID: 1BM0. (**b**) Locations of the seven binding sites common to fatty acid anions using palmitate as an example (*red*). PDB ID: 1e7h. (**c**) Binding site for warfarin (PDB ID: 2bxd) in site I (*yellow*) and for diazepam (PDB ID: 2bxf) in site II (*green*). (**d**) Lidocaine binding (*magenta*) in the central, interdomain crevice. PDB ID: 3JQZ

1.6.2 Cys34

Cys34 in subdomain IA (Fig. 1.3a) is located in a crevice on the surface of the protein and does not participate in any disulfide bridges. Despite a limited accessibility, the sulfhydryl group can bind Hg^{2+}, Ag^{+}, Au^{+}, and Pt^{2+} (Fanali et al. 2012). The residue can also interact with nitric oxide and 8-nitro-cGMP and thereby form *S*-nitrosothiol and *S*-cGMP-HSA, respectively (Ishima et al. 2012a, b). In the circulation, about half of HSA has Cys34 coupled to low molecular weight thiols such as cysteine, glutathione, and homocysteine. The thiol group can also interact with several drugs such as bucillamine derivatives, D-penicillamine, captopril, meso-2,3-dimercaptosuccinate, N-acetyl-L-cysteine, aurothiomalate, auranofin, and ethacrynate (Kragh-Hansen et al. 2002).

1.6.3 Subdomains IA and IIA

Palmitate has been reported to bind with a high affinity to an enclosed site located between subdomains IA and IIA (Simard et al. 2006). The methylene tail of the fatty acid anion binds in a nearly linear conformation within a narrow hydrophobic cavity formed by residues of the two subdomains, while the carboxyl forms specific salt bridge interactions with Tyr150, Arg257, and Ser287 (site 2 in Fig. 1.3b). Apparently, no other ligands, with the exception of other medium- and long-chain fatty acid anions, bind in this site.

HSA is the major Zn^{2+} transporter in plasma, and in the fatty acid-free protein, the metal ion binds primarily to an essentially preformed, 5-coordinate site located at the interface of domains I and II involving His67 and Asn99 from the former and His247 and Asp249 from the latter domain (Blindauer et al. 2009). The site is also a primary binding site for Cd^{2+} and a weak site for Cu^{2+} and Ni^{2+}. Therefore, the site is often called the multi-metal-binding site (Bal et al. 2013). Binding of a fatty acid to site 2 results in conformational changes which disrupt the site.

1.6.4 Subdomain IB

This subdomain houses a fairly large, L-shaped cavity with charged residues such as Arg117 and Arg186, capable of making hydrogen bondings, at its entrance (Zunszain et al. 2008). High-affinity binding of bilirubin and fusidic acid within this pocket causes only minor conformational changes in the site. By contrast, high-affinity binding of hemin and low-affinity binding of a fatty acid anion (site 1 in Fig. 1.3b) induce a significant conformational rearrangement of the subdomain. Wang et al. (2013) found that the subdomain is a major binding site for complex heterocyclic molecules such as the oncology agents camptothecin, 9-amino-camptothecin, etoposide, teniposide, bicalutamide, and idarubicin. The authors also observed that the large binding cavity has two access sites.

1.6.5 Subdomain IIA

The pioneering work of Sudlow et al. (1975), which was based on displacement of fluorescent probes, revealed that most drugs bind with a high affinity to one of two sites, called site I and site II. Of these, site I is placed in subdomain IIA (Fig. 1.3c).

The site is adaptable, because it can bind structurally very diverse ligands. However, usually they are dicarboxylic acids and/or bulky heterocyclic molecules with a negative charge localized in the middle of the molecule. The site is also large, because several examples have been found of independent binding of two different compounds (Kragh-Hansen et al. 2002). Several studies have shown that the site is composed of flexible subsites. Thus, Yamasaki et al. (1996) reported independent, high-affinity binding of acenocoumarol, dansyl-L-asparagine, and n-butyl p-aminobenzoate to three subsites which were named Ia, Ib, and Ic, respectively. Of these, Ia and Ib correspond to the warfarin and azapropazone binding regions, respectively.

X-ray crystallography has revealed many details about the site. Ghuman et al. (2005) found that it was a preformed binding pocket within the core of the subdomain including the lone tryptophan residue of the protein (Trp214). In the defatted protein, the interior of the pocket is predominantly apolar but contains two important clusters of polar residues, an inner one toward the bottom of the pocket (Tyr150, His242, and Arg257) and an outer cluster at the pocket entrance (Lys195, Lys199, Arg218, and Arg222). The large binding cavity is comprised of a central zone from which extend three distinct compartments. Warfarin, phenylbutazone, oxyphenbutazone, and the renal toxin 3-carboxy-4-methyl-5-propyl-2-furanpropanoic acid bind in the center of the pocket and partly into the compartments. Iophenoxic acid binds with a high affinity in the main chamber (central zone) and makes crucial hydrogen bonds to the side chains of Tyr150 and Arg257 (Ryan et al. 2011a). The ligand binds in a similar position and orientation to other compounds that are composed primarily of iodinated aromatic rings such as triiodobenzoic acid, diiodosalicylic acid, and iodipamide. The dansylated forms of L-asparagine, L-arginine, and L-glutamate are fluorescent marker ligands for the site, and they make essentially the same set of interactions with the binding pocket (Ryan et al. 2011b).

1.6.6 Subdomain IIIA

Sudlow's site II, also called the indole-benzodiazepine site, is situated in subdomain IIIA (Fig. 1.3c). Ligands binding to the site are often aliphatic or aromatic carboxylic acids with a negatively charged acidic group at one end of the molecule away from a hydrophobic center. Site II seems to be smaller, more selective, and less flexible than site I (Kragh-Hansen et al. 2002). Among the ligands known to bind to the site with a high affinity are L-tryptophan, diazepam, and several nonsteroidal antiinflammatory drugs. Octanoate and fatty acids with a longer chain length also bind to site II with a high affinity (Kragh-Hansen et al. 2006; Simard et al. 2006).

Crystallographic analyses have shown that, like site I, site II comprises a largely preformed hydrophobic cavity with distinct polar features in defatted HSA (Ghuman et al. 2005). However, differences between the sites also exist, among them being that site II is smaller than site I and without preformed sub-compartments. Furthermore, site II has only one main cluster of polar residues near the pocket entrance; it is composed of Arg410, Tyr411, Lys414, and Ser489. Diflunisal, diazepam, ibuprofen, and the renal toxin indoxyl sulfate were found to bind in the center of the binding pocket. Propofol also binds to the site but in a different conformation due to its hydroxyl group in the center of the molecule. The flexibility of the site is less pronounced than that of site I, because binding is often strongly affected by stereoselectivity (Kragh-Hansen et al. 2002). However, site II possesses some flexibility, because it can bind diazepam, which has a large, branched structure, and two molecules of myristate (Ghuman et al. 2005). Ryan et al. (2011b) have studied in detail the binding of the marker ligands dansyl-L-phenylalanine, dansyl-L-norvaline, and dansylsarcosine.

1.6.7 Subdomain IIIB

Palmitate and other medium- and long-chain fatty acids bind with a high affinity in a hydrophobic channel that spans the width of the subdomain (site 5 in Fig. 1.3b) (Bhattacharya et al. 2000; Simard et al. 2006; Kragh-Hansen et al. 2006). The carboxyl group of the fatty acids interacts with the side chain of Lys525, aided in most cases by Tyr401. So far, no drugs have been reported to bind to this subdomain with a high affinity. However, binding experiments with mutants of HSA and molecular dockings strongly suggest that L-thyroxine binds with a high affinity to a site in the subdomain and that Lys525 is also important for strong binding of this ligand (Kragh-Hansen et al. 2016).

1.6.8 Central, Interdomain Crevice

Only a very few ligands have been reported to bind in the large crevice between domains I and III in defatted HSA. However, two low-affinity sites exist for the binding of decanoate but not for fatty acids of a longer aliphatic chain: one close to the base of the crevice and one further up (Bhattacharya et al. 2000). The positively charged lidocaine binds to a superficially placed site in subdomain IB in the upper part of the interdomain cleft. Arg114, Lys190, and Asp187 are important residues for binding of the drug (Fig. 1.3d) (Hein et al. 2010).

1.6.9 Covalent Binding

Certain metal ions, nitric oxide, and several drugs and other organic compounds can bind covalently to Cys34 (see Sect. 1.6.2).

The metabolism of drugs with a carboxylic acid function involves usually the formation of acyl glucuronides. This type of metabolite is a reactive electrophilic species and can therefore react covalently with HSA. Several such examples exist, and serine, arginine, or lysine residues, especially Lys159 or Lys199, are involved (Kragh-Hansen et al. 2002).

D-glucose and other reducing monosaccharides can interact with HSA. The interaction is with positively charged amino acids, and especially Lys525 but also Lys199, Lys281, and Lys439 seem to be important in this respect (Nakajou et al. 2003). This nonenzymatic glycation starts by condensation of the carbonyl group in acyclic glucose with a free amino group forming a Schiff base or aldimine intermediate, which undergoes an Amadori rearrangement to form a stable ketoamine that can cyclize to a ring structure. The Amadori product may subsequently undergo oxidative and nonoxidative degradation or rearrangement, giving rise to a heterogeneous group of substances loosely described as advanced glycation end products. Because the ambient glucose concentration is the major determinant for the reaction, glycation is increased in diabetes, and glycoalbumin has been proposed as a useful test for short- and intermediate-term control (2–3 weeks). By contrast, glycohemoglobin (HbA1c) is the favorite test for monitoring long-term (2–3 months) control of blood sugar levels in diabetics. Glycoalbumin may also serve as a biomarker for complications of diabetes such as nephropathy, retinopathy, and cardiovascular disease (Cohen 2013; Furusyo and Hayashi 2013).

1.7 Coupling of Nonbinding Therapeutics to HSA

Several polypeptide and polynucleotide therapeutics are used in the clinic while others are promising in clinical trials. However, the usefulness of such therapeutics can be limited due to very short in vivo half-lives, in the order of minutes, due to, for example, rapid degradation and renal clearance. The short half-lives can be increased very much by coupling the therapeutics to albumin, and the use of albumin for this purpose is a rapidly growing area of research.

1.7.1 Reversible Binding of Modified Therapeutics

A well-known example to illustrate this possibility is Levemir® (Insulin detemir), which is an albumin-binding human insulin analog in which threonine in position B29 has been replaced by lysine to which myristic acid is covalently attached. The modified insulin binds reversibly to HSA, and its half-life is increased from 4–6 min to 5–7 h (Sleep 2014). More recently, Tresiba® (Insulin degludec) has been developed. In this insulin form, palmitic acid is conjugated to the mutated lysine in position B29 through a gamma-L-glutamyl spacer, and in this case the half-life has been increased to more than 24 h in the blood of patients (Sleep 2014). In principally the same way has the in vivo half-life of the glucagon-like peptide 1 (GLP-1) been extended. In liraglutide (Victoza® and Saxenda®) Lys37 has been substituted by arginine, and palmitic acid has been coupled to Lys26. Liraglutide has a half-life of 11–15 h in contrast to the endogenous hormone that has a half-life of ca. 2 min. Furthermore, semaglutide is a modified version of GLP-1 to which stearic diacid is bound, and this product has a half-life of 160 h allowing once-weekly dosing (Sleep 2014). Attachment of a fatty acid is also able to result in albumin binding of other therapeutics like cisplatin. In addition to fatty acids, also a small organic molecule based upon 4-(p-Iodophenyl) butanoic acid derivatives, called Albutag, and several peptides can imply albumin binding. For example, the binding of Albutag to a single-chain antibody has increased its half-life 40-fold in a tumor-bearing mouse model (Sleep 2014). For more examples, see the review of Bern et al. (2015).

Exogenous double-stranded siRNA can result in potent and sequence-specific, posttranslational gene silencing, but their in vivo use is impeded by a poor stability and short half-life. Bienk et al. (2015) observed that although cholesterol is not a known ligand for HSA, siRNA functionalized with two cholesteryl moieties exhibited high and specific affinity to the protein ($K_{ass} = 10^7\ M^{-1}$) and a sixfold increase in serum half-life in NMRI mice.

1.7.2 Covalent Binding of Modified Therapeutics

Polypeptide therapeutics can be modified by a maleimide- or thiol-containing compound which ensures conjugation to Cys34. This approach has been used in the case of, for example, aldoxorubicin (doxorubicin), insulin, methotrexate, exendin-4, protamine, growth hormone, the antiviral peptide PC-1505, the opioid agonist dynorphin A, YY peptide, and the granulocyte colony-stimulating factor (Elsadek and Kratz 2012; Sleep 2014; Bern et al. 2015). Usually, the half-lives of the therapeutics are increased very much to several hours or even days. Using this approach for increasing half-lives, it is necessary to release the prodrug or the drug itself from HSA. One way of obtaining this release is to use an acid-sensitive linker. In this way the compound can be set free extracellularly in the slightly acidic environment often present in, for example, tumor tissue or intracellularly in acidic endosomes or lysosomal compartments after cellular uptake. Alternatively, the linker can be hydrolyzed by making use of redox enzymes or albumin's enzymatic properties (see Sect. 1.11).

Using this way of extending half-lives and that mentioned in Sect. 1.7.1, one should be aware of the risk that binding of the modified therapeutic could displace albumin-bound drugs, or other ligands, simultaneously given to the patient. In addition, the bound therapeutic could interfere with albumin binding to receptors such as FcRn and thereby interfere with the half-life, distribution, or metabolism of itself and of albumin (see Sect. 1.3).

1.7.3 Genetic Fusion to HSA

If the gene of the therapeutic polypeptide or protein is available, then half-life extension can also be achieved by albumin fusion technology, where the therapeutic of interest is expressed together with albumin as a single polypeptide chain. If sufficient linker space is given, albumin and the therapeutic are able to fold correctly. A favorable feature of this technique is that it allows a simple one-step synthesis process with no need for in vitro chemical linking steps. The safe and general utility of albumin fusions can be seen by considering the number of fusions approved or in clinical development. Some examples can be mentioned here: coagulation factors (FVIIa, FIX, and von Willebrand factor), anticoagulants (hirudin, infestin, and barbourin), growth factors (erythropoietin and G-CSF), cytokines (IL-2, IL-1ra, interferon-α-2b, and interferon-β), GLP-1, insulin, and thioredoxin-1. For reviews, see Sleep (2014) and Bern et al. (2015).

The therapeutic molecule is fused either to the N- or C-terminus of albumin – or simultaneously to both ends. If fused to the C-terminus, the therapeutic can perhaps interfere with the pH-dependent binding of albumin to FcRn and thereby affect the half-life of the product (see Sect. 1.3). By contrast, nothing indicates that fusion to albumin potentiates the immune response to the therapeutic polypeptide or protein or to albumin itself.

1.8 Targeting

In addition to increasing the stability and half-life of drugs and therapeutics, native and modified HSA can act as a versatile and effective delivery tool for them and thereby increase their therapeutic effect and reduce unwanted side effects. The function can be exerted in three principally different manners, namely, by passive accumulation at disease sites, by adding a tag to the protein which can be recognized by specific receptors, or by mutating the protein.

1.8.1 Passive Targeting

HSA, with or without cargo, accumulates within tumor interstitium due to the enhanced permeation and retention effect (EPR). This effect occurs due to the extensive vascularization in these tissues and to the combined effects of a leaky, immature capillary network and an impaired lymphatic drainage. A supplementary explanation for albumin accumulation in tumors is based on albumin interaction with receptors. Thus, HSA could transcytose across the endothelium of the capillaries via the gp60 or FcRn receptor and bind to SPARC in the interstitium of the tumor (see Sects. 1.3 and 1.4). In addition to increased interstitial accumulation, tumor cells seem to have increased uptake and catabolism of albumin.

HSA can also accumulate in other types of diseased tissue such as in rheumatoid arthritis and in other inflamed tissues (Neumann et al. 2010). This type of accumulation is most probable also the result of a local EPR effect and the presence of SPARC (Sleep 2014).

The therapeutic potential of passive targeting of HSA in tumors can be illustrated by using *S*-nitrosated HSA as an example. Thus, Ishima et al. (2012a) found that *S*-nitrosated HSA accumulates in the tumor of C26 tumor-bearing mice due to the EPR effect and causes a pronounced cell death. Experiments with an *S*-nitrosated version of recombinant HSA dimer showed that both NO and the dimer itself enhance the EPR effect and that the modified dimer accumulates more than the corresponding monomer in the tumor. Furthermore, *S*-nitrosated HSA dimer delivers large amounts of cytotoxic NO into the tumor tissue. The antitumor effect of the *S*-nitrosated dimer can be further increased by adding more NO moieties to the protein (or by pegylation): binding of 13.5 mol NO/mol protein, using 2-iminothiolane as a linker, resulted in a ten times higher antitumor activity in the mice as compared to a preparation having only 1.5 mol NO/mol protein (Ishima et al. 2014). As mentioned, in addition to having a good antitumor effect, *S*-nitrosated HSA dimer can increase the EPR effect. This results in a higher concentration of the compound in the tumor. However, it can also increase the accumulation and thereby antitumor effect of other antitumor agents simultaneously given to the mice such as micelles like HPMA-ZnPP and liposomes like Doxil (Kinoshita et al. 2015).

Cell influx of NO mainly takes place by cell-surface protein disulfide isomerase, and apoptosis is mainly the result of ROS induction and activation of enzymes such as caspase-3 and heme oxygenase-1. The administration of *S*-nitrosated albumin is safe, because it has no effect on blood pressure, heart rate, or on several biochemical markers (Kinoshita et al. 2015).

1.8.2 Active Targeting

Targeting to cells and organs can be achieved by adding a targeting ligand to HSA. The ligand is targeted to a receptor in the wanted cell type, and by coupling a therapeutic to the albumin-ligand complex a treatment will become much more specific, and the toxic side effects will be reduced. In this respect, the liver has been one of the most desirable target organs. For example, Kupffer cells in the liver possess mannose receptors, and mannosylated HSA can selectively transfer NO to these cells and thereby be used in the treatment of ischemia/reperfusion injuries (Taguchi et al. 2015). The liver also contains hepatic stellate cells having receptors for mannose-6-phosphate, and HSA with this ligand can target antifibrotic drugs to these cells and thereby to the liver (Taguchi et al. 2015). Coupling of 5-fluorouracil to galactosylated HSA results in a potentially useful complex in the treatment of liver carcinoma, because it can bind to asialoglycoprotein receptors on the hepatocytes (Cai et al. 2006). A similar approach can be made by using lactosaminated HSA, because the outlying galactose moiety can bind to the same type of receptor. In this way can adenine arabinoside-AMP, fluorodeoxyuridine, and doxorubicin be targeted to the liver with good therapeutic results; see Fiume and Di Stefano (2010) who discuss the possible pros and cons of these treatments.

HSA nanocapsules can be surface modified with folic acid, and this complex can interact with folate receptor beta specifically expressed by activated macrophages. In this way, it should be possible to deliver co-bound therapeutics selectively to the pro-inflammatory cells playing a key role in the development of rheumatoid arthritis without causing toxicity and collateral damage to healthy cells (Rollett et al. 2012). Finally, cationized HSA itself and proteins and nanoparticles conjugated to it can cross the blood-brain barrier (Sleep 2014).

1.8.3 Mutated HSA

Not much work has been carried out for testing whether mutants of HSA can be used for targeting. Perhaps the work of Iwao et al. (2007, 2009) can give some hints in that respect. These authors studied the biodistribution of a large number of genetic variants of the protein in mice. They found that the mutations Cys177→Phe, Lys240→Glu, and Glu321→Lys and two examples of a shortening and modification of the C-terminal end resulted in a 2.2- to 20-fold increase in liver uptake and that the mutations Cys177→Phe, Lys313→Asn, and Lys541→Glu increased kidney clearance by a factor 2.2–4.4. Results like these could be useful when designing recombinant, therapeutic albumins or albumin products with a modified cell or organ uptake. Actually, the recombinant mutation of Arg410 to Ala resulted in a more than 27-fold increase of liver clearance, whereas uptake by kidneys, spleen, lungs, and heart were not significantly affected (Iwao et al. 2006).

1.9 Nanoparticles

The use of many drugs, for example, chemotherapy agents and agents for gene therapy, is hampered by low solubility in aqueous media, a pronounced toxicity, and/or fast in vivo degradation. One way of reducing or even solving these problems is to encapsulate the agents in nanoparticles. Albumin seems to be particularly useful in this respect, because it is stable, biodegradable, nontoxic, nonimmunogenic, and can readily bind various drugs. Nanoparticles made of albumin are easy to prepare under soft conditions, and a large variety of protocols for making them exist based on desolvation, emulsification, thermal gelation, nano spray drying, the nab technology, or self-assembly. Because functional groups exist on the nanoparticle surface, it is possible to harden the particles by cross-linking and to attach targeting ligands or compounds which imply a lower toxicity, because side effects are reduced. Among the two latter types of molecules are galactosamine, folate, polyethylene glycol, surfactants, cationic or thermosensitive polymers, and peptides like Arg-Gly-Asp-containing peptide; even proteins can be covalently bound to the surface of HSA nanoparticles for increasing brain uptake. The first commercial product based on protein nanoparticles in oncology was albumin-bound paclitaxel (Abraxane®) approved by the FDA in 2005. Since then, albumin nanoparticles encapsulating other active compounds have been fabricated, namely, doxorubicin, cisplatin, docetaxel, rapamycin, vinblastine sulfate, mitoxantrone, ferulate, iron oxide, EDTA, and octyl aldehyde. The purposes of making such particles are a potential use in the treatment of different cancer forms and vascular diseases as well as for imaging and chelation therapy. Many of these particles are currently in different phases of clinical trials. For reviews, see Elsadek and Kratz (2012), Elzoghby et al. (2012), and Bern et al. (2015).

1.10 Antioxidant Activities

Free radicals are normal components of cellular oxygen metabolism in mammals, and they often play a central role in signaling processes. However, radicals can be produced in surplus and thereby lead to oxidative stress resulting in reversible and irreversible modifications of sensitive macromolecules. Therefore, it is essential to have well-developed defense mechanisms against this toxicity. In extracellular fluids, a major antioxidant role is performed by albumin, and it can diminish the amounts of different reactive oxygen species (ROS) and reactive nitrogen species (RNS) (Quinlan et al. 2005; Roche et al. 2008; Anraku et al. 2013). In addition to directly interacting with ROS and RNS, albumin can to a certain extent prevent their formation.

HSA can act as a primary antioxidant and prevent ROS and RNS formation by binding transition metal ions. Most importantly, HSA binds copper ions with a high affinity, but the protein can also bind vanadium, cobalt, and nickel ions as well as heme and iron ions, the latter under conditions of iron overload or pronounced hemolysis (Quinlan et al. 2005; Roche et al. 2008). The activities of NO and sulfur-containing organic compounds like homocysteine are reduced through binding to Cys34 of HSA. Interestingly, albumin-bound bilirubin, but not free bilirubin, has antioxidant properties.

HSA can protect ligands against oxidation. For example, polyunsaturated fatty acids bind with a high affinity to the protein, and thereby they become protected from oxidant-mediated damage (Roche et al. 2008).

HSA can also act as a secondary antioxidant and scavenge preformed ROS and RNS. This activity can mainly be attributed to its single thiol group, Cys34, which accounts for ca. 80 % of the reduced thiols in human plasma (Anraku et al. 2013). The thiol group can react with several oxidizing species and become oxidized reversibly to sulfenic (SOH) or sulfinic acid (SO_2H) or irreversibly to sulfonic acid

(SO_3H). Actually, the redox state of Cys34 has been suggested as a useful biomarker for oxidative stress (Anraku et al. 2013). In addition to Cys34, also methionine residues and, probably to a lesser degree, tryptophan, arginine, tyrosine, lysine, threonine, proline, and valine residues contribute to the antioxidant properties of HSA. The involvement of the different amino acid residues depends on their molecular surroundings in the protein and on the type of oxidant (Iwao et al. 2012; Anraku et al. 2015).

1.11 Enzymatic Properties

HSA and some of its ligand complexes possess enzymatic properties which are useful both in vivo and in vitro (Kragh-Hansen 2013). The most pronounced of the many activities of HSA are different types of hydrolysis. Key examples are esterase-like activities involving Tyr411 (Fig. 1.4) or Lys199 and the thioesterase activity of Cys34. In the first case, hydrolysis involves water and both products are released, whereas in the latter cases one of the products is set free, and the other stays covalently bound to the protein. The protein has great impact on the metabolism of, for example, eicosanoids and xenobiotics. HSA is also useful in detoxification reactions, for activating prodrugs and for binding and activating drug conjugates.

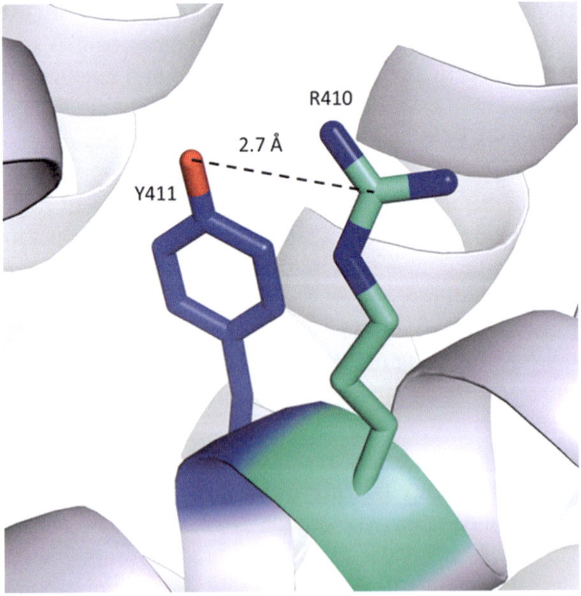

Fig. 1.4 A close-up of the nitrogen atoms of Arg410 and the phenolic oxygen of Tyr411 in subdomain IIIA, which are important for the esterase-like activity of HSA. The interresidue distance is 2.7 Å. PDB ID: 1BM0

The protein can be used to construct smart nanotubes with enzymatic properties useful for biomedical applications. Albumin with a metal ion-containing complex is capable of facilitating reactions involving reactive oxygen and nitrogen species and can be used for nanoparticle formation. Thus, even though the enzymatic properties of HSA are not always appreciated, they can be physiologically relevant and useful for biomedical and pharmaceutical purposes.

1.12 Other Functions

Physiologically, HSA may influence microvascular integrity, aspects of the inflammatory pathway, including neutrophil adhesion, and the activity of cell signaling moieties (Quinlan et al. 2005; Spinella et al. 2016). The protein also has an antithrombotic, anticoagulant effect due to its capacity to bind nitric oxide (Spinella et al. 2016). In addition, it possesses immune-modulating effects. Finally, albumin is a negative acute-phase protein. Perhaps one or more of these activities can be the object for pharmaceutical or pharmacological intervention in the future.

Acknowledgments I am indebted to Dr. Jan Terje Andersen, University of Oslo, Norway, for donating Fig. 1.2a and to Dr. Konrad Bienk, University of Aarhus, Denmark, for helping me in making the other figures.

References

Anraku M, Tsurusaki Y, Watanabe H, Maruyama T, Kragh-Hansen U, Otagiri M (2004) Stabilizing mechanisms in commercial albumin preparations: octanoate and N-acetyl-L-tryptophanate protect human serum albumin against heat and oxidative stress. Biochim Biophys Acta 1702:9–17

Anraku M, Chuang VTG, Maruyama T, Otagiri M (2013) Redox properties of serum albumin. Biochim Biophys Acta 1830:5465–5472

Anraku M, Shintomo R, Taguchi K, Kragh-Hansen U, Kai T, Maruyama T, Otagiri M (2015) Amino acids of importance for the antioxidant activity of human serum albumin as revealed by recombinant mutants and genetic variants. Life Sci 134:36–41

Bal W, Sokolowska M, Kurowska E, Faller P (2013) Binding of transition metal ions to albumin: sites, affinities and rates. Biochim Biophys Acta 1830:5444–5455

Bern M, Knudsen Sand KM, Nilsen J, Sandlie I, Andersen JT (2015) The role of albumin receptors in regulation of albumin homeostasis: implications for drug delivery. J Control Release 211:144–162

Bhattacharya AA, Grüne T, Curry S (2000) Crystallographic analysis reveals common modes of binding of medium and long-chain fatty acids to human serum albumin. J Mol Biol 303:721–732

Bienk K, Dagnæs-Hansen F, Wengel J, Kragh-Hansen U, Howard KA (2015) Albumin-mediated protection, reduced immunogenicity and extended circulatory half-life of cholesterol modified small interfering RNA. Presented at the 2015 Annual Meeting of Controlled Release Society, Edinburgh, Scotland, July 26–29. Abstract 844

Blindauer CA, Harvey I, Bunyan KE, Stewart AJ, Sleep D, Harrison DJ, Berezenko S, Sadler PJ (2009) Structure, properties and engineering of the major zinc binding site on human albumin. J Biol Chem 284:23116–23124

Bujacz A (2012) Structures of bovine, equine and leporine serum albumin. Acta Cryst D68:1278–1289

Cai C, Zhou K, Wu Y, Wu L (2006) Enhanced liver targeting of 5-fluorouracil using galactosylated human serum albumin as a carrier molecule. J Drug Target 14:55–61

Chen Z, He Y, Shi B, Yang D (2013) Human serum albumin from recombinant DNA technology: challenges and strategies. Biochim Biophys Acta 1830:5515–5525

Chuang VTG, Kragh-Hansen U, Otagiri M (2002) Pharmaceutical strategies utilizing recombinant human serum albumin. Pharm Res 19:569–577

Cohen MP (2013) Clinical, pathophysiological and structure/function consequences of modification of albumin by Amadori-glucose adducts. Biochim Biophys Acta 1830:5480–5485

Desai N, Trieu V, Damascelli B, Soon-Shiong P (2009) SPARC expression correlates with tumor response to albumin-bound Paclitaxel in head and neck cancer patients. Transl Oncol 2:59–64

Elsadek B, Kratz F (2012) Impact of albumin on drug delivery – new applications on the horizon. J Control Release 157:4–28

Elzoghby AO, Samy WM, Elgindy NA (2012) Albumin-based nanoparticles as potential controlled release drug delivery systems. J Control Release 157:168–182

Evans TW (2002) Review article: albumin as a drug – biological effects of albumin unrelated to oncotic pressure. Aliment Pharmacol Ther 16:6–11

Fanali G, Trezza V, Marino M, Fasano M, Ascenzi P (2012) Human serum albumin: from bench to bedside. Mol Asp Med 33:209–290

Ferrer ML, Duchowicz R, Carrasco B, de la Torre JG, Acuna AU (2001) The conformation of serum albumin in solution: a combined phosphorescence depolarization-hydrodynamic modelling study. Biophys J 80:2422–2430

Fiume L, Di Stefano G (2010) Lactosaminated human albumin, a hepatotropic carrier of drugs. Eur J Pharm Sci 40:253–262

Frahm GE, Smith DGS, Kane A, Lorbetskie B, Cyr TD, Girard M, Johnston MJW (2014) Determination of supplier-to-supplier and lot-to-lot variability in glycation of recombinant human serum albumin expressed in Oryza sativa. PLoS One 9:e109893

Furusyo N, Hayashi J (2013) Glycated albumin and diabetes mellitus. Biochim Biophys Acta 1830:5509–5514

Galis Z, Ghitescu L, Simionescu M (1988) Fatty acids binding to albumin increases its uptake and transcytosis by the lung capillary endothelium. Eur J Cell Biol 47:358–365

Gaze DC (2009) Ischemia modified albumin: a novel biomarker for the detection of cardiac ischemia. Drug Metab Pharmacokinet 24:333–341

Ghuman J, Zunszain PA, Petitpas I, Bhattacharya AA, Otagiri M, Curry S (2005) Structural basis of the drug-binding specificity of human serum albumin. J Mol Biol 353:38–52

He XM, Carter DC (1992) Atomic structure and chemistry of human serum albumin. Nature 358:209–215

Hein KL, Kragh-Hansen U, Morth JP, Jeppesen MD, Otzen D, Møller JV, Nissen P (2010) Crystallographic analysis reveals a unique lidocaine binding site on human serum albumin. J Struct Biol 171:353–360

Ishima Y, Chen D, Fang J, Maeda H, Minomo A, Kragh-Hansen U, Kai T, Maruyama T, Otagiri M (2012a) S-Nitrosated human serum albumin dimer is not only a novel anti-tumor drug but also a potentiator for anti-tumor drugs with augmented EPR effects. Bioconjug Chem 23:264–271

Ishima Y, Hoshino H, Shinagawa T, Watanabe K, Akaike T, Sawa T, Kragh-Hansen U, Kai T, Watanabe H, Maruyama T, Otagiri M (2012b) *S*-Guanylation of human serum albumin is a unique posttranslational modification and results in a novel class of antibacterial agents. J Pharm Sci 101:3222–3229

Ishima Y, Fang J, Kragh-Hansen U, Yin H, Liao L, Katayama N, Watanabe H, Kai T, Suenaga A, Maeda H, Otagiri M, Maruyama T (2014) Tuning of poly-S-nitrosated human serum albumin as superior antitumor nanomedicine. J Pharm Sci 103:2184–2188

Iwao Y, Anraku M, Yamasaki K, Kragh-Hansen U, Kawai K, Maruyama T, Otagiri M (2006) Oxidation of Arg-410 promotes the elimination of human serum albumin. Biochim Biophys Acta 1764:743–749

Iwao Y, Hiraike M, Kragh-Hansen U, Mera K, Noguchi T, Anraku M, Kawai K, Maruyama T, Otagiri M (2007) Changes of net charge and α-helical content affect the pharmacokinetic properties of human serum albumin. Biochim Biophys Acta 1774:1582–1590

Iwao Y, Hiraike M, Kragh-Hansen U, Kawai K, Suenaga A, Maruyama T, Otagiri M (2009) Altered chain-length and glycosylation modify the pharmacokinetics of human serum albumin. Biochim Biophys Acta 1794:634–641

Iwao Y, Ishima Y, Yamada J, Noguchi T, Kragh-Hansen U, Mera K, Honda D, Suenaga A, Maruyama T, Otagiri M (2012) Quantitative evaluation of the role of cysteine and methionine residues in the antioxidant activity of human serum albumin using recombinant mutants. IUBMB Life 64:450–454

Kinoshita R, Ishima Y, Ikeda M, Kragh-Hansen U, Fang J, Nakamura H, Chuang VTG, Tanaka R, Hit M, Kodama A, Watanabe H, Hir M, Otagiri M, Maruyama (2015) S-Nitrosated human serum albumin dimer as novel nano-EPR enhancer applied to macromolecular anti-tumor drugs such as micelles and liposomes. J Control Release 217:1–9

Kouno Y, Anraku M, Yamasaki K, Okayama Y, Iohara D, Ishima Y, Maruyama T, Kragh-Hansen U, Hirayama F, Otagiri M (2014) N-acetyl-L-methionine is a superior protectant of human serum albumin against photo-oxidation and reactive oxygen species compared to N-acetyl-L-tryptophan. Biochim Biophys Acta 1840:2806–2812

Kragh-Hansen U (1981) Molecular aspects of ligand binding to serum albumin. Pharmacol Rev 33:17–53

Kragh-Hansen U (2013) Molecular and practical aspects of the enzymatic properties of human serum albumin and of albumin-ligand complexes. Biochim Biophys Acta 1830:5535–5544

Kragh-Hansen U, Chuang VTG, Otagiri M (2002) Practical aspects of the ligand-binding and enzymatic properties of human serum albumin. Biol Pharm Bull 25:695–704

Kragh-Hansen U, Watanabe H, Nakajou K, Iwao Y, Otagiri M (2006) Chain length-dependent binding of fatty acid anions to human serum albumin studied by site-directed mutagenesis. J Mol Biol 363:702–712

Kragh-Hansen U, Minchiotti L, Galliano M, Peters T Jr (2013) Human serum albumin isoforms: genetic and molecular aspects and functional consequences. Biochim Biophys Acta 1830:5405–5417

Kragh-Hansen U, Minchiotti L, Coletta A, Bienk K, Galliano M, Schiøtt B, Iwao Y, Ishima Y, Otagiri M (2016) Mutants and molecular dockings reveal that the primary L-thyroxine binding site in human serum albumin is not the one which can cause familial dysalbuminemic hyperthyroxinemia. Biochim Biophys Acta 1860:648–660

Majorek KA, Porebski PJ, Dayal A, Zimmerman MD, Jablonska K, Stewart AJ, Chruszcz M, Minor W (2012) Structural and immunologic characterization of bovine, horse, and rabbit serum albumins. Mol Immunol 52:174–182

Mendez CM, McClain CJ, Marsano LS (2005) Albumin therapy in clinical practice. Nutr Clin Pract 20:314–320

Merlot AM, Kalinowski DS, Richardson DR (2014) Unraveling the mysteries of serum albumin –more than just a serum protein. Front Physiol 5:299, Article 299

Minchiotti L, Galliano M, Caridi G, Kragh-Hansen U, Peters T Jr (2013) Congenital analbuminaemia: molecular defects and biochemical and clinical aspects. Biochim Biophys Acta 1830:5494–5502

Minomo A, Ishima Y, Chuang VTG, Suwa Y, Kragh-Hansen U, Narisoko T, Morioka H, Maruyama T, Otagiri M (2013) Albumin domain II mutant with high bilirubin binding affinity has a great potential as serum bilirubin excretion enhancer for hyperbilirubinemia treatment. Biochim Biophys Acta 1830:2917–2923

Mitzner SR (2011) Extracorporeal liver support- albumin dialysis with the molecular adsorbent recirculating system (MARS). Ann Hepatol 10:S21–S28

Nakajou K, Watanabe H, Kragh-Hansen U, Maruyama T, Otagiri M (2003) The effect of glycation on the structure, function and biological fate of human serum albumin as revealed by recombinant mutants. Biochim Biophys Acta 1623:88–97

Neumann E, Frei E, Funk D, Becker MD, Schrenk H-H, Müller-Ladner U, Fiehn C (2010) Native albumin for targeted drug delivery. Expert Opin Drug Deliv 7:915–925

Oganesyan V, Damschroder MM, Cook KE, Li Q, Gao C, Wu H, Dall'Acqua WF (2014) Structural insights into neonatal Fc receptor-based recycling mechanisms. J Biol Chem 289:7812–7824

Olivieri JR, Craievich AF (1995) The subdomain structure of human serum albumin in solution under different pH conditions studied by small angle X-ray scattering. Eur Biophys J 24:77–84

Peters T Jr (1996) All about albumin: biochemistry, genetics and medical applications. Academic, San Diego

Quinlan GJ, Martin GS, Evans TW (2005) Albumin: biochemical properties and therapeutic potential. Hepatology 41:1211–1219

Roche M, Rondeau P, Singh NR, Tarnus E, Bourdon E (2008) The antioxidant properties of serum albumin. FEBS Lett 582:1783–1787

Rollett A, Reiter T, Nogueira P, Cardinale M, Loureiro A, Gomes A, Cavaco-Paulo A, Moreira A, Carmo AM, Guebitz GM (2012) Folic acid-functionalized human serum albumin nanocapsules for targeted drug delivery to chronically activated macrophages. Int J Pharm 427:460–466

Ryan AJ, Chung C-w, Curry S (2011a) Crystallographic analysis reveals the structural basis of the high-affinity binding of iophenoxic acid to human serum albumin. BMC Struct Biol 11:18

Ryan AJ, Ghuman J, Zunszain PA, Chung C-w, Curry S (2011b) Structural basis of binding of fluorescent, site-specific dansylated amino acids to human serum albumin. J Struct Biol 174:84–91

Sand KMK, Bern M, Nilsen J, Noordzij HT, Sandlie I, Andersen JT (2015) Unraveling the interaction between FcRn and albumin: opportunities for design of albumin-based therapeutics. Front Immunol 5:682, Article 682

Simard JR, Zunszain PA, Hamilton JA, Curry S (2006) Location of high and low affinity fatty acid binding sites on human serum albumin revealed by NMR drug-competition analysis. J Mol Biol 361:336–351

Sleep D (2014) Albumin and its application in drug delivery. Expert Opin Drug Deliv 12:793–812

Spinella R, Sawhney R, Jalan R (2016) Albumin in chronic liver disease: structure, functions and therapeutic implications. Hepatol Int 10:124–132

Sudlow G, Birkett DJ, Wade DN (1975) The characterization of two specific drug binding sites on human serum albumin. Mol Pharmacol 11:824–832

Sugio S, Kashima A, Mochizuki S, Noda M, Kobayashi K (1999) Crystal structure of human serum albumin at 2.5 Å resolution. Protein Eng 12:827–835

Taguchi K, Yamasaki K, Seo H, Otagiri M (2015) Potential use of biological proteins for liver failure therapy. Pharmaceutics 7:255–274

The Albumin Website: http://albumin.org. Accessed Sept 2015

Wang Z-m, Ho JX, Ruble JR, Rose J, Rüker F, Ellenburg M, Murphy R, Click J, Soistman E, Wilkerson L, Carter DC (2013) Structural studies of several clinically important oncology drugs in complex with human serum albumin. Biochim Biophys Acta 1830:5356–5374

Yamasaki K, Maruyama T, Kragh-Hansen U, Otagiri M (1996) Characterization of site I on human serum albumin: concept about the structure of a drug binding site. Biochim Biophys Acta 1295:147–157

Yamasaki K, Chuang VTG, Maruyama T, Otagiri M (2013) Albumin-drug interaction and its clinical implication. Biochim Biophys Acta 1830:5435–5443

Zunszain PA, Ghuman J, McDonagh AF, Curry S (2008) Crystallographic analysis of human serum albumin complexed with 4Z,15E-bilirubin-IXα. J Mol Biol 381:394–406

Chapter 2
Stability of Albumin and Stabilization of Albumin Preparations

Keishi Yamasaki and Makoto Anraku

Abstract Human serum albumin (HSA) is structurally stabilized by 17 disulfide bonds, and interactions between domains (or subdomains) of HSA also contribute to its stability. The effects of several other factors on the stability of HSA and pharmaceutical preparations that contain HSA have been widely investigated. When HSA is heated and in the presence of chemical denaturants, unfolding occurs through multiple steps, and the genetic variations of HSA and ligand binding to HSA have been shown to contribute to protecting HSA against such irreversible structural changes.

HSA is therapeutically used for treating shock, burns, hypoalbuminemia, surgery, trauma, cardiopulmonary bypass, acute respiratory distress, or hemodialysis, and its pharmaceutical preparations are supplied in the form of sterilized aqueous solutions. These solutions contain sodium octanoate and *N*-acetyl-L-tryptophanate to prevent the irreversible denaturation of HSA that occurs during pasteurization by heating at 60 °C for 10 h. Sodium octanoate has the greatest stabilizing effect against heat, whereas the presence of *N*-acetyl-L-tryptophanate diminishes the oxidation of HSA. Recently, *N*-acetyl-methioninate, as a new stabilizer, has been suggested to be superior to *N*-acetyl-L-tryptophanate with respect to scavenging reactive oxygen species and in protecting the protein against oxidation.

Nanoparticle and fusion proteins which contain HSA as a carrier for drug delivery are being actively developed. In addition to the use of sodium octanoate and *N*-acetyl-L-tryptophanate as a stabilizer for nanoparticles of paclitaxel, to stabilize HSA fusion proteins, ligand binding to HSA domains, genetic mutations of HSA domains, and lyophilization using sugars or surfactants are used.

Keywords Human serum albumin • Thermal stability • Chemical stability • Storage stability • Stabilizer

---

K. Yamasaki (✉) • M. Anraku
Faculty of Pharmaceutical Sciences, Sojo University,
4-22-1 Ikeda, Nishi-Ku, Kumamoto-shi 860-0082, Kumamoto, Japan
e-mail: kcyama@ph.sojo-u.ac.jp; anraku@ph.sojo-u.ac.jp

M. Otagiri, V.T.G. Chuang (eds.), *Albumin in Medicine*,
DOI 10.1007/978-981-10-2116-9_2

2.1 Introduction

Human serum albumin (HSA) is a major protein component of blood plasma and plays an important role in the regulation of colloidal osmotic pressure, the antioxidant capacity of human plasma, and the transport of numerous endogenous compounds such as fatty acids, hormones, toxic metabolites (eg, bilirubin), bile acids, amino acids, and metals (Peters 1995; Kragh-Hansen et al. 2002). Crystallographic data show that HSA contains three α-helical domains that are structurally similar: I (residues 1–195), II (196–383), and III (384–585), which are further divided into subdomains A and B (Carter and Ho 1994; Sugio et al. 1999; Curry et al. 1998). HSA contains 35 cysteine residues, and all of these except one, ^{34}Cys (in domain I), are involved in the formation of 17 disulfide bonds that serve to stabilize HSA (Fig. 2.1). The data also show that interdomain and intersubdomain interactions contribute significantly to the stability of the HSA molecule. HSA has been used as a therapeutic agent, and attempts to utilize it as a pharmaceutical additive have also been made. In all cases, the stability of HSA is an important aspect that should be considered especially at the stages of production and storage. The effects of heat, chemicals, genetic variation, and several other factors on the stability of HSA and its pharmaceutical preparations have been widely investigated.

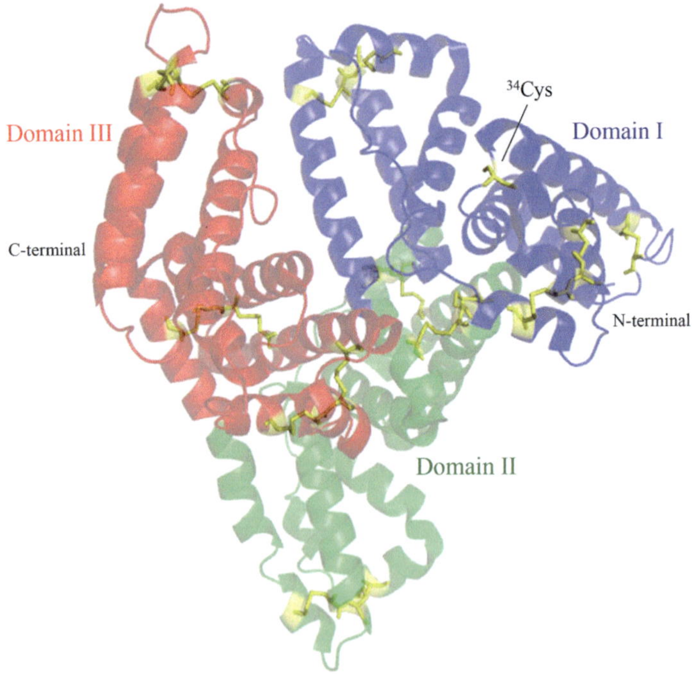

Fig. 2.1 Crystal structure of rHSA, showing the location of domains (I–III), and 35 cysteine residues (*yellow*). The illustration was made with PyMol on the basis of the atomic coordinates 1E78 available at the Brookhaven Protein Data Bank

2.2 Thermal Stability of HSA

Proteins are known to be denatured by a change in physical environment such as heating (Fersht 1999). The conformational responses of albumin against heating have also been studied in attempts to understand its structural features and stability.

2.2.1 *Fundamental Properties of HSA Structural Responses Against Heating*

An increase in temperature causes changes in the structure of HSA and intermolecular interactions, which can affect the thermal stability of HSA. The α-helical structure content in HSA which was estimated to be 66 % at 25 °C, based on circular dichroism (CD) spectra, decreased to 53 % at 65 °C (Moriyama and Takeda 2005). The 13 % loss in α-helical structure is not recovered after recooling to 25 °C.

Flora et al. investigated the effect of temperature on the structure of acrylodan-labeled HSA by steady-state and time-resolved fluorescence methods (Flora et al. 1998). Increasing the temperature to about 50 °C results in the reversible separation of domains I and II, as evidenced by fluorescent techniques. They also reported that heating to < 70 °C resulted in the irreversible unfolding of domain II, while increasing the temperature to 70 °C or higher resulted in the irreversible unfolding of domain I. It was suggested that this irreversible unfolding is accompanied by the unfolding of the pocket containing the free sulfhydryl group of 34Cys, which enables the formation of aggregates through disulfide bridges. Thus, they finally demonstrated that the unfolding of HSA that occurs at elevated temperatures involved multiple steps as shown in the following scheme:

$$N \leftrightarrow E \rightarrow I \rightarrow U, \tag{1}$$

where *N* is the native form of the protein, *E* is the expanded form, *I* is an intermediate where domain II is unfolded while domain I is intact, and *U* is the unfolded protein.

2.2.2 *Effect of Genetic Variation*

The effect of genetic variation on the thermal stability of HSA was studied by CD and differential scanning calorimetry (DSC) (Kragh-Hansen et al. 2005). The differences in Tm (midpoint of denaturation) and ΔHv (van't Hoff enthalpy) values from normal HSA (ΔTm and Δ(ΔHv)) were investigated. For the 33 generic variants examined, no clear relationship was found between thermal stability and the type of substitution, changes in protein charge, or hydrophobicity. The Tm of eight variants changed more than 5 °C as compared to normal HSA, where a positive ΔTm for three variants (Malmö, Maku, proalbumin Blenheim) and a negative ΔTm for five variants (Canterbury, Trieste, Verona, Venezia, Kénitra) were found (Table 2.1). Furthermore, variants mutated in domain I have a uniform effect (positive ΔTm and a negative Δ(ΔHv)), meaning that they denature more easily than normal HSA but do so at a higher temperature. They concluded that domain I is the most thermally unstable domain. Kosa et al. also suggested the importance of domain I on the stability of albumin based on results for the thermally induced denaturation of serum albumin from five mammalian species (human, bovine, dog, rabbit, and rat) (Kosa et al. 1998).

Watanabe et al. studied the thermal stabilities of residues in recombinant HSA domain II mutants (K199A, W214A, R218H, and H242Q) using DSC and CD techniques. Their results showed a minor increase in thermal stability for R218H and H242Q, a small decrease in stability for K199A, and a larger decrease in stability for W214A (Table 2.2) (Watanabe et al. 2001a). 214Trp is important for holding together the two halves of the heart-shaped albumin molecule, because this residue is an element in a major interdomain cluster of hydrophobic residues (Carter et al. 1994). Thus, 214Trp in domain II is thought to be an important amino acid residue that contributes to thermal stability as well as domain I. Furthermore, these data suggest that mutations of 218Arg and 242His in domain II are crucial in designing albumins with greater thermal stability. They also found only marginal changes in the thermodynamic parameters for the thermal denaturation of residues in mutants of domain III (R410A, Y411A, R410A/Y411A, Y411S, Y411F) (Table 2.2) (Watanabe et al. 2001a). However, a mutation in domain III also affects the thermal stability of the molecule, as observed in some variants, Ortonovo, Maku, Verona, Milano Fast, Catania, Venezia, Bazzano, or Kénitra. Thus, the contribution of domain III to thermal stability also should not be excluded.

2.2.3 *Effect of Ligand Binding*

Effects of the binding of ligands (fatty acids and some stabilizers) on the thermal stability of HSA have also been studied by many researchers. Ross and Shrake compared the stability of defatted and non-defatted HSA (Ross and Shrake 1988). They showed that the stability of defatted HSA (reflected by its denaturation temperature, Td), monitored from 50 to 80 °C by DSC, decreased with increasing HSA concentration from 1.43 to 74.1 mg/mL. This was shown to be compatible with a simple model in which the reversible polymerization of the denatured HSA monomer promotes unfolding. The denaturation of the non-defatted HSA monomer, which is subsaturated with a high-affinity endogenous long-chain fatty acid (LCFA), was a biphasic process. The Td for the first endotherm, associated with the denaturation of the LCFA-poor species, decreased with increasing protein concentration, similar to that for the defatted monomer, whereas the Td for the second endotherm, which is associated with the denaturation of LCFA-rich species, was found to be independent of concentration. The magnitude of the concentration

Table 2.1 Thermodynamic parameters for thermal denaturation of albumin variants, proalbumin variants, albumins modified at the C-terminal end, and glycosylated variants (Kragh-Hansen et al. 2005)

Variant	Mutation	ΔT_m (K)	$\Delta(\Delta H_v)$ (kJ/mol)	Variant	Mutation	ΔT_m (K)	$\Delta(\Delta H_v)$ (kJ/mol)
Blenheim	1 Asp→Val	1.94	−132.61	proAlb Lille	−2 Arg→His	4.64	−13.82
Malmö-95	63 Asp→Asn	6.07	−163.15	proAlb Blenheim	1 Asp→Val	7.10	−118.49
Vibo Valentia	82 Glu→Lys	2.03	−35.48	Arg-Alb	Albumin with −1 Arg	0.23	12.65
Tregasio	122 Val→Glu	0.57	26.85				
Hawkes Bay	177 Cys→Phe	−1.59	−17.66	Catania	580–582: substituted	0.13	−57.27
Tradate-2	225 Lys→Glu	−4.86	44.13		579–585: deleted		
Herborn	240 Lys→Glu	−2.74	−71.99	Venezia	572–578: substituted	−5.74	99.43
Niigata	269 Asp→Gly	3.67	−79.90		579–585: deleted		
Caserta	276 Lys→Asn	4.87	13.42	Bazzano	567–582: substituted	4.67	−8.54
Canterbury	313 Lys→Asn	−7.16	6.84		583–585: deleted		
Brest	314 Asp→Val	−0.38	24.09	Kénitra	575–585: substituted	−5.30	15.72
Roma	321 Glu→Lys	1.42	28.98		extended with 586–601		
Sondrio	333 Glu→Lys	−2.56	−21.89				
Trieste	359 Lys→Asn	−6.56	−13.91	Malmö-95	63 Asp→Asn	4.06	−10.41
Parklands	365 Asp→His	0.89	58.06		glycosylated at 63 Asn		
Milano Slow	375 Asp→His	−0.09	−94.33				
Kashmir	501 Glu→Lys	0.13	−1.52	Redhill	−1 Arg retained	1.93	−9.52
Ortonovo	505 Glu→Lys	1.87	−83.36		320 Ala→Thr		
Maku	541 Lys→Glu	6.12	−58.32		glycosylated at 318 Asn		
Church Bay	560 Lys→Glu	0.70	15.23				

(continued)

Table 2.1 (continued)

Variant	Mutation	ΔT_m (K)	$\Delta(\Delta H_v)$ (kJ/mol)	Variant	Mutation	ΔT_m (K)	$\Delta(\Delta H_v)$ (kJ/mol)
Paris-2	563 Asp→Asn	4.17	−154.35	Casebrook	494 Asp→Asn	−1.11	54.36
Verona	570 Glu→Lys	−6.53	53.92		glycosylated at 494 Asn		
Milano Fast	573 Lys→Glu	2.08	41.45				

Table 2.2 Thermodynamic parameters for thermal denaturation of recombinant HSA mutants (Watanabe et al. 2001a)

Recombinant Domain I mutants	Mutation	ΔT_m (K)	$\Delta(\Delta H_v)$ (kJ/mol)	Recombinant Domain III mutants	Mutation	ΔT_m (K)	$\Delta(\Delta H_v)$ (kJ/mol)
K199A	199 Lys→Ala	−1.84	−53.5	R410A	410 Arg→Ala	−0.5	−8.0
W214A	214 Trp→Ala	−3.0	−178.7	Y411A	411 Tyr→Ala	−0.8	−13.8
R218H	218 Arg→His	1.6	68.0	R410A/Y411A	410 Arg→Ala	−0.9	−18.2
H242Q	242 His→Gln	1.5	45.0		411 Tyr→Ala		
				Y411S	411 Tyr→Ser	0.3	3.9
				Y411F	411 Tyr→Phe	−0.7	−10.8

dependence of Td is directly related to the extent of polymerization of denatured monomer, which decreases with increasing level of bound ligand.

Anraku et al. showed that, after preheating HSA for 30 min at 60 °C, followed by cooling, it was impossible to obtain an ordinary DSC thermogram (Anraku et al. 2004). This can be attributed to the irreversible denaturation that occurs during preheating. The presence of sodium octanoate (Oct) and *N*-acetyl-L-tryptophanate (*N*-AcTrp) prevented the irreversible protein denaturation that occurred during preheating. The high-affinity binding of Oct also has a greater stabilizing effect against heat as indicated by the obvious increase in Td and calorimetric enthalpy, while N-AcTrp, which is mainly present in the unbound form, diminishes the oxidation of HSA. Furthermore, *N*-acetyl-methioninate (*N*-AcMet) was proposed as a substitute for *N*-AcTrp which has possible side effects in intracerebral disease (Anraku et al. 2007; Kouno et al. 2014).

2.3 Chemical Stability of Human Serum Albumin

A protein can be denatured by changing its chemical environment. The most common methods involve adding a chemical denaturant, such as guanidinium hydrochloride (GdnHCl) or urea (Fersht 1999). These compounds are strong denaturants that act by disrupting hydrogen bonding and thereby causing many proteins to adopt a highly unfolded and less compact conformation in solution.

2.3.1 Fundamental Properties of HSA Structural Responses Against Chemical Denaturant

Flora et al. investigated the effect of GdnHCl-induced denaturation on the spatial relationship between 214Trp of domain II and 34Cys of domain I, labeled with acrylodan by CD and fluorescence spectroscopy (Flora et al. 1998). They demonstrated that denaturation of HSA by GdnHCl occurs via a pathway involving at least the following three distinct steps: (1) initial reversible separation of domains I and II at GdnHCl concentrations less than 1.0 M, (2) the irreversible unfolding of domain II at 2.0 M GdnHCl, and (3) the irreversible unfolding of domain I at higher concentrations of GdnHCl. Other reports have proposed contradictory opening sequences of the domains: a sequence of domain III → II → I and the presence of a molten globule-like state of domain III around 1.8 M GdnHCl (Ahmad et al. 2005) and a sequence involving domain II → I → III (Santra et al. 2005).

González-Jiménez and Cortijo, using fluorescence and CD measurements, concluded that HSA is denatured using urea in a single two-state transition with a midpoint at about 6 M urea due to the unfolding of domain II (Gonzalez-Jimenez and Cortijo 2002). Even at a urea concentration of 8 M, some residual structure remains in domain I, where denaturation only takes place when a stronger denaturant, GdnHCl, is added. Different conclusions for the urea-induced denaturation were obtained by Muzammil et al. (2000). They suggested that domains I and/or III are involved in the formation of an intermediate at a urea concentration of 2.0–4.6 M, followed by denaturation at higher concentrations of urea, whereas no changes occurred in domain II at urea concentrations of 0.0–5.4 M. Furthermore, the urea-induced transition underwent a single-step cooperative transition in the presence of anions, indicating that anions were chiefly responsible for stabilizing HSA.

2.3.2 Effect of Genetic Variation

Watanabe et al. compared the stability of single and double residue mutants at 410Arg and 411Tyr (R410A, Y411A, Y411S, Y411F, R410A/Y411A) against GndHCl-induced denaturation of rHSA (Watanabe et al. 2000). No obvious differences between the mutants and rHSA were observed, suggesting that a limited substitution of one or two amino acid residues in domain III (subdomain IIIA) has no effect on the stability of the molecule. Kosa et al. compared the effects of GdnHCl-induced denaturation between mammalian (human, bovine, dog, rabbit, and rat) albumins using fluorescence and CD spectroscopies (Kosa et al. 1998). Dog albumin, which possesses a different homology in terms of the amino acid sequences in domain I, showed the highest stability against GdnHCl-induced denaturation. Thus, in addition to Cys residues that are involved in formation of disulfide bonds, several other amino acid residues located in domain I might also contribute to the stability of the molecule (Kosa et al. 1998; Kragh-Hansen et al. 2005).

2.3.3 Effect of Ligand Binding

Effects of ligand binding (fatty acids and some stabilizers) on the chemical denaturation of HSA have been also investigated. Recently, detailed investigations of the chemical denaturation of fatted and defatted HSA were carried out using small-angle X-ray scattering, dynamic light scattering, fluorescence, and CD (Galantini et al. 2008; Leggio et al. 2009). Multistep unfolding pathways involving two intermediates were proposed for both the GdnHCl- and urea-induced denaturation of defatted HSA. The opening sequences of the domains were evidenced as III → II → I and I → II → III for the GdnHCl- and urea-induced denaturation, respectively (Figs. 2.2a and 2.3a). Fatty acids have a somewhat protective effect on GdnHCl-induced denaturation (Fig. 2.2b), whereas urea-induced denaturation is strongly inhibited by the presence of a fatty acid (palmitic acid) (Fig. 2.3b), which remains in the native form up to high urea concentrations. In this case, the unfolding process of fatty acid-containing HSA by urea can be characterized by a single-step mechanism.

The unfolding of native HSA and recombinant wild-type HSA (rHSA) by GdnHCl was followed by changes in their far-UV CD spectra (Watanabe et al. 2001b). The transition curve for both albumins exhibited an apparent two-state denaturation behavior. Oct with *N*-AcTrp had a greater stabilizing effect against the GdnHCl-induced denaturation of both HSA than the other fatty acids examined (palmitate and oleate). The stability of HSA against urea-induced denaturation is enhanced upon the binding of ligands such as ibuprofen, diazepam, ketoprofen, and sodium dodecyl sulfate (SDS) (Galantini et al. 2010; Manoharan et al. 2015; Duggan and Luck 1948).

The binding of surfactants such as SDS is known to provide marginal structural rigidity to the native state of HSA at low SDS concentrations (Anand et al. 2015). In contrast, higher concentrations of SDS cause the unfolding of HSA. Anand et al. investigated the effect of β-cyclodextrin (β-CyD) on the unfolding process of bovine serum albumin (BSA) in which SDS causes a structural change in a similar manner to HSA. Although β-CyD itself induced a marginal structural loss of BSA, it contributed to the recovery of the structure of unfolded BSA by SDS (Anand and

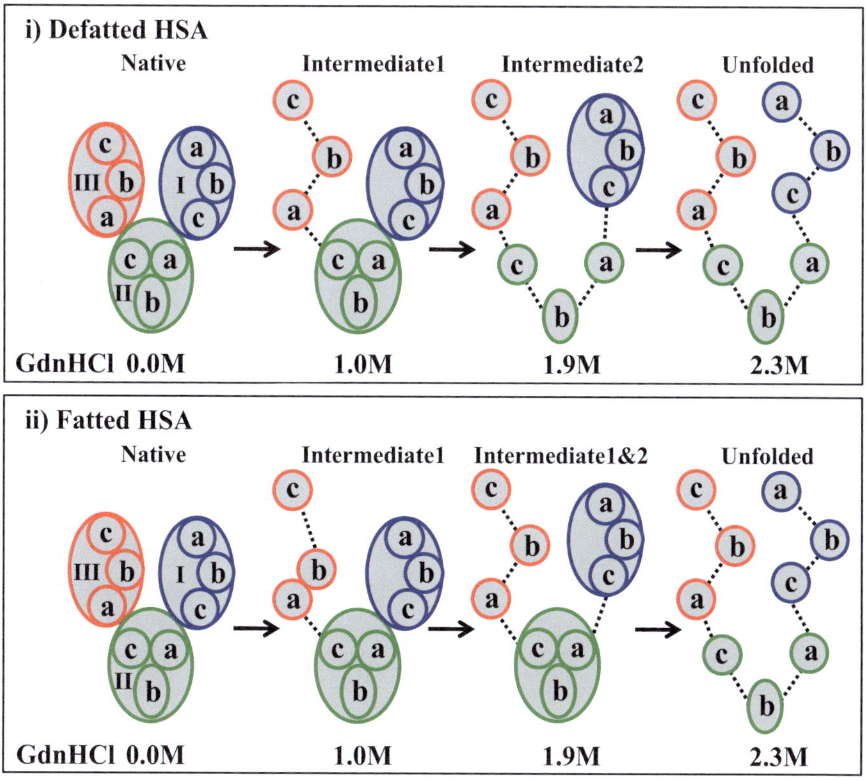

Fig. 2.2 Summarizing scheme of GdnHCl-induced denaturation process for defatted (**i**) and fatted HSA (**ii**) "*a*," "*b*," and "*c*": three loops for each domain (Galantini et al. 2008)

Mukherjee 2013). Interestingly, the effect of β-CyD on protein structure was not observed for BSA that was denatured by GdnHCl. The refolding of BSA by β-CyD is attributed to the inclusion or removal of SDS molecules from BSA (Anand and Mukherjee 2013). Thus, endogenous and exogenous compounds (eg, fatty acids, drugs, surfactants, and cyclodextrins) may stabilize HSA against denaturation.

2.4 Stabilization of Marketed Albumin

HSA has been used as therapeutic agent for treating shock, burns, hypoalbuminemia, surgery or trauma, cardiopulmonary bypass, acute respiratory distress, and hemodialysis (Peters 1995; Mendez et al. 2005). HSA products in the USA and Japan are supplied in the form of sterilized aqueous solutions at concentrations of 5, 20, or 25% (Table 2.3). In Japan, most HSA solutions are produced using domestically donated blood. The commercial purification of HSA from donated human blood is carried out based on the Cohn fractionation method (Cohn et al.

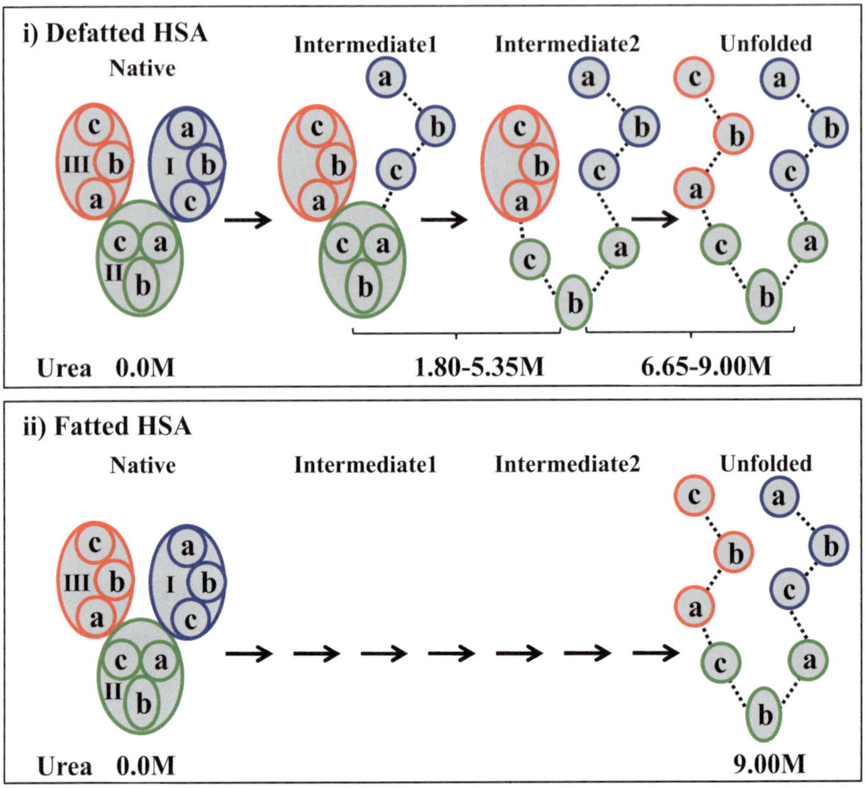

Fig. 2.3 Summarizing scheme of urea-induced denaturation process for defatted (**i**) and fatted HSA (**ii**) "*a*," "*b*," and "*c*": three loops for each domain (Leggio et al. 2009)

1946). By controlling the pH, temperature, and cold ethanol content, the precipitate obtained after the final step is Fraction V which is comprised of more than 96 % HSA. After removing the ethanol, the pH and ionic strength are adjusted for commercial use. For the production of HSA, a final pasteurization of HSA is carried out to prevent the risk of transmitting pathogenic viruses, such as those causing hepatitis or HIV. This process involves heating at 60 °C for 10 h, therefore Oct (0.08 mmol/g HSA) and *N*-AcTrp (0.08 mmol/g HSA) are used to maintain the stability of HSA (Bertolini et al. 2012). To avoid the potential spread of blood pathogens, an alternative to blood-derived HSA, recombinant HSA (rHSA), has been successfully produced using *Pichia pastoris*, *Saccharomyces cerevisiae*, or *Oryza sativa*. Solution of rHSA produced by *Pichia pastoris* is licensed as Medway® in Japan for the treatment of hypoalbuminemia and hemorrhagic shock. This solution also includes Oct and *N*-AcTrp as stabilizers.

Highly pure commercial HSA solutions (almost 100 %) can be achieved with the current production system. The shelf life of most HSA solutions is 3 years at 37 °C or 5 years under refrigeration in glass vials. Although it seems to be unnecessary to

Table 2.3 Marketed albumin solution in the USA and Japan

Country	Origin	Brand name	Company	Contents (%)	Volume (mL)	Container	Stabilizers
US	Donated human blood	Albumin	Octapharma	5 %	100, 250, 500 mL	Infusion bottle	OCT(0.08 mmol/g HSA) and *N*-AcTrp (0.08 mmol/g HSA)
				25 %	50, 100 mL		
		Albuked™	Kedrion	5 %	50, 250 mL	Glass vial	
				25 %	20, 50,100 mL		
		Plasbumin	Grifols	5 %	50, 250, 500 mL	Glass vial	
					20, 50,100 mL		
		AlbuRx®	CSL Behring	5 %	250, 500 mL	Glass vial	
				25 %	50, 100 mL		
		Albutein®	Grifols	5 %	250, 500 mL	Glass vial	
				25 %	50, 100 mL		
		Buminate	Baxalta	5 %	250, 500 mL	Glass bottle	
				25 %	20, 50, 100 mL		
		Flexbumin	Baxalta	5 %	250 mL	Plastic container	
				25 %	50, 100 mL		
Japan	Domestically donated human blood	Kenketsu-Kaketsuken	Kaketsuken	20 %	20, 50 mL	Glass vial	
				25 %	50 mL		
		Kenketsu-Benesis	Japan Blood Products Org.	5 %	100, 250 mL		
				25 %	20, 50 mL		
		Kenketsu-Sekijuji	Japan Blood Products Org.	20 %	20, 50 mL		
				25 %	50 mL		
		Kenketsu-Nichiyaku	Nihon Pharmaceutical	5 %	250 mL		
				20 %	20, 50 mL		
				25 %	50 mL		
	Donated human blood	Albuminar	CSL Behring	5 %	250 mL		
				20 %	50 mL		
				25 %	50 mL		
	Recombinant	Medway®	Mitsubishi Tanabe	25 %	50 mL		

improve the stability of HSA under production and storage, there are some issues yet to be settled. For example, the mechanism responsible for stabilization by Oct and *N*-AcTrp is not fully understood. Furthermore, *N*-AcTrp has a possible side effect of intracerebral disease (Aguilera et al. 2001), and alternative stabilizers are required for ensuring safety as well as stability.

2.4.1 Stabilizing Mechanisms in Marketed Albumin

Since the only source of HSA for clinical application is donated human blood, the risks of transmitting pathogenic viruses, such as those causing hepatitis, HIV, and some other unidentified diseases, exists. Pasteurization of HSA is carried out by heating at 60 °C for several hours with Oct and *N*-AcTrp as commonly used stabilizers (Shrake et al. 1984), a process that usually destroys the viruses present. These commonly used additives effectively protect HSA by increasing the melting temperature as determined by DSC and by decreasing the formation of aggregates after heating (Arakawa and Kita 2000a, b). Also, Oct is probably five- to tenfold more effective than *N*-AcTrp in reducing aggregation during the thermal unfolding of albumin (Arakawa and Kita 2000a, b). Thus, the mechanism of stabilization of *N*-AcTrp is not fully elucidated. We found that *N*-AcTrp was much better than Oct to protect HSA against general oxidation caused by prolonged exposure to reactive oxygen species (ROS) (Anraku et al. 2004). In contrast to *N*-acetyl-L-cysteinate (*N*-AcCys), L-tryptophanate (L-Trp) also had this effect. The protection brought about by *N*-AcTrp was pronounced and more immediate in the case of the sulfhydryl group of HSA. This protective role of *N*-AcTrp can be of practical importance, because the 34Cys of HSA represents the largest fraction of free thiols in blood (Kragh-Hansen et al. 2002). In albumin products, Oct has the greatest stabilizing effect against heat, whereas the presence of *N*-AcTrp diminishes oxidation of HSA.

2.4.2 Important Amino Acids for the Antioxidant Activity of Marketed Albumin

N-AcTrp has a possible side effect of causing intracerebral disease (Aguilera et al. 2001). In Trp metabolism, 3-hydroxykynurenine is known to have particularly strong neurotoxic properties (Topczewska-Bruns et al. 2003), and the accumulation of Trp metabolites in the nervous tissue due to HSA product administration may be involved in the pathogenesis of several neurological disorders in uremia. To provide safe and risk-free albumin preparations, it is important to find new stabilizing reagents in place of *N*-AcTrp.

All amino acid residues of proteins are susceptible to oxidative modification by one or more forms of ROS (Vogt 1995; Davies 2005). The oxidative modifications of sulfur-containing amino acids such as Cys and Met could serve as antioxidants via their cyclic oxidation and reduction. Among the 585 amino acid residues in HSA, Bourdon et al. reported that Met and Cys accounted for 40–80 % of the total antioxidant activity of HSA. They concluded that Cys functions largely as a free radical scavenger, whereas Met mainly functions as a metal chelator (Bourdon et al. 2005). Carballal et al. also concluded that thiol groups of albumin may well be an important scavenger of ROS (Carballal et al. 2003). We also quantitatively evaluated the role of Cys and Mets in the antioxidative activity of HSA using recombinant mutants, in which ^{34}Cys and/or the six Met residues had been mutated to alanine using site-directed mutagenesis (Iwao et al. 2012). Our results showed that the level of contribution of ^{34}Cys and the Met residues to the anti-oxidative activity of HSA were 61 % and 29 % against $O_2^{\bullet -}$, 68 % and 61 % against H_2O_2, 38 % and 6 % against $HO^\bullet$, 36 % and 13 % against HOCl, and 51 % and 1 % against $^\bullet$NO, respectively.

Recently, Miyamura et al., using ESI-TOFMS analysis, reported that the fraction of ^{34}Cys-cysteinylated HSA (Cys-^{34}Cys-HSA) is different between commercially available HSA solutions (Miyamura et al. 2016). Furthermore, they showed that the antioxidative and ligand-binding activities of marketed HSA solutions were significantly negatively correlated with the Cys-^{34}Cys-HSA fraction. Thus, free ^{34}Cys is important for maintaining the antioxidative function of HSA solutions. Furthermore, in our recent study, the substitution of a glutamic acid at positions 122 or 541, but not at 240 or 560, improved the antioxidant effect, perhaps by allowing the Met residues in their vicinity, ^{123}Met and ^{548}Met, respectively, more accessible for the oxidant (Fig. 2.4) (Anraku et al. 2015). On the other hand, no effect on oxidation was found for *N*-AcCys in our previous studies, although Cys residues of proteins are susceptible to oxidation (Anraku et al. 2004). Thus, a sulfur-containing amino acid having mercapto groups, *N*-AcMet could be the most suitable antioxidant for use as a stabilizer in albumin products.

2.4.3 Role of N-Acetyl-Methioninate as a New Stabilizer for Marketed Albumin

To provide safe and risk-free albumin products, we validated *N*-AcMet as a new stabilizer for albumin products. As a result, we suggested that the use of *N*-AcMet in albumin preparations was more safe and free of the risk of side effects than the use of *N*-AcTrp (Anraku et al. 2007). We also investigated the roles of Oct, *N*-AcTrp, and *N*-AcMet for the photostability of albumin. Interestingly, we found that *N*-AcMet is superior to *N*-AcTrp with respect to scavenging ROS and for protecting the protein against oxidation. Furthermore, *N*-AcMet, with or without Oct, has a good stabilizing effect on the structure of albumin. By contrast, *N*-AcTrp promotes photooxidative degradation of the protein (Fig. 2.5) (Kouno et al. 2014). Thus, *N*-AcMet should be useful as a new stabilizer and antioxidant for albumin preparations. HSA is isolated by fractionating human plasma, which exposes it to

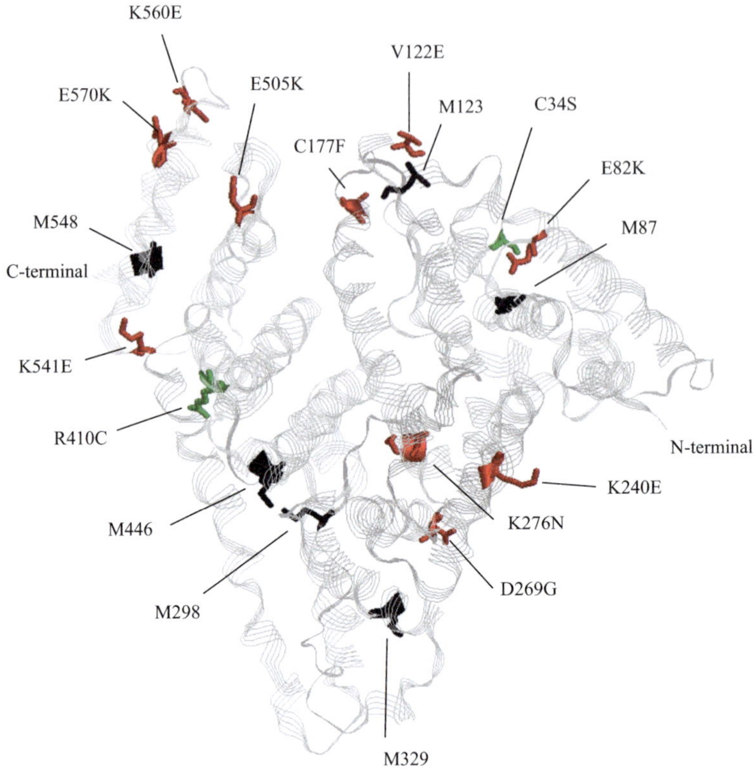

Fig. 2.4 The crystal structure of HSA showing the locations of the six methionine residues (*black*) and the amino acids substituted in the two recombinant mutants (*green*) and the ten genetic variants (*red*) used in this study (Anraku et al. 2015). The illustration was made with RasMol on the basis of the atomic coordinates 1E78 available at the Brookhaven Protein Data Bank

possible contamination by viruses or prions. As an alternative to blood-derived HSA, recombinant HSA (rHSA) has been successfully produced using yeast cells. rHSA produced by the methylotrophic yeast *Pichia pastoris* is identical to blood-derived HSA in neo-antigenicity (Kobayashi 2006). The preparation has been shown to have comparable safety, tolerability, pharmacokinetics, and pharmacodynamics. In the near future, highly stable and inexpensive rHSA stabilized by *N*-AcMet may be available for clinical applications.

2.5 Stability and Stabilization of Novel Pharmaceutical Products That Contain HSA

HSA is also used as excipient for certain types of pharmaceutical products to protect the drug from degradation or from being adsorbed by the container and to be a carrier of drugs. Recent research and development of pharmaceutical products that

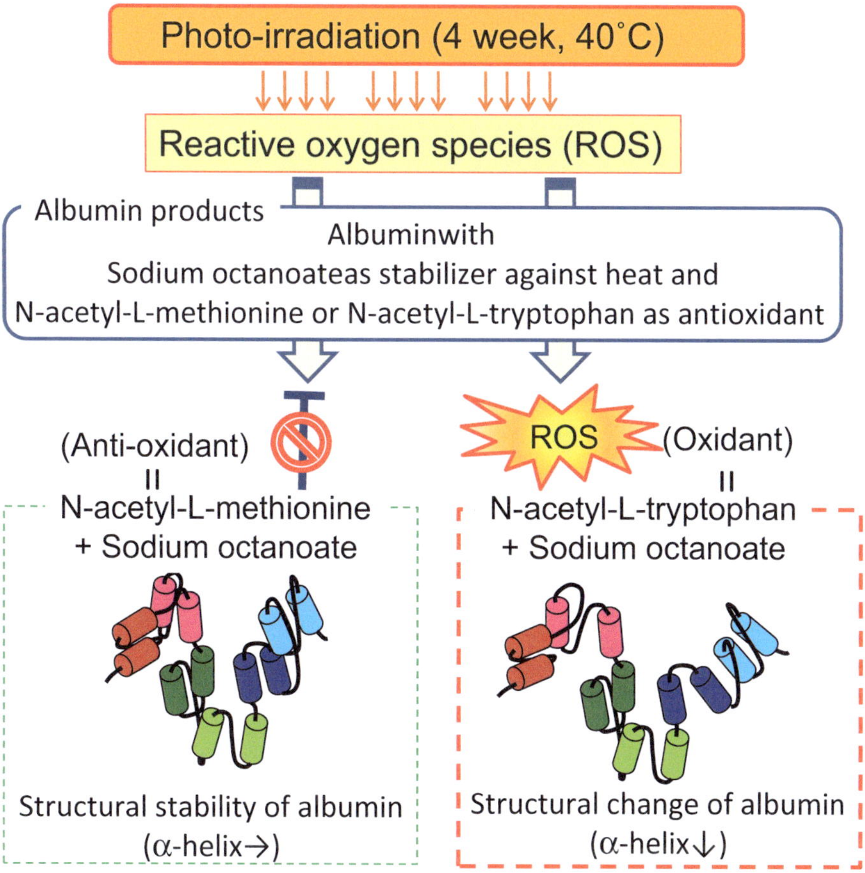

Fig. 2.5 Protective mechanism of N-acetyl-L-methionine against photooxidation of albumin (Kouno et al. 2014)

contain HSA as a carrier of drugs are currently being actively pursued (Table 2.4). These products also should be formulated so that they are stable during their production and storage until used in the clinic.

2.5.1 Nab-Paclitaxel

Nanoparticle albumin-bound (nab) paclitaxel has been evaluated clinically for tumor treatment (Kratz 2008; Hawkins et al. 2008). Abraxane® which is nab-paclitaxel was approved by the FDA in 2005 for the treatment of metastatic breast cancer (Desai et al. 2006). Paclitaxel is mixed with HSA in an aqueous solution, and the mixture is then passed through a high-pressure homogenization process to form nanoparticles with an approximate diameter of 130 nm. Abraxane® is supplied as a

Table 2.4 Example of pharmaceutical products containing HSA as DDS carrier

Active	Characteristics	Use	Brand name	Form	Container	Status	About Stability
Nab-Paclitaxel (Hawkins et al. 2008)	HSA-bound Paclitaxel (nanoparticle)	Metastatic breast cancer	Abraxane®	Lyophilized Powder	Grass vial	Marketed	36 months (20–25 °C) Stabilizers involved
Albiglutide (Sharma et al. 2016)	Long acting GLP-1-HSA fusion	Type 2 diabetes	Tanzeum®	Lyophilized powder	Prefilled Syringe	Marketed	2 years (2–8 °C)
IFN-α2b-HSA (Subramanian et al. 2007; Bain et al. 2006)	Long-acting IFN-α2b- HSA fusion	Hepatitis C	Albuferon™	Lyophilized Powder	Vial	Phase III (withdrawn BLA)	Effects of surfactant and substitution of ^{34}Cys (Zhao et al. 2009)
HSA-hGH (Osborn et al. 2002)	Long-acting hGH-HSA fusion	Growth hormone deficiency	Albutropin™	Lyophilized Powder	Vial	Phase II	Effects of surfactants (Chou et al. 2005) and octanoic acid binding (Cordes et al. 2012b)
Balugrastim (Volovat et al. 2014)	Long acting GCSF -HSA fusion	Neutropenia in chemotherapy	–	Lyophilized Powder	Vial	Phase III (withdrawn BLA, MAA)	Effects of fusion and octanoic acid binding (Cordes et al. 2012a)
IL-2-HSA (Melder et al. 2005)	Long acting IL-2-HSA fusion	Solid tumor	Albuleukin™	–	–	Phase I	–
rVIIa-FP (Schulte 2014, 2013)	Long acting factor VIIa- HSA fusion	Hemophilia A and B	–	Lyophilized Powder	Vial	Phase I Phase II/III (enrolled)	–
rIX-FP (Schulte 2014, 2013)	Long acting factor IX- HSA fusion	Hemophilia B	–	Lyophilized Powder	Vial	Phase III (submitted BLA)	–

CJC-1131 (Tiessen et al. 2008; Christensen and Knop 2010)	GLP-1 analog chemically coupled to HSA (long acting)	Type 2 diabetes	–	–	–	Phase II (discontinued)	–
Albenatide (CJC-1134) (Christensen and Knop 2010)	Exendin-4 chemically coupled to HSA (long acting)	Type 2 diabetes	–	–	–	Phase II	–
MTX-HSA (Burger et al. 2001)	Methotrexate chemically conjugated with HSA	Metastatic renal cell carcinoma	–	–	–	Phase II (discontinued)	–
sEphB4-HSA (Shi et al. 2012)	Long acting sEphB4-HSA fusion	Vascular proliferative disease, cancer	–	–	–	Preclinical	Stabilization by sucrose and trehalose during lyophilization (Shi et al. 2012)
HSA-Trx (Ikuta et al. 2010; Tanaka et al. 2013)	Long acting thioredoxin-HSA fusion	Idiopathic pulmonary fibrosis	–	–	–	Preclinical	–
HSA dimer (Taguchi et al. 2010, 2015)	Chemically or genetically cross-linked HSA	DDS carrier or plasma expander	–	–	–	Preclinical	–

(continued)

Table 2.4 (continued)

Active	Characteristics	Use	Brand name	Form	Container	Status	About Stability
SNO-HSAs (Ishima et al. 2010; Kinoshita et al. 2015)	S-nitrosated HSA analogs	Antitumor	–		–	Preclinical	–
Man-rHSAs (Hirata et al. 2010)	Genetically engineered mannosylated-HSAs	DDS carrier for liver-selective therapeutics	–	–	–	Preclinical	–

white to yellow, sterile, lyophilized powder for reconstitution with 0.9 % sodium chloride. This lyophilized powder shows long-term storage characteristics (36 months) in sealed glass vials packaged in the original carton at 20–25 °C. The original carton can prevent the paclitaxel from photo-degradation. It is recommended that the reconstituted suspension in the vial or the infusion bag be used immediately, in order to ensure that the nanoparticle is stable as well as photo-stable. Thus, further research and development are expected to improve particle stability and photo-stability of lyophilized or reconstituted products. Abraxane® contains Oct and *N*-AcTrp as excipients as well as commercial HSA products. If ROS generated by the exposure of lyophilized or reconstituted products to light contributes to the degradation of paclitaxel or the instability of nanoparticles, *N*-AcMet, a superior ROS scavenger, might be substituted for *N*-AcTrp as an effective stabilizer.

2.5.2 *Albiglutide*

Albiglutide is a once-weekly injectable glucagon-like peptide 1 (GLP-1) agonist that was approved in 2014 by FDA for treatment of type 2 diabetes (Sharma et al. 2016). Albiglutide was first marketed as a recombinant albumin-fused peptide, consisting of two molecules of modified human GLP-1 that had been genetically fused to HSA, and is produced in *Saccharomyces cerevisiae* cells. This is commercially supplied as a pen-type injection system, in which lyophilized powder and water for injection are separately prefilled. This product is stable for 2 years at 2–8 °C or for up to 4 weeks at room temperature. The pen should be used within 8 h after reconstitution of the commercial product. Lyophilization is known to be a suitable technique for improving the storage stability of proteins. However, dehydration during the lyophilization process sometimes causes protein denaturation and aggregation, resulting in a loss of biological activity. Sugar excipients have been widely used to stabilize the protein during lyophilization and storage (Anhorn et al. 2008). Sucrose and trehalose have also been demonstrated to be suitable stabilizers against the aggregation of rHSA (Han et al. 2007) and the other HSA fusion protein (sEphB4-HSA; discussed at Sect. 2.5.3) (Shi et al. 2012). A commercial albiglutide product is also formulated using excipients, sodium dihydrogen phosphate monohydrate, disodium phosphate anhydrous, trehalose dehydrate, mannitol, and polysorbate 80, some of which might be used as stabilizer.

2.5.3 *IFN-α2b-HSA*

Interferon-α2b and HSA fusion protein (IFN-α2b-HSA) has been developed as a promising long-acting formulation of IFN-α2b for the treatment of hepatitis C (Subramanian et al. 2007; Bain et al. 2006). The product, which was evaluated in clinical trials, was supplied as lyophilized form by Human Genome Sciences, Inc.

Zhao et al. separately developed IFN-α2b-HSA and investigated the stability of IFN-α2b-HSA solutions against mechanical and thermal stress (Zhao et al. 2009). Under these stress conditions, IFN-α2b-HSA was prone to undergo disulfide-linked aggregation. The addition of Tween 80 attenuated the aggregation caused by agitation but did not attenuate the aggregation caused by heating. Tween 80, a nonionic surfactant, is not effective for stabilizing the product during storage. They substituted the unpaired cysteine residue (^{34}Cys) of the HSA domain in the fusion protein with serine by site-directed mutagenesis. This new fusion protein, IFN-α2b-HSA (C34S), possessed a significant higher stability over IFN-α2b-HSA against mechanical and thermal stability when the free sulfhydryl group in the HSA moiety of IFN-α2b-HSA was eliminated.

2.5.4 HSA-hGH

Human growth hormone (hGH) genetically fused to HSA (HSA-hGH) was developed for the long term treatment of growth hormone deficiency, because of its longer half-life than hGH alone (Osborn et al. 2002). This fusion protein was produced and developed under the trade name of Albutropin™ by Human Genome Science, Inc. Chow et al. studied the effects of nonionic surfactants (Tween 20 and Tween 80) on the stability of Albutropin™ against agitation (Chou et al. 2005). Both surfactants protected Albutropin™ against agitation-induced aggregation, even at concentrations below the critical micelle concentration, possibly due to binding to the native protein. They suggested that surfactants be employed as a rational approach to stabilizing protein pharmaceuticals such as Albutropin™. Recently, Cordes et al. investigated the effect of selective HSA domain stabilization by octanoic acid on the aggregation of HSA-hGH (Cordes et al. 2012b). They demonstrated that the reduction of aggregation under solution conditions is due to an increased colloidal stability resulting from the binding of octanoate to the HSA domain, but not due to an increased conformational stability of the HSA domain.

2.5.5 Balugrastim

Balugrastim is a granulocyte colony-stimulating factor (GCSF) that is fused to the C-terminus of HSA (HSA-GCSF), which allows for the once-per-cycle administration without pegylation (Volovat et al. 2014). Phase III clinical trials for the treatment of neutropenia in chemotherapy demonstrated that balugrastim is an effective and safe alternative to pegfilgrastim, a pegylated GCSF analog (Volovat et al. 2014). Cordes et al. reported that the fusion of GCSF to HSA reduces the rate of aggregation of HSA-GCSF under solution conditions as compared with that of GCSF itself (Cordes et al. 2012a). Because the aggregation of HSA-hGH is reduced by octanoic acid binding to an HSA domain despite its structural similarity between

GCSF and hGH, Cordes et al. suggested that each HSA fusion protein requires individually tailored conditions to reduce aggregation.

2.5.6 *sEphB4-HSA*

Shi et al. investigated the biophysical properties and stabilization of recombinant HSA fusion with the extracellular domain (sEphB4) of EphB4, a member of the Eph family of tyrosine kinase receptors (Shi et al. 2012). This protein (sEphB4-HSA) was designed by fusing HSA to the C-terminal of sEphB4 and is thought to be a promising therapeutic candidate for the treatment of vascular proliferative disease and cancer. They demonstrated that sEphB4-HSA is most stable at pH ≥5 and at temperatures lower than 50 °C. Elevated temperatures caused the aggregation of sEphB4-HSA. The screening of stabilizers including amino acids and sugars, sucrose, and trehalose indicated that they can be effective stabilizers against sEphB4-HSA aggregation. As mentioned in Sect. 2.5.2, sucrose and trehalose have also been reported to be suitable stabilizers against the aggregation of rHSA (Han et al. 2007) and HSA nanoparticles (Anhorn et al. 2008) during the lyophilization and long-term storage.

There are the other exciting avenues for medical applications of HSA that have not been fully explored (Table 2.4). For the formulations of these new products, more research directed at identifying factors that affect the stability of such molecules will be highly desirable.

2.6 Conclusion and Future Prospects

The identification of factors that affect the structures of HSA (ie, heat, chemicals, and genetic variations) helps not only to explain the dynamic structure of this protein but also to identify the important positions on the molecule (ie, domains or amino acid residues) that affect its stability, which finally serves to predict the stability of HSA at the point of production and storage. Developments of more stable proteins, new stabilizers, suitable production processes, and containers will also contribute to the stabilization of novel pharmaceutical formulations that contain HSA. Of course, the maintenance of the pharmacological and pharmacokinetic functions of HSA and its pharmaceutical preparations should be considered during the introduction of stabilizers.

Acknowledgment We wish to acknowledge Dr. Victor Tuan Giam Chuang, Curtin University, Australia, for reviewing our article.

References

Aguilera A, Selgas R, Diez JJ, Bajo MA, Codoceo R, Alvarez V (2001) Anorexia in end-stage renal disease: pathophysiology and treatment. Expert Opin Pharmacother 2:1825–1838

Ahmad B, Ahmed MZ, Haq SK, Khan RH (2005) Guanidine hydrochloride denaturation of human serum albumin originates by local unfolding of some stable loops in domain III. Biochim Biophys Acta 1750:93–102

Anand U, Mukherjee S (2013) Reversibility in protein folding: effect of beta-cyclodextrin on bovine serum albumin unfolded by sodium dodecyl sulphate. Phys Chem Chem Phys 15:9375–9383

Anand U, Ray S, Ghosh S, Banerjee R, Mukherjee S (2015) Structural aspects of a protein-surfactant assembly: native and reduced states of human serum albumin. Protein J 34:147–157

Anhorn MG, Mahler HC, Langer K (2008) Freeze drying of human serum albumin (HSA) nanoparticles with different excipients. Int J Pharm 363:162–169

Anraku M, Tsurusaki Y, Watanabe H, Maruyama T, Kragh-Hansen U, Otagiri M (2004) Stabilizing mechanisms in commercial albumin preparations: octanoate and N-acetyl-L-tryptophanate protect human serum albumin against heat and oxidative stress. Biochim Biophys Acta 1702:9–17

Anraku M, Kouno Y, Kai T, Tsurusaki Y, Yamasaki K, Otagiri M (2007) The role of N-acetyl-methioninate as a new stabilizer for albumin products. Int J Pharm 329:19–24

Anraku M, Shintomo R, Taguchi K, Kragh-Hansen U, Kai T, Maruyama T, Otagiri M (2015) Amino acids of importance for the antioxidant activity of human serum albumin as revealed by recombinant mutants and genetic variants. Life Sci 134:36–41

Arakawa T, Kita Y (2000a) Stabilizing effects of caprylate and acetyltryptophanate on heat-induced aggregation of bovine serum albumin. Biochim Biophys Acta 1479:32–36

Arakawa T, Kita Y (2000b) Protection of bovine serum albumin from aggregation by Tween 80. J Pharm Sci 89:646–651

Bain VG, Kaita KD, Yoshida EM, Swain MG, Heathcote EJ, Neumann AU, Fiscella M, Yu R, Osborn BL, Cronin PW, Freimuth WW, McHutchison JG, Subramanian GM (2006) A phase 2 study to evaluate the antiviral activity, safety, and pharmacokinetics of recombinant human albumin-interferon alfa fusion protein in genotype 1 chronic hepatitis C patients. J Hepatol 44:671–678

Bertolini J, Goss N, Curling J (2012) Production of plasma proteins for therapeutic use. Hoboken: Wiley

Bourdon E, Loreau N, Lagrost L, Blache D (2005) Differential effects of cysteine and methionine residues in the antioxidant activity of human serum albumin. Free Radic Res 39:15–20

Burger AM, Hartung G, Stehle G, Sinn H, Fiebig HH (2001) Pre-clinical evaluation of a methotrexate-albumin conjugate (MTX-HSA) in human tumor xenografts in vivo. Int J Cancer J Int Cancer 92:718–724

Carballal S, Radi R, Kirk MC, Barnes S, Freeman BA, Alvarez B (2003) Sulfenic acid formation in human serum albumin by hydrogen peroxide and peroxynitrite. Biochemistry 42:9906–9914

Carter DC, Ho JX (1994) Structure of serum albumin. Adv Protein Chem 45:153–203

Chou DK, Krishnamurthy R, Randolph TW, Carpenter JF, Manning MC (2005) Effects of Tween 20 and Tween 80 on the stability of albutropin during agitation. J Pharm Sci 94:1368–1381

Christensen M, Knop FK (2010) Once-weekly GLP-1 agonists: how do they differ from exenatide and liraglutide? Curr Diab Rep 10:124–132

Cohn EJ, Strong LE, Hughes WL, Mulford DJ, Ashworth JN, Melin M, Taylor HL (1946) Preparation and properties of serum and plasma proteins. IV. A system for the separation into fractions of the protein and lipoprotein components of biological tissues and fluids1a, b, c, d. J Am Chem Soc 68:459–475

Cordes AA, Carpenter JF, Randolph TW (2012a) Selective domain stabilization as a strategy to reduce human serum albumin-human granulocyte colony stimulating factor aggregation rate. J Pharm Sci 101:2009–2016
Cordes AA, Platt CW, Carpenter JF, Randolph TW (2012b) Selective domain stabilization as a strategy to reduce fusion protein aggregation. J Pharm Sci 101:1400–1409
Curry S, Mandelkow H, Brick P, Franks N (1998) Crystal structure of human serum albumin complexed with fatty acid reveals an asymmetric distribution of binding sites. Nat Struct Biol 5:827–835
Davies MJ (2005) The oxidative environment and protein damage. Biochim Biophys Acta 1703:93–109
Desai N, Trieu V, Yao Z, Louie L, Ci S, Yang A, Tao C, De T, Beals B, Dykes D, Noker P, Yao R, Labao E, Hawkins M, Soon-Shiong P (2006) Increased antitumor activity, intratumor paclitaxel concentrations, and endothelial cell transport of cremophor-free, albumin-bound paclitaxel, ABI-007, compared with cremophor-based paclitaxel. Clin Cancer Res 12:1317–1324
Duggan EL, Luck JM (1948) The combination of organic anions with serum albumin; stabilization against urea denaturation. J Biol Chem 172:205–220
Fersht A (1999) Structure and mechanism in protein science: a guide to enzyme catalysis and potein folding. W. H Freeman, San Francisco
Flora K, Brennan JD, Baker GA, Doody MA, Bright FV (1998) Unfolding of acrylodan-labeled human serum albumin probed by steady-state and time-resolved fluorescence methods. Biophys J 75:1084–1096
Galantini L, Leggio C, Pavel NV (2008) Human serum albumin unfolding: a small-angle X-ray scattering and light scattering study. J Phys Chem B 112:15460–15469
Galantini L, Leggio C, Konarev PV, Pavel NV (2010) Human serum albumin binding ibuprofen: a 3D description of the unfolding pathway in urea. Biophys Chem 147:111–122
Gonzalez-Jimenez J, Cortijo M (2002) Urea-induced denaturation of human serum albumin labeled with acrylodan. J Protein Chem 21:75–79
Han Y, Jin BS, Lee SB, Sohn Y, Joung JW, Lee JH (2007) Effects of sugar additives on protein stability of recombinant human serum albumin during lyophilization and storage. Arch Pharm Res 30:1124–1131
Hawkins MJ, Soon-Shiong P, Desai N (2008) Protein nanoparticles as drug carriers in clinical medicine. Adv Drug Deliv Rev 60:876–885
Hirata K, Maruyama T, Watanabe H, Maeda H, Nakajou K, Iwao Y, Ishima Y, Katsumi H, Hashida M, Otagiri M (2010) Genetically engineered mannosylated-human serum albumin as a versatile carrier for liver-selective therapeutics. J Control Release 145:9–16
Ikuta S, Chuang VT, Ishima Y, Nakajou K, Furukawa M, Watanabe H, Maruyama T, Otagiri M (2010) Albumin fusion of thioredoxin – the production and evaluation of its biological activity for potential therapeutic applications. J Control Release 147:17–23
Ishima Y, Hiroyama S, Kragh-Hansen U, Maruyama T, Sawa T, Akaike T, Kai T, Otagiri M (2010) One-step preparation of S-nitrosated human serum albumin with high biological activities. Nitric Oxide Biol Chem Off J Nitric Oxide Soc 23:121–127
Iwao Y, Ishima Y, Yamada J, Noguchi T, Kragh-Hansen U, Mera K, Honda D, Suenaga A, Maruyama T, Otagiri M (2012) Quantitative evaluation of the role of cysteine and methionine residues in the antioxidant activity of human serum albumin using recombinant mutants. IUBMB Life 64:450–454
Kinoshita R, Ishima Y, Ikeda M, Kragh-Hansen U, Fang J, Nakamura H, Chuang VT, Tanaka R, Maeda H, Kodama A, Watanabe H, Maeda H, Otagiri M, Maruyama T (2015) S-Nitrosated human serum albumin dimer as novel nano-EPR enhancer applied to macromolecular anti-tumor drugs such as micelles and liposomes. J Control Release 217:1–9
Kobayashi K (2006) Summary of recombinant human serum albumin development. Biologicals 34:55–59
Kosa T, Maruyama T, Otagiri M (1998) Species differences of serum albumins: II. Chemical and thermal stability. Pharm Res 15:449–454

Kouno Y, Anraku M, Yamasaki K, Okayama Y, Iohara D, Ishima Y, Maruyama T, Kragh-Hansen U, Hirayama F, Otagiri M (2014) N-acetyl-L-methionine is a superior protectant of human serum albumin against photo-oxidation and reactive oxygen species compared to N-acetyl-L-tryptophan. Biochim Biophys Acta 1840:2806–2812

Kragh-Hansen U, Chuang VT, Otagiri M (2002) Practical aspects of the ligand-binding and enzymatic properties of human serum albumin. Biol Pharm Bull 25:695–704

Kragh-Hansen U, Saito S, Nishi K, Anraku M, Otagiri M (2005) Effect of genetic variation on the thermal stability of human serum albumin. Biochim Biophys Acta 1747:81–88

Kratz F (2008) Albumin as a drug carrier: design of prodrugs, drug conjugates and nanoparticles. J Control Release 132:171–183

Leggio C, Galantini L, Konarev PV, Pavel NV (2009) Urea-induced denaturation process on defatted human serum albumin and in the presence of palmitic acid. J Phys Chem B 113:12590–12602

Manoharan P, Wong YH, Tayyab S (2015) Stabilization of human serum albumin against urea denaturation by diazepam and ketoprofen. Protein Pept Lett 22:611–617

Melder RJ, Osborn BL, Riccobene T, Kanakaraj P, Wei P, Chen G, Stolow D, Halpern WG, Migone TS, Wang Q, Grzegorzewski KJ, Gallant G (2005) Pharmacokinetics and in vitro and in vivo anti-tumor response of an interleukin-2-human serum albumin fusion protein in mice. Cancer Immunol Immunother: CII 54:535–547

Mendez CM, McClain CJ, Marsano LS (2005) Albumin therapy in clinical practice. Nutr Clin Pract Off Publ Am Soc Parenter Enter Nutr 20:314–320

Miyamura S, Imafuku T, Anraku M, Taguchi K, Yamasaki K, Tominaga Y, Maeda H, Ishima Y, Watanabe H, Otagiri M, Maruyama T (2016) Comparison of post-translational modification and the functional impairment of human serum albumin in commercial preparations. J Pharm Sci 105:1043–1049

Moriyama Y, Takeda K (2005) Protective effects of small amounts of bis(2-ethylhexyl) sulfosuccinate on the helical structures of human and bovine serum albumins in their thermal denaturations. Langmuir 21:5524–5528

Muzammil S, Kumar Y, Tayyab S (2000) Anion-induced stabilization of human serum albumin prevents the formation of intermediate during urea denaturation. Proteins 40:29–38

Osborn BL, Sekut L, Corcoran M, Poortman C, Sturm B, Chen G, Mather D, Lin HL, Parry TJ (2002) Albutropin: a growth hormone-albumin fusion with improved pharmacokinetics and pharmacodynamics in rats and monkeys. Eur J Pharmacol 456:149–158

Peters T (1995) All about albumin: biochemistry, genetics, and medical applications. San Diego: Academic Press

Ross PD, Shrake A (1988) Decrease in stability of human albumin with increase in protein concentration. J Biol Chem 263:11196–11202

Santra MK, Banerjee A, Rahaman O, Panda D (2005) Unfolding pathways of human serum albumin: evidence for sequential unfolding and folding of its three domains. Int J Biol Macromol 37:200–204

Schulte S (2013) Innovative coagulation factors: albumin fusion technology and recombinant single-chain factor VIII. Thromb Res 131:S2–S6

Schulte S (2014) Challenges for new haemophilia products from a manufacturer's perspective. Thromb Res 134(Suppl 1):S72–S76

Sharma AK, Thanikachalam PV, Rajput SK (2016) Albiglutide: is a better hope against diabetes mellitus? Biomed Pharmacother 77:120–128

Shi S, Liu J, Joshi SB, Krasnoperov V, Gill P, Middaugh CR, Volkin DB (2012) Biophysical characterization and stabilization of the recombinant albumin fusion protein sEphB4-HSA. J Pharm Sci 101:1969–1984

Shrake A, Finlayson JS, Ross PD (1984) Thermal stability of human albumin measured by differential scanning calorimetry. I. Effects of caprylate and N-acetyltryptophanate. Vox Sang 47:7–18

Subramanian GM, Fiscella M, Lamouse-Smith A, Zeuzem S, McHutchison JG (2007) Albinterferon alpha-2b: a genetic fusion protein for the treatment of chronic hepatitis C. Nat Biotechnol 25:1411–1419

Sugio S, Kashima A, Mochizuki S, Noda M, Kobayashi K (1999) Crystal structure of human serum albumin at 2.5 a resolution. Protein Eng 12:439–446

Taguchi K, Urata Y, Anraku M, Watanabe H, Kawai K, Komatsu T, Tsuchida E, Maruyama T, Otagiri M (2010) Superior plasma retention of a cross-linked human serum albumin dimer in nephrotic rats as a new type of plasma expander. Drug Metab Dispos 38:2124–2129

Taguchi K, Chuang VT, Yamasaki K, Urata Y, Tanaka R, Anraku M, Seo H, Kawai K, Maruyama T, Komatsu T, Otagiri M (2015) Cross-linked human serum albumin dimer has the potential for use as a plasma-retaining agent for the fatty acid-conjugated antidiabetic drugs. J Pharm Pharmacol 67:255–263

Tanaka R, Watanabe H, Kodama A, Chuang VT, Ishima Y, Hamasaki K, Tanaka K, Mizushima T, Otagiri M, Maruyama T (2013) Long-acting human serum albumin-thioredoxin fusion protein suppresses bleomycin-induced pulmonary fibrosis progression. J Pharmacol Exp Ther 345:271–283

Tiessen RG, Castaigne JP, Dreyfus JF, Nemansky M, Kruizinga HH, van Vliet AA (2008) Pharmacokinetics and tolerability of a novel long-acting glucagon-like peptide-1 analog, CJC-1131, in healthy and diabetic subjects. Int J Clin Pharmacol Ther 46:443–452

Topczewska-Bruns J, Pawlak D, Tankiewicz A, Chabielska E, Buczko W (2003) Kynurenine metabolism in central nervous system in experimental chronic renal failure. Adv Exp Med Biol 527:177–182

Vogt W (1995) Oxidation of methionyl residues in proteins: tools, targets, and reversal. Free Radic Biol Med 18:93–105

Volovat C, Gladkov OA, Bondarenko IM, Barash S, Buchner A, Bias P, Adar L, Avisar N (2014) Efficacy and safety of balugrastim compared with pegfilgrastim in patients with breast cancer receiving chemotherapy. Clin Breast Cancer 14:101–108

Watanabe H, Tanase S, Nakajou K, Maruyama T, Kragh-Hansen U, Otagiri M (2000) Role of arg-410 and tyr-411 in human serum albumin for ligand binding and esterase-like activity. Biochem J 349(Pt 3):813–819

Watanabe H, Kragh-Hansen U, Tanase S, Nakajou K, Mitarai M, Iwao Y, Maruyama T, Otagiri M (2001a) Conformational stability and warfarin-binding properties of human serum albumin studied by recombinant mutants. Biochem J 357:269–274

Watanabe H, Yamasaki K, Kragh-Hansen U, Tanase S, Harada K, Suenaga A, Otagiri M (2001b) In vitro and in vivo properties of recombinant human serum albumin from Pichia pastoris purified by a method of short processing time. Pharm Res 18:1775–1781

Zhao HL, Xue C, Wang Y, Sun B, Yao XQ, Liu ZM (2009) Elimination of the free sulfhydryl group in the human serum albumin (HSA) moiety of human interferon-α2b and HSA fusion protein increases its stability against mechanical and thermal stresses. Eur J Pharm Biopharm 72:405–411

Chapter 3
Albumin as a Biomarker

Hiroshi Watanabe and Toru Maruyama

Abstract Recent advancements in technology have given rise to the emergence of data regarding posttranslational modifications of human serum albumin (HSA) such as oxidation, glycation, truncation, dimerization, and carbamylation in disease conditions. We developed a simple and rapid analytical method that allows the redox state of Cys-34 of HSA to be quantitatively and qualitatively evaluated with a high degree of sensitivity using an ESI-TOFMS technique. An increase in the level of oxidized HSA accompanied by a decrease in the level of reduced HSA was observed in cases of chronic liver disease, chronic renal disease, and diabetes mellitus, although the redox state of HSA is not associated with colloid osmotic pressure. The degree of oxidation of Cys-34 was correlated with ligand binding and the antioxidative functions of HSA. Since there is no oxidized form of HSA immediately after its secretion from liver cells, the oxidized species could constitute a potential marker of the extent of oxidative stress and of the scavenging activity of Cys-34. Evidence has accumulated to demonstrate that monitoring of the redox state of Cys-34 could not only be a useful marker for evaluating the progression of oxidative stress-related disease and the development of its complications but also in predicting therapeutic efficacy. In addition, the usefulness of determining the redox state of Cys-34 as an index of the quality of HSA preparations was also shown. These data strongly suggest that monitoring the posttranslational modifications of HSA can be important, since HSA function is related not only to its serum concentration but also to the preservation of its structural integrity under disease conditions.

Keywords Albumin • Biomarker • Oxidation • Cysteine-34 • Liver disease • Kidney disease • Diabetes

---

H. Watanabe (✉) • T. Maruyama
Department of Biopharmaceutics, Graduate School of Pharmaceutical Sciences, Kumamoto University, 5-1, Oe-honmachi, Chuo-ku, Kumamoto 862-0973, Japan

Center for Clinical Pharmaceutical Sciences, School of Pharmacy, Kumamoto University, 5-1, Oe-honmachi, Chuo-ku, Kumamoto 862-0973, Japan
e-mail: hnabe@kumamoto-u.ac.jp

M. Otagiri, V.T.G. Chuang (eds.), *Albumin in Medicine*,
DOI 10.1007/978-981-10-2116-9_3

3.1 Introduction

Posttranslationally modified proteins can be used as biomarkers for the diagnosis of diseases or for the assessment of therapeutic responses. A prime example of this is the quantification of glycated hemoglobin and glycoalbumin for the diagnosis and treatment of diabetes mellitus (Association 2013; Furusyo and Hayashi 2013; American Diabetes Association 2013).

Human serum albumin (HSA) is the most abundant protein in plasma, constituting 60–65 % of the total protein (Peters 1996). It is synthesized in the liver (10–15 g/day: 3.5 % of the total HSA pool) and released into the intravascular space. An equivalent mass of HSA is catabolized in the muscle, skin, liver, and kidney. In the whole body of the HSA pool, 30–40 % is located in the intravascular space, whereas the remaining 60–70 % is in the interstitial space. The transfer rate of HSA from the intravascular space to the interstitial space is 4–5 % per hour, and approximately the same percentage is returned to the intravascular space through the lymphatic system (Peters 1996). It has been proposed that an important link exists between the intravascular HSA level and disease (Peters 1996). In addition, due to the long plasma half-life of HSA (approximate 19 days), the occurrence of posttranslational modifications, resulting from oxidation, glycation, and truncation involving Cys-34 and other sites, contributes to the microheterogeneity of circulating HSA, which may alter its biological activity under disease conditions. Therefore, not only its quantitative variation but also posttranslational modifications of HSA are thought to serve as an indicator of the severity of pathophysiological conditions and could result in an altered protein function and structure. This chapter explores the available evidences for the use of HSA as biomarker in diseases.

3.2 Cys-34 Is an Important Target for Oxidants

In addition to its function as a determinant of colloid osmotic pressure, HSA has other biological functions such as a carrier for endogenous and exogenous substances as well as radical scavenging properties. HSA is comprised of a single, nonglycosylated chain of 585 amino acids containing 17 disulfide bridges. Of the 35 cysteine residues, only that at position 34 (Cys-34) remains free, representing the main antioxidant site of the molecule (Fig. 3.1) (Peters 1996), because the thiol group is an effective reducing agent and nucleophile; it can react by one- and two-electron mechanisms, and they are susceptible to reversible and irreversible oxidative modification (Turell et al. 2008, 2013, 2014; Torres et al. 2012).

In healthy subjects, 70–80 % of the HSA circulates as human mercaptalbumin (HMA: reduced form), characterized by a reduced Cys-34 with preserved antioxidant and scavenging activities, whereas in 20–30 % of the population, the Cys-34 residue is reversibly oxidized and binds small thiol molecules such as cysteine, homocysteine, or glutathione (human nonmercaptalbumin 1 (HNA1): oxidized

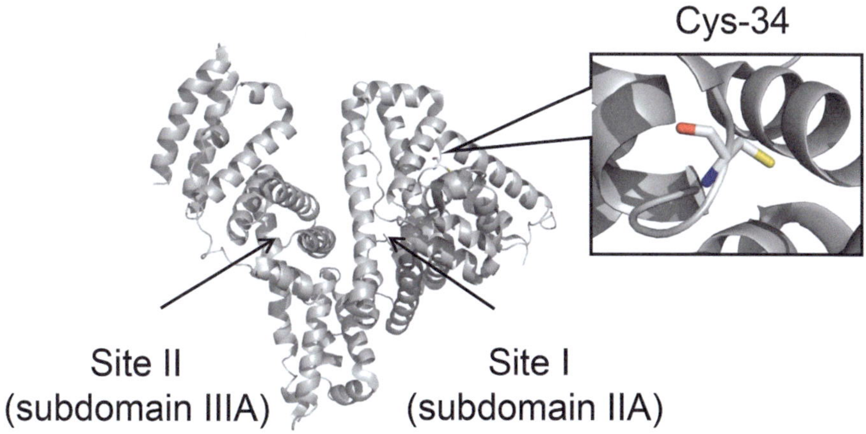

Fig. 3.1 The location of Cys-34 and ligand binding sites (Site I and Site II) in HSA. PDB ID: 1BM0

form). The remaining ~5 % of the total HSA circulates as nonmercaptalbumin 2 (HNA2: oxidized form), with the Cys-34 residue irreversibly oxidized to sulfinic or sulfonic acids, thus leading to the permanent loss of its function (Turell et al. 2008; Anraku et al. 2013). Furthermore, the antioxidant activity is also exerted by the NH_2-terminal region of HSA, which can chelate metal ions.

In cells, the thiol group is present at millimolar concentrations and is almost completely in the reduced form. The intracellular total glutathione concentration is ~15 mM (Requejo et al. 2010), and 90 % of the glutathione is reduced (Hansen et al. 2009). In addition, protein thiols are more abundant (10–50 mM) than glutathione in cells (Hansen et al. 2009). In contrast, in the extracellular compartment, particularly in the plasma compartment, the thiol groups are more oxidized, and their concentration is lower. The concentration of total reduced thiol in plasma is ~0.6 mM. Therefore, Cys-34 in HSA represents the largest fraction (>80 %) of free thiol groups in plasma, thus representing a potent scavenger of reactive oxygen and nitrogen species.

3.3 Detection of the Redox State of Cys-34

Recent studies have demonstrated an association between chronic diseases and oxidative stress. There is growing interest in developing diagnostic tools for monitoring the extent of oxidative damage in tissue and organs, and the use of novel antioxidants for the prevention or treatment of oxidative stress-disease including liver, kidney, heart, vessel, cancer and neurodegenerative disease (Poli 2000; Descamps-Latscha and Witko-Sarsat 2001; Pergola et al. 2011). However, at present, a rapid and sensitive clinical laboratory testing method for the rapid assessment of oxidative stress in human has not been available.

Era's group and our laboratory have developed a high-performance liquid chromatography (HPLC) method for monitoring the redox status of Cys-34 in HSA. This HPLC analytical method has been shown to be useful for assessing the level of oxidative stress in disease states and for evaluating the anti-oxidative activity of therapeutic agents (Sogami et al. 1985; Terawaki et al. 2004; Anraku et al. 2004). However, this HPLC analytical method does not provide any molecular structural information regarding the oxidized form of Cys-34, as mentioned above. To address this need, we developed a simple and rapid analytical method that allows the quantitative and qualitative evaluation of the redox state of Cys-34 with a high degree of sensitivity using electrospray ionization time-of-flight mass spectrometer (ESI-TOFMS) (Kawakami et al. 2006; Kubota et al. 2009; Nagumo et al. 2014). We found that cysteinylation at Cys-34 (Cys-Cys34-HSA) accounts for the majority of the oxidized forms of Cys-34 (Fig. 3.2). A significant positive correlation was observed between the Cys-Cys34-HSA fraction determined by EIS-TOFMS and the degree of oxidized Cys-34-HSA determined by HPLC of plasma samples from 229

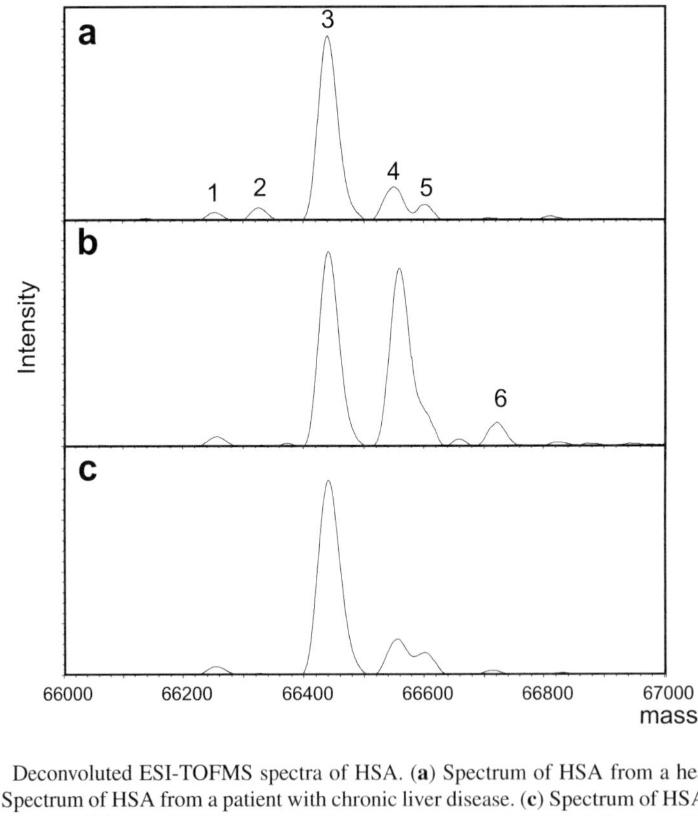

Fig. 3.2 Deconvoluted ESI-TOFMS spectra of HSA. (**a**) Spectrum of HSA from a healthy subject. (**b**) Spectrum of HSA from a patient with chronic liver disease. (**c**) Spectrum of HSA from the same patient that (*b*) after DTT treatment. The peaks correspond to the following: (*1*) Asp-Ala truncation from N-terminal of HSA, (*2*) Leu truncation from C-terminal of HSA, (*3*) reduced HSA, (*4*) Cys-Cys34-HSA, (*5*) glycated HSA and (*6*) glycated Cys-Cys34-HSA (Nagumo et al. 2014)

patients with chronic liver disease, chronic kidney disease, and diabetes mellitus. Compared to HPLC, ESI-TOFMS is a high throughput method that can be used to analyze large numbers of samples rapidly and sensitively and thus may be suitable for use in clinical laboratory testing. Furthermore, the ESI-TOFMS method offers an additional advantage of having the capability of characterizing the different oxidized forms of Cys-34 that are produced by oxidation by a variety of endogenous substances. These data permitted the fraction of Cys-Cys34-HSA determined by ESI-TOFMS to be used as a novel marker for oxidative stress in the systemic circulation (Nagumo et al. 2014).

3.4 Alterations of Cys-34 in Pathological Conditions

Recent advanced technology has given rise to the collection of data concerning the redox modifications of HSA in several disease conditions, where it was observed a decrease in the amount of the reduced form of HSA is accompanied by an increase in the amount of oxidized form, although redox state of HSA is not associated with colloid osmotic pressure (Sakata et al. 2010). Colombo et al. referred to as "redox albuminomics" (Colombo et al. 2012). Oxidation of HSA occurs in vivo, and its process is correlated to organ dysfunction. Since the oxidized form of HSA is not present immediately after secretion from liver cells, the oxidized species could constitute potential markers of the involvement of oxidative stress and of the scavenging activity of the Cys-34. So far, there are several reports regarding the detection of oxidized form of HSA in different conditions such as chronic liver disease (Nagumo et al. 2014; Klammt et al. 2007; Jalan et al. 2009; Oettl et al. 2013), chronic kidney disease (Anraku et al. 2004; Musante et al. 2006, 2007; Terawaki et al. 2004), diabetes mellitus (Suzuki et al. 1992; Boisvert et al. 2010), aging (Era et al. 1995), etc.

3.4.1 Cirrhosis

3.4.1.1 Redox State of Cys-34 in Cirrhosis

Since the plasma HSA concentration is diminished in patients with liver disease, the HSA level is related to the prognosis in liver disease. Recent studies have shown that not only the HSA concentration but also the HSA function is reduced in liver disease (Nagumo et al. 2014; Klammt et al. 2007; Jalan et al. 2009; Oettl et al. 2013). Oxidative stress is believed to play an important role in liver disease (Sen et al. 2002) and is reflected by the oxidative modification of HSA.

We recently reported on the effect of disease severity or branched chain amino acid (BCAA) treatment on the Cys-Cys34-HSA fraction measured by ESI-TOFMS in 139 patients with chronic liver disease (cirrhosis) (Nagumo et al. 2014). The relationship between the Cys-Cys34-HSA fraction and the Child-Pugh classifica-

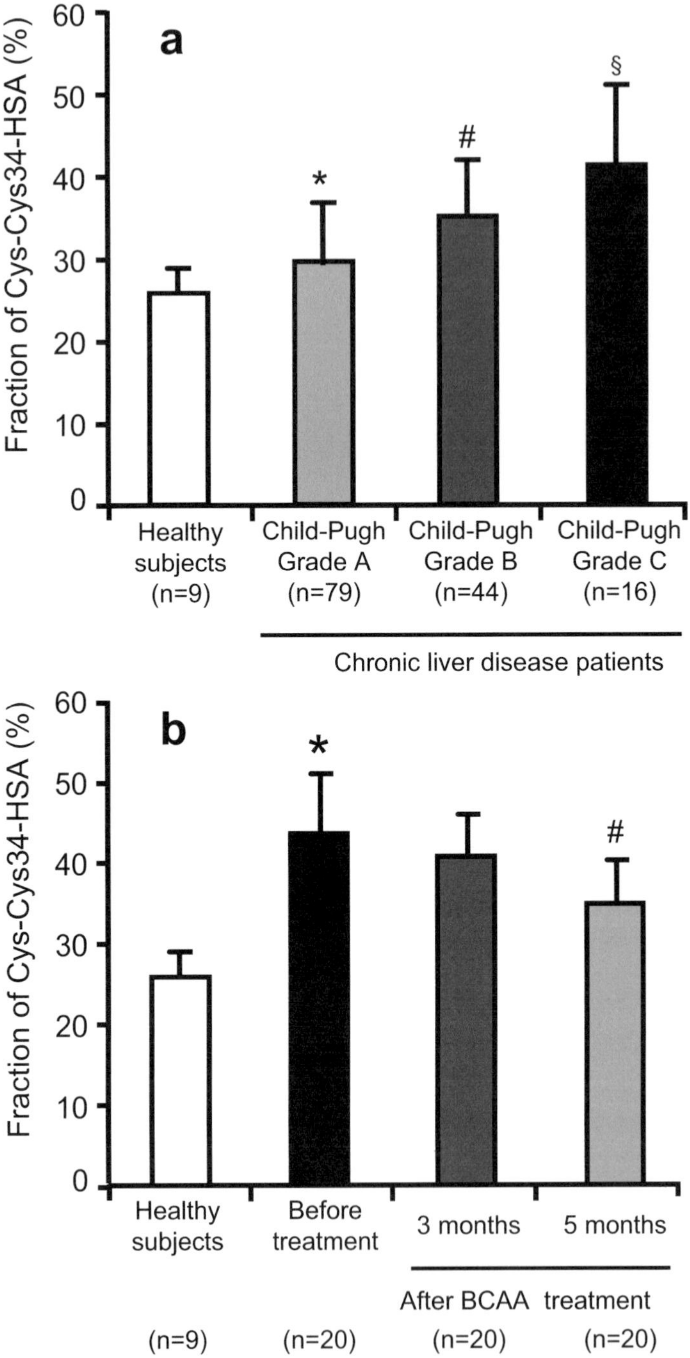

Fig. 3.3 (**a**) Effect of disease progression on the Cys-Cys34-HSA fraction in chronic liver disease (cirrhosis) patients. The Cys-Cys34-HSA fraction was measured by ESI-TOFMS. Values are expressed as the mean ± SD (n=9–79). *P<0.05 as compared with healthy subjects, #P<0.05 as compared with Child-Pugh Grade A, and §P<0.05 as compared with Child-Pugh Grade B. (**b**) Effect of BCAA treatment on the Cys-Cys34-HSA fraction in chronic liver disease patients. Values are expressed as the mean ± SD (n=20). *P<0.01 as compared with healthy subjects, #P<0.05 as compared with before treatment (Nagumo et al. 2014)

tion in patients with chronic liver disease was examined. As shown in Fig. 3.3, an increase in the severity of the disease was associated with an increase in the Cys-Cys34-HSA fraction, suggesting an association between the progression of chronic liver disease and oxidative stress. The effect of BCAA treatment on the Cys-Cys34-HSA fraction in patients with chronic liver disease was also examined. As shown in Fig. 3.3, the Cys-Cys34-HSA fraction was decreased as the result of the BCAA treatment, accompanied by a significant improvement compared to the pretreatment level.

We further investigated that the effect of the cysteinylation of Cys-34 on the ligand-binding ability of HSA purified from the plasma of chronic liver disease patients. HSA has two major ligand-binding sites, so-called Site I and Site II, which are localized in subdomains IIA and IIIA on HSA, respectively (Fig. 3.1). To investigate the effect of cysteinylation on ligand-binding properties, binding experiments using an ultrafiltration technique were carried out using endogenous and exogenous substances. Bilirubin and L-tryptophan, which preferentially bind to Site I and Site II, respectively, were selected as endogenous ligands related to liver disease. On the other hand, warfarin and diazepam, representative drugs that bind to Site I and Site II, respectively, were used as exogenous ligands. The increases in the unbound fraction of all of the four ligands were associated with the progression of liver disease. The decreased ligand-binding capacity of HSA observed in patients with chronic liver disease was significantly improved as the result of the BCAA treatment. Interestingly, a moderate correlation was observed for the unbound fraction of warfarin and diazepam, and a causal relationship was found in the case of the unbound fraction of bilirubin and L-tryptophan with the Cys-Cys34-HSA fraction (Fig. 3.4), suggesting that the cysteinylation of Cys-34 affected the microenvironment of the ligand-binding sites. Since the decreased ligand-binding properties of HSA against these endogenous substances may contribute to an acceleration or inhibition of the progression of a disease, the correlation between the Cys-Cys34-HSA fraction and the fraction of ligand binding to HSA suggests that the Cys-Cys34-HSA fraction might also serve as a marker for estimating changes in the physiological functions of HSA. These data suggest that the Cys-Cys34-HSA fraction is, in fact, an appropriate marker for a decreased HSA function caused by the oxidative stress-related progression of liver disease or in response to a therapeutic treatment.

In a previous study, we also reported that Cys-34 accounted for approximately 40 % of the total radical scavenging activity of HSA (Iwao et al. 2012; Anraku et al. 2011). The observed changes in the Cys-Cys34-HSA fraction with increasing severity of liver disease or BCAA treatment are likely to reflect an alteration in available SH groups resulting from the cysteinylation or de-cysteinylation of Cys-34. Thus, such changes in the cysteinylation of Cys-34 could influence the anti-oxidative capacity of the protein in the systemic circulation, which may contribute to the progression of chronic liver disease.

In patients with advanced liver disease such acute on liver failure, Oettl K et al. investigated the oxidative modification of HSA and its relation to physiological properties such as the binding function of HSA (Oettl et al. 2008). The distribution of HMA, HNA1, and HNA2 was determined by an HPLC method. They reported

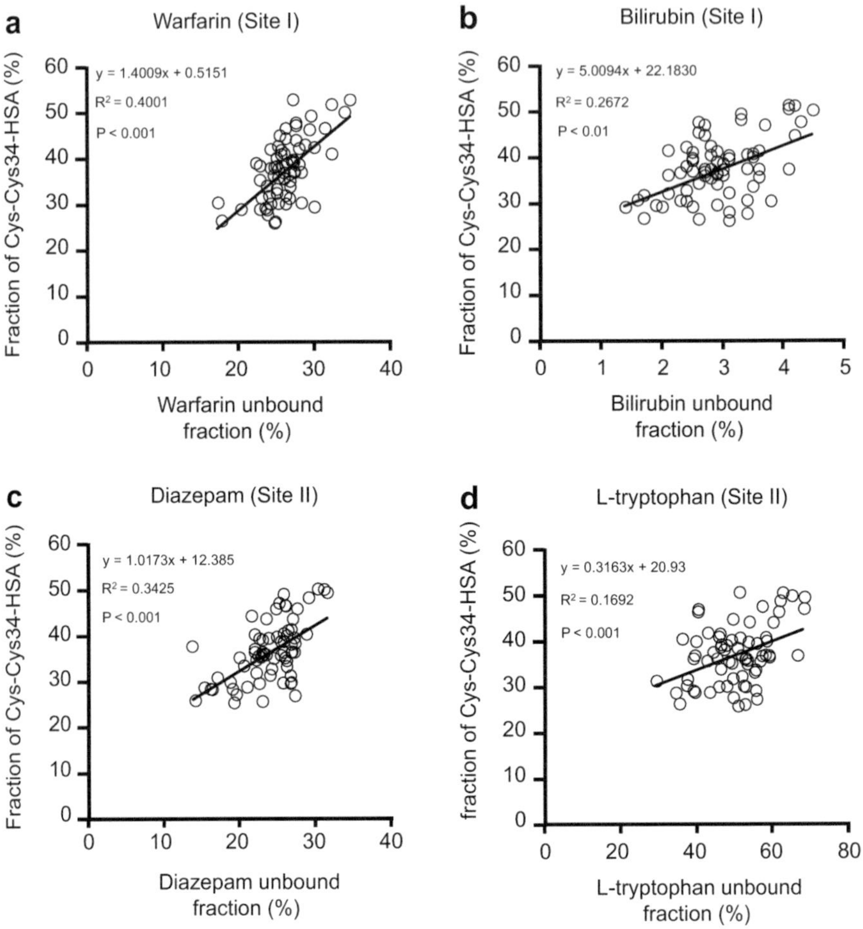

Fig. 3.4 Relationship between the Cys-Cys34-HSA fraction and the ligand unbound fraction in chronic liver disease (cirrhosis) patients. Correlation between the Cys-Cys34-HSA fraction and the unbound fraction of (**a**) warfarin, (**b**) bilirubin, (**c**) diazepam and (**d**) L-tryptophan to purified HSA in chronic liver disease (Nagumo et al. 2014)

that, in acute chronic liver failure, the extent of oxidation of Cys-34 increased with increasing severity of liver failure, while HMA was significantly decreased in cirrhosis and acute-on-chronic liver failure. Bilirubin-binding affinity (Site I) was in the order HMA>HNA1>HNA2, suggesting that the decreased percentages of HMA as well as the lower total HSA concentrations in plasma of patients provide the pathophysiological basis for impaired HSA function in advanced liver disease.

These findings were consistent with their later findings. They also recently observed marked alterations of the redox state of HSA and a significant impairment of HSA-binding capacity (Site II) in patients with chronic liver disease as well as in septic patients without liver disease (Oettl et al. 2013; Stauber et al. 2014). In fact, a significant increase of HNA1 or HNA2 accompanied by a decrease in HMA and impaired dansylsarcosine binding (Site II) were observed in these patients. In patients with cirrhosis, HNA1 levels were found to be significantly correlated with the MELD score, serum bilirubin levels, INR, and CRP, suggesting that both liver disease and inflammation have an influence on the generation of HNA1 and HNA2. Receiver operating characteristic (ROC) analysis also revealed the diagnostic accuracy of HNA2 for 30-day and 90-day survival in patients with cirrhosis (Oettl et al. 2013). In a multivariate analysis of prognostic variables, HNA2 was the only remaining predictor of 90 day mortality (Stauber et al. 2014). The diagnostic accuracy of HNA2 tended to be superior to that of the MELD score. In addition, a Kaplan-Meier analysis demonstrated a significantly higher 90 day mortality for cirrhotic patients with baseline HNA2 levels >12 % (Oettl et al. 2013). Therefore, they concluded that the irreversible oxidation of HSA is associated with a poor prognosis and thus may represent a novel biomarker for advanced chronic liver disease such as an acute liver failure.

Patients with cirrhosis display an increased tendency to and mortality from infections due to impaired innate immune systems. O'Brien et al. reported evidence that eicosanoid prostaglandin E_2 drives cirrhosis-associated immunosuppression (O'Brien et al. 2014). Importantly, they also demonstrated that HSA which reduces prostaglandin E_2 bioavailability in cirrhosis patients was decreased, and decreased HSA level contributes to immunosuppression and increasing the risk of infection by increasing the bioavailability of prostaglandin E_2. These data suggest that the decreased ligand-binding properties of HSA due to its redox change also contribute to an impaired innate immune system in cirrhosis patients.

The clinical relevance of HSA dysfunction in liver disease is supported by studies showing a reduction in the severity of hepatic encephalopathy and improved survival of cirrhotic patients as the result of an infusion of HSA (Jalan and Kapoor 2004; Sort et al. 1999).

3.4.1.2 Cysteinylated and N-Terminal Truncated Isoform of HSA in Cirrhosis

Domenicali et al. attempted to identify the structural alterations of HSA in 168 cirrhosis patients with stable condition or with acute clinical complications and determined their relationship with specific clinical complications and patient survival (Domenicali et al. 2014). Using HPLC/ESI-TOFMS, they identified seven HSA isoforms, including (1) truncation of the last two amino acid residues at the N-terminal portion (HSA-DA), (2) truncation of the last amino acid residue at the C-terminal portion (HSA-L), (3) cysteinylation of the Cys-34 residue (Cys-Cys34-HSA), (4) sulfinylation of the Cys-34 residue (HSA-SO_2H), (5) glycosylation (HSA+GLYC) in addition to two combinations of (6) cysteinylated with the N-terminal truncated form (Cys-Cys34-HSA-DA) or the (7) glycosylated form (Cys-Cys34-HSA+GLYC). In patients with cirrhosis, the unchanged form of HMA was significantly reduced, while the levels of cysteinylated (Cys-Cys34-HSA, Cys-Cys34-HSA-DA, and Cys-Cys34-HSA+GLYC) and glycosylated (HSA+GLYC) forms were increased. No changes in other isoforms (HSA-DA, HSA-SO_2H, and HSA-L) were observed between healthy and cirrhosis patients. Among these isoforms, they demonstrated that the unchanged forms HMA and Cys-Cys34-HSA-DA were predictors of 1-year survival, with greater prognostic accuracy than the total HSA concentration. From these data, they proposed that the concept of the "effective albumin concentration" is an important value which implies that HSA function is related not only to its serum concentration but also to the preservation of its structural integrity in patients with cirrhosis.

3.4.1.3 HSA Dimer in Cirrhosis

Naldi et al. reported that the HSA dimer/monomer ratio was also increased in plasma from cirrhosis patients compared to that from healthy subjects (Naldi et al. 2015), suggesting the dimer/monomer ratio might be useful as a biomarker for liver disease. The dimeric form of HSA was characterized by using ESI-TOF and MALDI-TOFMS techniques. The N- or C-terminal truncated HSA and native HSA undergo dimerization to form homo- and heterodimeric forms of HSA. The dimerization site was shown to be at Cys-34, involving the formation of a disulfide bridge between two HSA molecules.

3.4.1.4 Ischemia-Modified Albumin in Cirrhosis

Ischemia-modified albumin (IMA) has been developed and licensed for routine clinical applications as a cardiac biomarker in the EU and US. IMA is a test that measures the cobalt-binding capacity of HSA. The strong binding of metal ions such as Co^{2+}, Cu^{2+}, Ni^{2+}, and Al^{3+} is associated with the three N-terminal amino acid residues (Asp1 (D)-Ala2 (A)-His3 (H)) in HSA (Sadler et al. 1994). However, the N-terminal portion of HSA is susceptible to biochemical degradation and is less stable than native HSA (Chan et al. 1995). Therefore, IMA has the potential for assessing patients with conditions associated with oxidative stress. Jalan et al. demonstrated that the functional capacity of HSA is altered in patients with cirrhosis (Jalan et al. 2009). They showed that the IMA/albumin ratio (IMAR) was significantly higher in acute chronic liver failure than in healthy or cirrhosis patients, and a Kaplan-Meier analysis confirmed an increased mortality in the group with an IMAR>0.02. Thus, IMAR is correlated with disease severity and may have prognostic use in acute chronic liver failure.

3.4.2 *Chronic Kidney Disease*

3.4.2.1 Redox State of Cys-34 in Chronic Kidney Disease

Oxidative stress has been proposed to play an important role in the progression of chronic kidney disease and the development of its complication such a cardiovascular disease. Terawaki et al. measured the redox state of Cys-34 by HPLC methods in 55 non-dialysis patients with chronic kidney disease (Terawaki et al. 2004). They showed that the fraction of total HNA (HNA1+HNA2) and HNA2 was increased with decreasing renal function, and a significant positive correlation with serum creatinine was found, demonstrating that the oxidative stress is correlated with the degree of renal function, even before dialysis. In the same year, we also reported that increased HNA1 and HNA2 levels were observed in hemodialysis patients, while the level of HMA was decreased in the same patients (Anraku et al. 2004). In patients with focal segmental glomerulosclerosis, Musante et al. reported that HSA had undergone massive oxidation, in which Cys-34 was oxidized to a sulfonic group (HSA-SO_3^-) (Musante et al. 2006, 2007). Recently, Suzuki et al. showed that an increase in total cysteine in plasma and a decrease of HMA were accompanied by a decrease in renal function in nondiabetic chronic kidney disease patients (Suzuki et al. 2014). These data suggest that an analysis of the redox state of Cys-34 is a potentially useful method for the quantitative and qualitative evaluation of oxidative stress and the degree of renal dysfunction in pre- or post-hemodialysis patients.

In our animal experiments, the treatment of AST-120 (Kremezin®), an oral carbonaceous adsorbent prescribed for pre-dialysis patients, reduced the level of oxidized Cys-34 in 5/6-nephrectomized rats as a consequence of the removal of uremic toxins from the systematic circulation (Shimoishi et al. 2007). In hemodialysis patients, we also showed that the oral administration of AST-120 reduced the oxidized HSA ratio in addition to a continuous reduction of some uremic toxins (Yamamoto et al. 2015). Furthermore, olmesartan and termisartan, angiotensin II type 1 receptor antagonists, effectively reduced the extent of oxidized Cys-34 in hemodialysis patients beyond their blood-lowering actions (Kadowaki et al. 2007). We also determined the effect of a HSA-thioredoxin fusion protein, a long-acting redox active form of thioredoxin, on oxidized Cys-34 in mice with a rhabdomyolysis-induced kidney injury. The level of oxidized Cys-34 in the plasma fraction was significantly increased in this model but was restored by a HSA-thioredoxin treatment (Nishida et al. 2015). Thus, monitoring of the redox state of Cys-34 could not only be a useful marker for evaluating the progression of oxidative stress in chronic kidney disease but also in predicting therapeutic efficacy (Fig. 3.5).

Terawaki et al. also evaluated whether changes in the redox state of Cys-34 influence the incidence of cardiovascular disease in 86 patients who were undergoing hemodialysis (Terawaki et al. 2010, 2011). During their 2-year follow-up periods, 20 patients had experienced cardiovascular event. As a result, the value for the fraction of HMA in patients who experienced a cardiovascular event was significantly

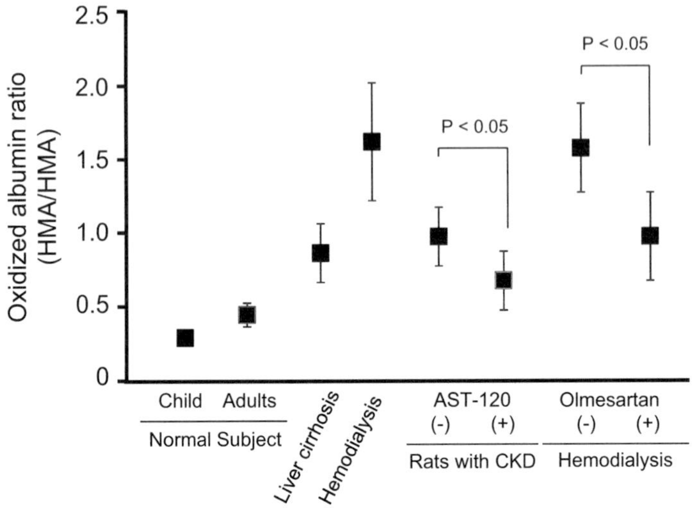

Fig. 3.5 Change in oxidized HSA ratio in healthy subjects and patients (Anraku et al. 2008, 2013; Shimoishi et al. 2007; Kadowaki et al. 2007). Administration of AST-120 to 5/6 nephrectomized (CKD) rats reduced the oxidized HSA ratio. Administration of olmesartan lowered the extent of oxidized HSA ratio in hemodialysis patients

lower than that in the absence of a cardiovascular event, and the lower HMA value was indicative of an increased cardiovascular mortality, demonstrating that the redox state of Cys-34 is closely related to serious cardiovascular incidents and mortality among patients who are undergoing hemodialysis.

In patients with continuous ambulatory peritoneal dialysis, the total HNA fraction was higher than that in healthy subjects (Terawaki et al. 2007). Lim et al. recently investigated the relationship between the redox state of Cys-34 from serum and the transport type of the peritoneal membrane in a cohort of 80 patients with end-stage renal disease receiving peritoneal dialysis (Lim et al. 2015). Very interestingly, serum HMA levels were significantly higher in patients with a high transport status than those with a low transport status, indicating that higher serum HMA levels appear to be associated with high peritoneal membrane transport characteristics in peritoneal dialysis patients.

3.4.2.2 Carbamylation of HSA in Chronic Kidney Disease

Protein carbamylation is an unavoidable consequence of the presence of excess urea. Berg et al. measured the carbamylation of HSA and tested whether HSA carbamylation was correlated with outcomes in patients with end-stage renal disease (Berg et al. 2013). As a result, they identified HSA carbamylation as a risk factor for mortality in patients with end-stage renal disease. Afterward, Drechsler et al. also reported that, in diabetic patients with end-stage renal disease, HSA carbamylation was strongly associated with a 1-year adjusted risk of cardiovascular mortality, sudden cardiac death, and a 4-year risk of death from congestive heart failure (Drechsler et al. 2015).

3.4.3 *Diabetes Mellitus*

Suzuki et al. first reported that the reduced form of Cys-34 HSA was significantly lower in patients with diabetes mellitus (Suzuki et al. 1992). In addition, the reduced form of Cys-34 HSA was lower in poorly controlled patients than in well-controlled patients, suggesting the presence of a rapidly altered oxidative change in HSA due to hyperglycemia (Suzuki et al. 1992). Moreover, recent analyses using ESI-TOFMS indicate that a significant amount of Cys-Cys34-HSA is present in the amniotic fluid of patients with gestational diabetes mellitus (Boisvert et al. 2010). Their results demonstrated that HSA in the amniotic fluid is in a highly oxidized state and that the increased oxidative stress associated with gestational diabetes mellitus alters the HSA in the amniotic fluid toward the formation of irreversibly oxidized isoforms. Bar-Or et al. also found that maternal serum HSA during gestational diabetes mellitus was predominantly oxidized as the cysteinylated isoform (Bar-Or et al. 2005b). Thus, direct measurements of maternal serum HSA oxidation might provide an even better, and less invasive, method of assessing pregnancy-associated oxidative stress and may have the potential for use in the detection of gestational diabetes mellitus.

3.5 Posttranslational Modifications and the Functional Alterations of Commercial HSA Preparations: Cys-Cys34-HSA as a Predictive Marker for the Functional Impairment of Albumin Preparations

HSA preparations have long been used as a multipurpose plasma substitute in clinical practice, such as the emergency treatment for shock, restoring blood volume, and acute management of burns, and in clinical situations associated with hypoproteinemia (Peters 1996; Garcia-Martinez et al. 2015). HSA is subject to a variety of posttranslational modifications in vivo, as mentioned above. Thus, albumin products prepared from such pooled plasma sources can be posttranslationally modified (Bar-Or et al. 2005a). In addition, during its preparation from pooled plasma and subsequent storage, HSA can also be modified posttranslationally. Such modifications often compromise albumin's functions other than its osmotic pressure maintaining function.

We estimated the degree of posttranslational modifications and the differences in the function of HSA molecules derived from five kinds of albumin preparations that are currently commercially available in Japan, in order to examine the relationship between posttranslational modifications and the functional impairment of HSA (Fig. 3.6) (Miyamura et al. 2016). We used ESI-TOFMS to evaluate the degree of posttranslational modification of the entire HSA molecule and found that the fraction of Cys-Cys34-HSA varied substantially among the albumin preparations. Meanwhile, no remarkable difference was found in the degree of glycated or

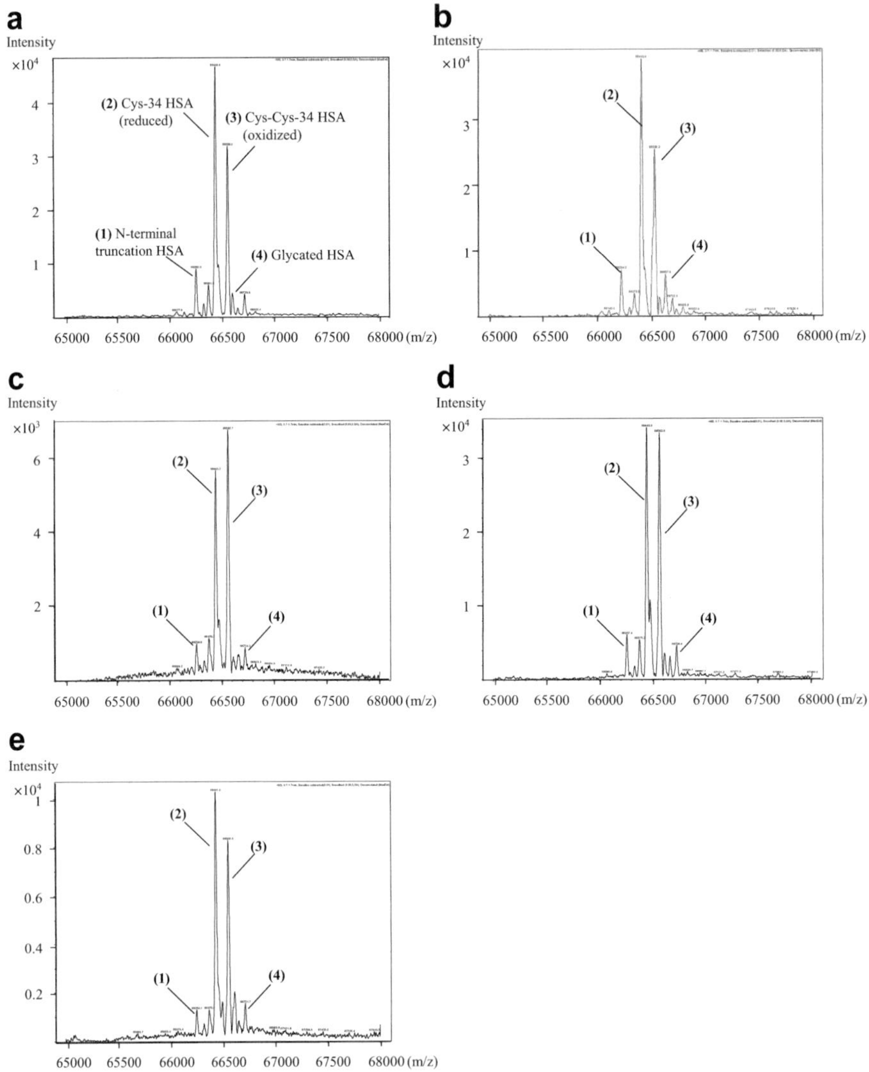

Fig. 3.6 Representative deconvoluted ESI-TOFMS spectra of HSA in five types of commercial albumin preparations. (**a**) KAKETSUKEN; (**b**) BENESIS: (**c**) CSL Behring; (**d**) NICHIYAKU; (**e**) JBPO (Miyamura et al. 2016)

N-terminal truncated HSA among the preparations tested. The non-osmotic pressure maintenance functions of HSA, such as its anti-oxidative and ligand-binding activities, significantly differed among the preparations. Interestingly, the alterations of these functions showed a significantly negative correlation only with the Cys-Cys34-HSA fraction, indicating that lower levels of Cys-Cys34-HSA were more likely to retain these functions. These observations suggest the usefulness of determining the Cys-Cys34-HSA fraction by ESI-TOFMS as a quality measure to estimate the "effective albumin concentration" in preparations. Preparations with a low degree of cysteinylation would be preferable for use in clinical practice (Miyamura et al. 2016).

3.6 Conclusion

This chapter summarizes the available evidence for the use of HSA as biomarker, especially the impact of the posttranslational modification of HSA such as oxidation, glycation, truncation, dimerization, and carbamylation under chronic liver disease, chronic renal disease, and diabetes mellitus. We developed a simple and rapid analytical method that permits the redox state of Cys-34 to be quantitatively and qualitatively evaluated with a high degree of sensitivity using an ESI-TOFMS technique. Monitoring the redox state of Cys-34 could not only be a useful marker for evaluating the progression of oxidative stress-related diseases and subsequent complications but also in predicting therapeutic efficacy. In addition, usefulness of determining the redox state of Cys-34 as a quality measure of HSA preparations was also shown. These data strongly point to the importance of monitoring the posttranslational modifications of HSA, which implies that HSA function is related, not only to its serum concentration but also to the preservation of its structural integrity in disease conditions.

Acknowledgment We are grateful to Dr. Masaki Otagiri, Faculty of Pharmaceutical Sciences, Sojo University, Japan; Dr. Victor T. G. Chuang, School of Pharmacy, Faculty of Health Sciences, Curtin Health Innovation Research Institute, Curtin University, Australia; and Dr. Makoto Anraku, Faculty of Pharmaceutical Sciences, Sojo University, Japan, for their valuable advices and kind consideration.

References

American Diabetes Association (2013) Diagnosis and classification of diabetes mellitus. Diabetes Care 36(Suppl 1):S67–S74

Anraku M, Kitamura K, Shinohara A, Adachi M, Suenga A, Maruyama T, Miyanaka K, Miyoshi T, Shiraishi N, Nonoguchi H, Otagiri M, Tomita K, Suenaga A (2004) Intravenous iron administration induces oxidation of serum albumin in hemodialysis patients. Kidney Int 66:841–848

Anraku M, Kitamura K, Shintomo R, Takeuchi K, Ikeda H, Nagano J, Ko T, Mera K, Tomita K, Otagiri M (2008) Effect of intravenous iron administration frequency on AOPP and inflammatory biomarkers in chronic hemodialysis patients: a pilot study. Clin Biochem 41:1168–1174
Anraku M, Takeuchi K, Watanabe H, Kadowaki D, Kitamura K, Tomita K, Kuniyasu A, Suenaga A, Maruyama T, Otagiri M (2011) Quantitative analysis of cysteine-34 on the anitioxidative properties of human serum albumin in hemodialysis patients. J Pharm Sci 100:3968–3976
Anraku M, Chuang VT, Maruyama T, Otagiri M (2013) Redox properties of serum albumin. Biochim Biophys Acta 1830:5465–5472
Association AD (2013) Diagnosis and classification of diabetes mellitus. Diabetes Care 36(Suppl 1):S67–S74
Bar-Or D, Bar-Or R, Rael LT, Gardner DK, Slone DS, Craun ML (2005a) Heterogeneity and oxidation status of commercial human albumin preparations in clinical use. Crit Care Med 33:1638–1641
Bar-Or D, Heyborne KD, Bar-Or R, Rael LT, Winkler JV, Navot D (2005b) Cysteinylation of maternal plasma albumin and its association with intrauterine growth restriction. Prenat Diagn 25:245–249
Berg AH, Drechsler C, Wenger J, Buccafusca R, Hod T, Kalim S, Ramma W, Parikh SM, Steen H, Friedman DJ, Danziger J, Wanner C, Thadhani R, Karumanchi SA (2013) Carbamylation of serum albumin as a risk factor for mortality in patients with kidney failure. Sci Transl Med 5:175ra129
Boisvert MR, Koski KG, Skinner CD (2010) Increased oxidative modifications of amniotic fluid albumin in pregnancies associated with gestational diabetes mellitus. Anal Chem 82:1133–1137
Chan B, Dodsworth N, Woodrow J, Tucker A, Harris R (1995) Site-specific N-terminal autodegradation of human serum albumin. Eur J Biochem 227:524–528
Colombo G, Clerici M, Giustarini D, Rossi R, Milzani A, Dalle-Donne I (2012) Redox albuminomics: oxidized albumin in human diseases. Antioxid Redox Signal 17:1515–1527
Descamps-Latscha B, Witko-Sarsat V (2001) Importance of oxidatively modified proteins in chronic renal failure. Kidney Int Suppl 78:S108–S113
Domenicali M, Baldassarre M, Giannone FA, Naldi M, Mastroroberto M, Biselli M, Laggetta M, Patrono D, Bertucci C, Bernardi M, Caraceni P (2014) Posttranscriptional changes of serum albumin: clinical and prognostic significance in hospitalized patients with cirrhosis. Hepatology 60:1851–1860
Drechsler C, Kalim S, Wenger JB, Suntharalingam P, Hod T, Thadhani RI, Karumanchi SA, Wanner C, Berg AH (2015) Protein carbamylation is associated with heart failure and mortality in diabetic patients with end-stage renal disease. Kidney Int 87:1201–1208
Era S, Kuwata K, Imai H, Nakamura K, Hayashi T, Sogami M (1995) Age-related change in redox state of human serum albumin. Biochim Biophys Acta 1247:12–16
Furusyo N, Hayashi J (2013) Glycated albumin and diabetes mellitus. Biochim Biophys Acta 1830:5509–5514
Garcia-Martinez R, Noiret L, Sen S, Mookerjee R, Jalan R (2015) Albumin infusion improves renal blood flow autoregulation in patients with acute decompensation of cirrhosis and acute kidney injury. Liver Int 35:335–343
Hansen RE, Roth D, Winther JR (2009) Quantifying the global cellular thiol-disulfide status. Proc Natl Acad Sci U S A 106:422–427
Iwao Y, Ishima Y, Yamada J, Noguchi T, Kragh-Hansen U, Mera K, Honda D, Suenaga A, Maruyama T, Otagiri M (2012) Quantitative evaluation of the role of cysteine and methionine residues in the antioxidant activity of human serum albumin using recombinant mutants. IUBMB Life 64:450–454
Jalan R, Kapoor D (2004) Reversal of diuretic-induced hepatic encephalopathy with infusion of albumin but not colloid. Clin Sci (Lond) 106:467–474
Jalan R, Schnurr K, Mookerjee RP, Sen S, Cheshire L, Hodges S, Muravsky V, Williams R, Matthes G, Davies NA (2009) Alterations in the functional capacity of albumin in patients with decompensated cirrhosis is associated with increased mortality. Hepatology 50:555–564

Kadowaki D, Anraku M, Tasaki Y, Kitamura K, Wakamatsu S, Tomita K, Gebicki JM, Maruyama T, Otagiri M (2007) Effect of olmesartan on oxidative stress in hemodialysis patients. Hypertens Res 30:395–402

Kawakami A, Kubota K, Yamada N, Tagami U, Takehana K, Sonaka I, Suzuki E, Hirayama K (2006) Identification and characterization of oxidized human serum albumin. A slight structural change impairs its ligand-binding and antioxidant functions. FEBS J 273:3346–3357

Klammt S, Mitzner S, Stange J, Brinkmann B, Drewelow B, Emmrich J, Liebe S, Schmidt R (2007) Albumin-binding function is reduced in patients with decompensated cirrhosis and correlates inversely with severity of liver disease assessed by model for end-stage liver disease. Eur J Gastroenterol Hepatol 19:257–263

Kubota K, Nakayama A, Takehana K, Kawakami A, Yamada N, Suzuki E (2009) A simple stabilization method of reduced albumin in blood and plasma for the reduced/oxidized albumin ratio measurement. Int J Biomed Sci 5:293–301

Lim PS, Chen HP, Chen CH, Wu MY, Wu CY, Wu TK (2015) Association between redox status of serum albumin and peritoneal membrane transport properties in patients on peritoneal dialysis. Blood Purif 40:243–249

Miyamura S, Imafuku T, Anraku M, Taguchi K, Yamasaki K, Tominaga Y, Maeda H, Ishima Y, Watanabe H, Otagiri M, Maruyama T (2016) Comparison of post-translational modification and the functional impairment of human serum albumin in commercial preparations. J Pharm Sci 105:1043–1049

Musante L, Bruschi M, Candiano G, Petretto A, Dimasi N, Del Boccio P, Urbani A, Rialdi G, Ghiggeri GM (2006) Characterization of oxidation end product of plasma albumin 'in vivo'. Biochem Biophys Res Commun 349:668–673

Musante L, Candiano G, Petretto A, Bruschi M, Dimasi N, Caridi G, Pavone B, Del Boccio P, Galliano M, Urbani A, Scolari F, Vincenti F, Ghiggeri GM (2007) Active focal segmental glomerulosclerosis is associated with massive oxidation of plasma albumin. J Am Soc Nephrol 18:799–810

Nagumo K, Tanaka M, Chuang VT, Setoyama H, Watanabe H, Yamada N, Kubota K, Matsushita K, Yoshida A, Jinnouchi H, Anraku M, Kadowaki D, Ishima Y, Sasaki Y, Otagiri M, Maruyama T (2014) Cys34-cysteinylated human serum albumin is a sensitive plasma marker in oxidative stress-related chronic diseases. PLoS One 9:e85216

Naldi M, Baldassarre M, Nati M, Laggetta M, Giannone FA, Domenicali M, Bernardi M, Caraceni P, Bertucci C (2015) Mass spectrometric characterization of human serum albumin dimer: a new potential biomarker in chronic liver diseases. J Pharm Biomed Anal 112:169–175

Nishida K, Watanabe H, Ogaki S, Kodama A, Tanaka R, Imafuku T, Ishima Y, Chuang VT, Toyoda M, Kondoh M, Wu Q, Fukagawa M, Otagiri M, Maruyama T (2015) Renoprotective effect of long acting thioredoxin by modulating oxidative stress and macrophage migration inhibitory factor against rhabdomyolysis-associated acute kidney injury. Sci Rep 5:14471

O'Brien AJ, Fullerton JN, Massey KA, Auld G, Sewell G, James S, Newson J, Karra E, Winstanley A, Alazawi W, Garcia-Martinez R, Cordoba J, Nicolaou A, Gilroy DW (2014) Immunosuppression in acutely decompensated cirrhosis is mediated by prostaglandin E2. Nat Med 20:518–523

Oettl K, Stadlbauer V, Petter F, Greilberger J, Putz-Bankuti C, Hallström S, Lackner C, Stauber RE (2008) Oxidative damage of albumin in advanced liver disease. Biochim Biophys Acta 1782:469–473

Oettl K, Birner-Gruenberger R, Spindelboeck W, Stueger HP, Dorn L, Stadlbauer V, Putz-Bankuti C, Krisper P, Graziadei I, Vogel W, Lackner C, Stauber RE (2013) Oxidative albumin damage in chronic liver failure: relation to albumin binding capacity, liver dysfunction and survival. J Hepatol 59:978–983

Pergola PE, Krauth M, Huff JW, Ferguson DA, Ruiz S, Meyer CJ, Warnock DG (2011) Effect of bardoxolone methyl on kidney function in patients with T2D and Stage 3b-4 CKD. Am J Nephrol 33:469–476

Peters TJ (1996) All about albumin: biochemistry, genetics, and medical applications. Academic, San Diego
Poli G (2000) Pathogenesis of liver fibrosis: role of oxidative stress. Mol Aspects Med 21:49–98
Requejo R, Hurd TR, Costa NJ, Murphy MP (2010) Cysteine residues exposed on protein surfaces are the dominant intramitochondrial thiol and may protect against oxidative damage. FEBS J 277:1465–1480
Sadler PJ, Tucker A, Viles JH (1994) Involvement of a lysine residue in the N-terminal Ni2+ and Cu2+ binding site of serum albumins. Comparison with Co2+, Cd2+ and Al3+. Eur J Biochem 220:193–200
Sakata M, Kawaguchi T, Taniguchi E, Nakayama A, Ishizaki S, Sonaka I, Maganuma M, Nakamura T, Itou M, Oriishi T, Abe M, Yanagimoto C, Koga H, Harada M, Sakamoto T, Oda S, Sata M (2010) Redox state of albumin is not associated with colloid osmotic pressure. Mol Med Rep 3:685–687
Sen S, Williams R, Jalan R (2002) The pathophysiological basis of acute-on-chronic liver failure. Liver 22(Suppl 2):5–13
Shimoishi K, Anraku M, Kitamura K, Tasaki Y, Taguchi K, Hashimoto M, Fukunaga E, Maruyama T, Otagiri M (2007) An oral adsorbent, AST-120 protects against the progression of oxidative stress by reducing the accumulation of indoxyl sulfate in the systemic circulation in renal failure. Pharm Res 24:1283–1289
Sogami M, Era S, Nagaoka S, Kuwata K, Kida K, Shigemi J, Miura K, Suzuki E, Muto Y, Tomita E (1985) High-performance liquid chromatographic studies on non-mercapt in equilibrium with mercapt conversion of human serum albumin. II. J Chromatogr 332:19–27
Sort P, Navasa M, Arroyo V, Aldeguer X, Planas R, Ruiz-del-Arbol L, Castells L, Vargas V, Soriano G, Guevara M, Ginès P, Rodés J (1999) Effect of intravenous albumin on renal impairment and mortality in patients with cirrhosis and spontaneous bacterial peritonitis. N Engl J Med 341:403–409
Stauber RE, Spindelboeck W, Haas J, Putz-Bankuti C, Stadlbauer V, Lackner C, Oettl K (2014) Human nonmercaptalbumin-2: a novel prognostic marker in chronic liver failure. Ther Apher Dial 18:74–78
Suzuki E, Yasuda K, Takeda N, Sakata S, Era S, Kuwata K, Sogami M, Miura K (1992) Increased oxidized form of human serum albumin in patients with diabetes mellitus. Diabetes Res Clin Pract 18:153–158
Suzuki Y, Suda K, Matsuyama Y, Era S, Soejima A (2014) Close relationship between redox state of human serum albumin and serum cysteine levels in non-diabetic CKD patients with various degrees of renal function. Clin Nephrol 82:320–325
Terawaki H, Yoshimura K, Hasegawa T, Matsuyama Y, Negawa T, Yamada K, Matsushima M, Nakayama M, Hosoya T, Era S (2004) Oxidative stress is enhanced in correlation with renal dysfunction: examination with the redox state of albumin. Kidney Int 66:1988–1993
Terawaki H, Matsuyama Y, Era S, Matsuo N, Ikeda M, Ogura M, Yokoyama K, Yamamoto H, Hosoya T, Nakayama M (2007) Elevated oxidative stress measured as albumin redox state in continuous ambulatory peritoneal dialysis patients correlates with small uraemic solutes. Nephrol Dial Transplant 22:968
Terawaki H, Takada Y, Era S, Funakoshi Y, Nakayama K, Nakayama M, Ogura M, Ito S, Hosoya T (2010) The redox state of albumin and serious cardiovascular incidence in hemodialysis patients. Ther Apher Dial 14:465–471
Terawaki H, Era S, Nakayama M, Hosoya T (2011) Decrease in reduced-form albumin among chronic kidney disease patients: new insights in cardiovascular complications. Ther Apher Dial 15:156–160
Torres MJ, Turell L, Botti H, Antmann L, Carballal S, Ferrer-Sueta G, Radi R, Alvarez B (2012) Modulation of the reactivity of the thiol of human serum albumin and its sulfenic derivative by fatty acids. Arch Biochem Biophys 521:102–110
Turell L, Botti H, Carballal S, Ferrer-Sueta G, Souza JM, Durán R, Freeman BA, Radi R, Alvarez B (2008) Reactivity of sulfenic acid in human serum albumin. Biochemistry 47:358–367

Turell L, Radi R, Alvarez B (2013) The thiol pool in human plasma: the central contribution of albumin to redox processes. Free Radic Biol Med 65:244–253
Turell L, Botti H, Bonilla L, Torres MJ, Schopfer F, Freeman BA, Armas L, Ricciardi A, Alvarez B, Radi R (2014) HPLC separation of human serum albumin isoforms based on their isoelectric points. J Chromatogr B Analyt Technol Biomed Life Sci 944:144–151
Yamamoto S, Kazama JJ, Omori K, Matsuo K, Takahashi Y, Kawamura K, Matsuto T, Watanabe H, Maruyama T, Narita I (2015) Continuous reduction of protein-bound uraemic toxins with improved oxidative stress by using the oral charcoal adsorbent AST-120 in haemodialysis patients. Sci Rep 5:14381

Chapter 4
Albumin Fusion Protein

Toru Maruyama, Victor Tuan Giam Chuang, and Masaki Otagiri

Abstract Albumin fusion technology can improve the blood retention properties of protein-based pharmaceutics that have a short circulation half-life. The procedure consists of fusing the HSA gene with the gene for a therapeutic protein or peptide, thus permitting the existing recombinant protein therapeutics to be reengineered using albumin as a suitable and safe carrier. Improving the pharmacokinetic properties of biopharmaceutics by albumin fusion technology could lead to greatly extending the pharmacological activity of such pharmaceutics over a much long period of time and would permit a bolus intravenous administration with less dosing frequency, making the treatment more convenient and cost-effective. Albumin fusion technology also has benefits including improving the stability, versatility, safety, and ease of manufacture for various biopharmaceutics. Moreover, recent studies have extended the boundaries of albumin fusion technology in which a fusion protein is delivered to targeted organs and cells or to permit the efficient intracellular delivery of partner biopharmaceutics.

Keywords Albumin fusion technology • Blood circulation • Active targeting

4.1 Introduction

To avoid rapid degradation in the digestive tract, the majority of biopharmaceuticals are administered by injection. However, a number of these proteins are also rapidly cleared from the blood. Therefore, relatively high doses and/or frequent injections

T. Maruyama (✉)
Department of Biopharmaceutics, Graduate School of Pharmaceutical Sciences, Kumamoto University, 5-1, Oe-honmachi, Chuo-ku, Kumamoto 862-0973, Japan

Center for Clinical Pharmaceutical Sciences, School of Pharmacy, Kumamoto University, 5-1, Oe-honmachi, Chuo-ku, Kumamoto 862-0973, Japan
e-mail: tomaru@kumamoto-u.ac.jp

V.T.G. Chuang
School of Pharmacy, Faculty of Health Sciences, Curtin University, GPO Box U1987, Perth 6845, WA, Australia

M. Otagiri
Faculty of Pharmaceutical Sciences, Sojo University, 4-22-1 Ikeda, Kumamoto 860-0082, Japan

M. Otagiri, V.T.G. Chuang (eds.), *Albumin in Medicine*,
DOI 10.1007/978-981-10-2116-9_4

are typically required to maintain therapeutic blood levels, potentially resulting in increased side effects and treatment costs, as well as a significant reduction in treatment compliance and the patient's quality of life.

A number of attempts have been made to improve the retention of biologically active peptides in the blood, such as nanoparticulation with polylactic acid, macromolecularization via micelle formation, peglytization (PEG) with polyethylene glycol, and other different types of drug delivery systems (Werle and Bernkop-Schnurch 2006; Chapman 2002; Sinclair and Elliott 2005). The advancements in recombinant DNA technology allow novel proteins with desirable properties to be engineered and produced, e.g., the production of two genetically fused protein molecules. When a protein possesses favorable blood retention properties as a carrier, the use of genetic fusion can improve the plasma half-life of a protein drug.

Recombinant HSA (rHSA) has been successfully produced using a cell culture system, and its safety profile has been established (Chuang and Otagiri 2007). Since HSA has an elimination half-life in the blood circulation of about 20 days in the human body, due to its large molecular size (MW: 66.5 kDa) and negative electrical potential which inhibits glomerular filtration, this results in vascular permeability or decreased lymphatic clearance and FcRn-mediated recycling in endothelium (Chuang and Otagiri 2007). Therefore, recombinant HSA would also be expected to be useful as a carrier for drug delivery systems. In this chapter, we introduce a novel approach for prolonging the half-life of protein drugs in the blood using rHSA, a process that is referred to as "albumin fusion technology" and its potential for clinical applications.

4.2 Principles of Albumin Fusion Technology

4.2.1 Basis of Albumin Fusion Technology and Development of Innovative Biopharmaceuticals

Albumination, in which albumin molecule is attached to a biologics, has been applied to proteins or peptides in case where the pharmacokinetics need to be improved, especially blood retention as a biotherapeutics. Albumin fusion technology which consists of fusing the HSA gene with the gene for a therapeutic protein or peptide, may allow existing recombinant protein therapeutics to be re-engineered using albumin as a suitable and safe carrier (Chuang et al. 2002). This fused gene can then be used as a signal to produce an albumin fusion protein, using suitable expression systems. It is noteworthy that albumin fusion technology has received considerable attention for prolonging the plasma half-life of biologically active peptides or low molecular weight proteins with poor blood retention. With a more stable plasma concentration that falls within the therapeutic range, the use of a fusion protein has the potential to improve the efficacy of therapy with less dosing frequencies. This could result in numerous benefits, including improving the safety of a medication, reduced medical wastes, etc. and, most importantly, the cost-effectiveness of the treatment. In addition to the extended blood retention, other benefits of albumin fusion technology include improving the stability, versatility, safety, and ease of manufacture for a biopharmaceutics.

4.2.2 Albumin Fusion Protein in Clinical Trial

As of this writing, more than 40 successful attempts to use albumin fusion technology for delivering small-sized proteins or peptides have been reported. Table 4.1 summarizes a list of albumin fusion proteins that have been considered and potential therapeutic applications. Most of them are still in the preclinical stage, but some have proceeded in the clinical trial stage. For instance, albinterferon-α2b, a long-acting interferon (IFN) formulation, was developed for the treatment of chronic hepatitis C (Subramanian et al. 2007; Rustgi 2009). This agent exhibits a prolonged half-life and duration of antiviral activity, indicating that it could be administered with dosing intervals of 2–4 weeks (Zeuzem et al. 2010). Phase 2 clinical trials in prior IFN nonresponders and IFN-naïve patients with genotype 1 or 2/3 chronic hepatitis C have shown antiviral activity and an acceptable safety/tolerance for albinterferon-α2b.

In 2014, albiglutide (Eperzan™ and Tanzeum™), the first therapeutics that used albumin fusion technology, was approved by the FDA and the European Medicines Agency. Albiglutide, a glucagon-like peptide-1(GLP-1) receptor agonist, is administered subcutaneously at weekly intervals and is marketed by GlaxoSmithKline for the treatment of type 2 diabetes. In albiglutide, HSA is fused with two tandem copies of a modified GLP-1 which confers resistance to dipeptidyl peptidase IV-mediated proteolysis (Tomkin 2009). It has been demonstrated that weekly albiglutide administration significantly improved glycemic control and elicited weight loss in type 2 diabetic patients, with a favorable safety and tolerability profile in a randomized, multicenter, double-blind, parallel-group study (Pratley et al. 2014).

HSA-coagulation factor IX (HSA-IX: CSL654) has been developed for use as a long-acting coagulant for the treatment of hemophilia B by CSL Behring (Nolte et al. 2012; Santagostino et al. 2012). A global phase 3 trial was conducted by the PROLONG-9FP Investigators Study Group using male patients who had been previously treated for severe hemophilia B at two dosing regimens, 40 IU/kg weekly and 75 IU/kg biweekly (Santagostino et al. 2016). This result indicated that HSA-IX was safe and effective for preventing and treating bleeding episodes in patients with hemophilia B at both dosing regimens. Moreover, HSA-human growth hormone (hGH) (TV-1106) for treating a growth hormone deficiency, HSA-human interleukin (IL)-2 (Albuleukin) for malignancy, HSA-human granulocyte colony-stimulating factor (G-CSF: GW003) for neutropenia, HSA-coagulation factor VIIa (CSL689) for hemophilia A or B, HSA-IFN-α2a for viral infection, and Her3-HSA-Her2 bispecific antibody fusion (MM-111) for malignancy are also in phase 1 or 2 clinical trial.

Table 4.1 A list of albumin fusion proteins and their potential therapeutic applications

Albumin fusion products	Potential application	Features	References
HSA-growth hormone (GH) (TV-1106, Albutropin™)	Growth hormone deficiency	Clinical trial (phase 2)	Osborn et al. (2002)
HSA-mutant GH	Obesity	Animal model	Wang et al. (2013)
HSA-granulocyte colony-stimulating factor (G-CSF)	Neutropenia	Clinical trial (phase 3 completed: withdraw after phase 3) (phase 1)	Halpern et al. (2002)
(Egranli™, GW003)		Animal model	Xu et al. (2015)
HSA-mutated G-CSF			Huang et al. (2014)
HSA domain III-G-CSF			Zhao et al. (2013a, b)
HSA-interferon (IFN)-α2b (albinterferon -α2b: Albuferon)	Chronic hepatitis C	Clinical trial (phase 3 completed: withdraw after phase 2b)	Subramanian et al. (2007)
		Animal model	Rustgi et al. (2009)
		(Releasable partner protein)	Zeuzem et al. (2010)
			Zhao et al. (2012a, b)
HSA-IFNγ	Malignancy	Animal model (gene delivery)	Miyakawa et al. (2011)
HSA-interleukin (IL)-2 (Albuleukin)	Malignancy	Animal model	Melder et al. (2005)
HSA-IL-1 receptor antagonist	Rheumatoid arthritis	Animal model (selective delivery to inflamed joint)	Liu et al. (2012)
HSA-(glucagon-like peptide (GLP) 1)2 (albiglutide)	Type 2 diabetes mellitus	Clinical trial (phase 3)	Tomkin (2009)
			Pratley et al. (2014)
HSA-atrial natriuretic peptide (ANP)	Hypertension	Animal model	de Bold et al. (2012)
HSA-B-type natriuretic peptide (BNP)	Congestive heart failure		
	Acute myocardial infraction		Wang et al. (2004)
HSA-tumor necrosis factor (TNF) ligands (TNF-α, TWEAK, TRAIL)	Malignancy	Animal model	Muller et al. (2012)

(continued)

Table 4.1 (continued)

Albumin fusion products	Potential application	Features	References
HSA-thymosin-α	Hepatitis B and C	Animal model	Chen et al. (2010)
	Malignancy		
HSA-vascular endothelial growth factor (VEGF)165b	Malignancy	Animal model	Zhu et al. (2012)
HSA-onconase	Malignancy	Animal model (releasable partner protein)	Zhao et al. (2012a, b)
HSA-tissue inhibitor of metalloproteinase (TIMP)-2	Malignancy	Animal model (tumor targeting by cRGD labeling)	Lee et al. (2011, 2012)
Cyclic arginine-glycine-aspartate (cRGD)-labeled HSA-TIMP-2	Tumor-imaging probe for SPECT and PET		Choi et al. (2012)
123I- and 68Ga-labeled cRGD-HSA-TIMP-2			
HSA-coagulation factor VIIa (CSL689)	Hemophilia A or B	Animal model	Weimer et al. (2008)
		Clinical trial (phase 1 completed)	Zollner et al. (2014)
			Golor et al. (2013)
HSA-coagulation factor IX (CSL654)	Hemophilia B	Animal model	Sheffield et al. (2004)
		Clinical trial (phase 3 completed)	Nolte et al. (2012)
			Santagostino et al. (2012, 2016)
HSA domain III-retinol-binding protein	Liver fibrosis	Animal model (stellate cell targeting)	Park et al. (2012)
			Choi et al. (2012)
			Lee et al. (2015)
HSA-Tregitope	Flaring autoimmune disease (alternative to IVIG treatment)	EpiVax-Novozyme	
HSA-exendin-4	Type 2 diabetes	Animal model	Huang et al. (2008)
			Zhang et al. (2014)
			Hou et al. (2015)
HSA-IL-28B	Chronic hepatitis C	Animal model	Zhao et al. (2013a, b)
HSA-platelet aggregation inhibitor (hirudin, barbourin)	Thrombosis	Animal model	Sheffield et al. (2007)
			Marques et al. (2001)

(continued)

Table 4.1 (continued)

Albumin fusion products	Potential application	Features	References
HSA-butyrylcholinesterase	Cocaine toxicity, addiction	Animal model	Gao et al. (2008)
HSA-insulin analog (Albulin)	Diabetes	Animal model	Duttaroy et al. (2005)
HSA-Me (consensus sequence of influenza A virus)	Influenza A virus vaccine	Animal model	Mu et al. (2016)
HSA-thioredoxin-1	Oxidative stress and inflammatory diseases	Animal model	Detail in Table 4.2
HSA-monovalent FcrRIIIA-specific antibody	Immune thrombocytopenia (autoimmune disease)	Animal model	Yu et al. (2016)
TAT-HSA-α-melanocyte-stimulating hormone	Central nerve disorder (brain delivery)	Animal model	Wang et al. (2016)
HSA-p53i or PMI	Malignancy (intracellular delivery)	Animal model	Joshi et al. (2013)
HSA-N-terminal fragment (ATF) of urokinase	Malignancy (tumor targeting carrier)	Animal model	Li et al. (2014a, b)
HSA-mutant CCL2	Inflammation	Animal model	Gerlza et al. (2015)
HSA-IFNβ	Multiple sclerosis	Animal model	Zhang et al. (2009)
HSA-parathyroid hormone (1–34)	Osteoporosis	Animal model	Wu et al. (2013)
HSA-scFv637	Immunosuppressive (myasthenia gravis)	Animal model	Li et al.(2014a, b)
HSA-Vβ	Toxic shock syndrome	Animal model	Yuan et al. (2016)
HSA-erythropoietin	Anemia	Animal model	Joung et al. (2009)
HSA-IFN-α2a	Antiviral	Clinical trial (phase 1/2)	
Her3-HSA-Her2 (bispecific antibody fusion, MM-111)	Malignancy (breast cancer, gastric cancers, esophageal)	Animal model	McDonagh et al. (2012)
			Richards et al. (2014)
		Clinical trial (phase 2)	Kontermann and Brinkmann (2015)
HSA-(thrombopoietin mimetic peptide)2	Immune thrombocytopenic purpura	Animal model	Wang et al. (2016)
HSA-somatostatin (14) or (28)	Acromegaly	Animal model	Ding et al. (2013, 2014)
	Cushing's disease		

Table 4.2 Therapeutic impact of HSA-Trx against disease model animals

Disease model	Features	References
Bleomycin-induced pulmonary fibrosis (mouse)	Lung protection by pre- and post-administration	Tanaka et al. (2013)
	Improve survival	
Ovalbumin-induced lung injury (mouse)	Lung protection	Furukawa et al. (2011)
Influenza virus-induced lung injury (mouse)	Lung protection by post-administration without affecting virus titer	Tanaka et al. (2014)
	Improve survival	
Endotoxic shock (mouse)	Improve survival	Ikuta et al. (2010)
Acetaminophen-induced hepatitis (mouse)	Hepatoprotection by pre- and post-administration	Tanaka et al. (2014)
	Improve survival	
Contrast-induced nephropathy (rat)	Prevention	Kodama et al. (2013)
	Renoprotection	
Cisplatin-induced nephropathy (mouse)	Prevention	Kodama et al. (2014)
	Renoprotection	
Rhabdomyolysis-associated acute kidney injury (mouse)	Renoprotection by pre- and post-administration	Nishida et al. (2015)
	Improve survival	

4.2.3 Albumin Fusion Protein in Preclinical Stage

In the preclinical stage, various proteins and peptides with diverse functions have been used in albumin fusion technology. Liu et al. develop a HSA-interleukin (IL)-1 receptor antagonist (HSA-IL-1Ra) (Liu et al. 2012). HSA-IL-1Ra markedly extended the circulatory half-life of IL-1Ra and resulted in a specific accumulation in arthritic paws of mice and consequently improved the therapeutic effect of the preparation in experimental arthritis model mice.

Albumin fusion technology also has been applied to create a vaccine against the influenza A virus. Mu X et al. fused the C-terminal domain of HSA with M2e consensus sequence of influenza A viruses (HSA-M2e) (Mu et al. 2016). HSA-M2e induced strong anti-M2e-specific humoral immune responses and, hence, reduced the viral load in the mice lungs and provided significant protection against a lethal challenge with an H1N1 or an H3N2 virus.

Thioredoxin-1 (Trx) is a redox-active protein that plays an important role in the maintenance of homeostasis (Arner and Holmgren 2000; Nakamura et al. 2009). Trx exerts various biological activities with antioxidative, anti-inflammatory, and antiapoptotic effects. It also interacts with macrophage migration inhibitory factor (MIF), which functions as a pleiotropic protein, participating in inflammatory and immune responses, and consequently inhibits its activities. Because of this, attempts have been made to develop it as a biopharmaceutics for the treatment of various diseases related to oxidative stress and inflammation. However, its plasma elimination half-life is only limited to within several hours. In order to overcome such inferior pharmacokinetic property of Trx, a long-acting Trx, HSA-Trx, was developed using albumin fusion technology (Ikuta et al. 2010). The physicochemical and structural properties of HSA and the fusion protein are comparable. The pharmacokinetic properties, such as plasma concentration and organ distribution profiles, were similar to that for HSA in both healthy animals and disease models. In vitro studies revealed that the biological activity of the fusion protein was somewhat reduced but was largely retained compared to Trx. The therapeutic efficacy of HSA-Trx has been demonstrated using various disease models as shown in Table 4.2. HSA-Trx suppresses various types of lung injuries bleomycin-induced pulmonary fibrosis (Tanaka et al. 2013), ovalbumin-induced lung injury (Furukawa et al. 2011), influenza virus-induced lung injury (Tanaka et al. 2014a, b), endotoxic shock (Ikuta et al. 2010), and acetaminophen-induced hepatitis (Tanaka et al. 2014a, b) by pre- and post-administration owing to its extended effects of modulating oxidative stress and MIF-induced inflammation. HSA-Trx also effectively prevented acute kidney injuries (AKI) and contrast-induced and cisplatin-induced nephropathy (Kodama et al. 2013, 2014). In addition, HSA-Trx showed a superior renoprotective effect against rhabdomyolysis-associated AKI and improved the survival rate, even though it was post-administered (Nishida et al. 2015). Thus, HSA-Trx has some potential as a biotherapeutics for the treatment of oxidative and inflammatory-related diseases.

Albumin fusion technology has also been applied to the development of therapeutic antibodies such as various types of single-chain variable fragment (scFv), a bispecific monoclonal antibody (bsAb), etc., to extend their half-life and the targeting of drugs to their specific recognition site (Muller et al. 2007; Leung 2008). For example, Fcr receptor (FcrR)-specific antibodies are expected to improve immune thrombocytopenia in refractory human patients because this type of thrombocytopenia is induced by the interaction between antiplatelet antibodies and FcrR. However, FcrR-specific antibodies induced inflammatory responses that could be related to Fc function. To overcome such an adverse reaction of FcrR-specific antibodies, Yu et al. designed a novel FcrR that is a fusion protein between the monovalent FcrRIIIA-specific antibody and HSA (Yu et al. 2016). This fusion protein effectively inhibited therapeutic responses against an antibody-dependent ITP model and reduced inflammatory responses induced by FcrR-specific antibodies. Thus, it is a potential candidate for the immunosuppressive therapy of Fcr receptor-mediated autoimmune disease.

4.2.4 *Effect of Linkers on the Properties of Albumin Fusion Proteins*

The expression efficacy and biological activity of a fusion protein may be influenced by both the type and length of the linker. Yang et al. attempted to clarify this issue using HSA-L(n)-onconase, a member of the pancreatic ribonuclease A superfamily, with linkers having four different lengths: L0, no linker; L1, (GGGGS)1; L2, (GGGGS)2; and L3, (GGGGS)3 (Yang et al. 2015). They found that linker length was not associated with the amount of fusion protein expressed, while its cytotoxic effect against tumor cells in vitro was enhanced with increasing linker length; the maximal effect was obtained in the case of the longest linker length. Therefore, it is possible that the biological activity of a fusion protein is influenced by the length of the linker.

Albumin fusion technology sometimes causes the intrinsic biological activities of fusion systems to be suppressed, due to the steric hindrance effect of HSA. To overcome such a disadvantage associated with conventional techniques, a cleavable fusion technology in which the linker switched from non-cleavable to cleavable was developed. This permits the intact fusion partner protein to be released with full activity. The cleavage of the linker is achieved by the introduction of protease cleavage sites or a disulfide linkage in the linkers between the two proteins. For instance, Zhao et al. successfully designed an IFN-α2b releasable albumin fusion protein by connecting two proteins with a cleavable peptide linker, and the results indicated that this new type of fusion protein improves the balanced pharmacokinetics and pharmacodynamics of IFN-α2b (Zhao et al. 2012a, b). The same research group also applied this technique to onconase and to tethering onconase to albumin by a cleavable disulfide linker to increase its potential as an antitumor agent (Zhao et al. 2012a, b).

4.2.5 *Immunogenicity of Albumin Fusion Protein*

Based on extensive in vivo studies and clinical trials, it is generally thought that HSA fusion proteins do not induce immune responses against endogenous HSA, even for HSA fused to immunostimulating cytokines, such as IFN-α2b. However, a recent animal study using cynomolgus monkeys revealed that a neutralizing antibody against G-CSF was induced after a multidose exposure of HSA-G-CSF (GW003) (Xu et al. 2015). In addition, Zhao et al. found that HSA-IFN-α2b was prone to disulfide-linked aggregation during accelerated mechanism and thermal stress tests, and such aggregates was associated with an increase in immunogenicity in mice (Zhao et al. 2009). This can be attributed by the intermolecular disulfide bond between Cys34 and HSA-IFN-α2b. To solve this, they produced HSA(C34S)-IFN-α2b in which the Cys34 in HSA was replaced with serine to prevent disulfide bond formation. HSA(C34S)-IFN-α2b showed improved stability against mechanical and thermal stress, with similar pharmacokinetic properties as HSA-IFN-α2b. Therefore, removing the free sulfhydryl group in Cys34 of an HSA fusion protein appears to be a better approach for improving the stability of a fusion protein.

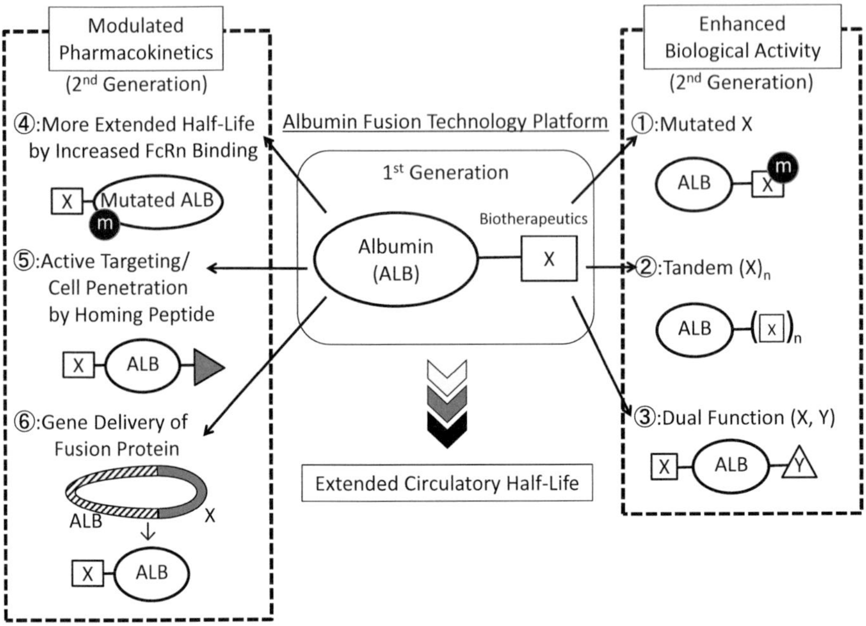

Fig. 4.1 Diversity of albumin fusion technology

4.3 Diversity of Albumin Fusion Technology Platform

Attempts have been made to establish a second generation of albumin fusion technology to create better biopharmaceuticals with improved pharmacokinetics and pharmacodynamics as shown in Fig. 4.1.

4.3.1 Improving Biological Activity of Albumin Fusion Protein

To improve the biological activity of the fusion protein, three approaches have been tested: (1) using a mutated protein with enhanced biological activity as a partner protein, (2) increasing the number of protein copies fused to HSA, and (3) fusing two different partner proteins to the N- and C-terminal of HSA to obtain dual functional properties.

4.3.1.1 Mutated Protein with Enhanced Biological Activity as a Partner Protein

Kungl's group produced a chemokine CCL2 mutant that improved the glycosaminoglycans on the endothelium of inflamed tissue. They further designed an HSA-CCL2 mutant fusion protein to extend the circulatory half-life of mutant (Gerlza et al. 2015). Interestingly, this fusion protein enhanced the affinity of the product to glycosaminoglycans as compared to unfused CCL2 mutant.

Wang et al. replaced the amino terminus of human growth hormone (hGH) to retain its biological lipolytic function. Moreover, they further produced HSA-mutated hGH to extend in vivo half-life of mutated hGH as an antiobesity agent (Wang et al. 2013).

4.3.1.2 Multi-copies of Partner Protein

Increasing the number of therapeutic protein copies in a fusion protein is an attractive approach for enhancing the biological activity of the therapeutic protein. In fact, albiglutide is composed of two tandem copies of GLP-1. Ding et al. examined the effect of the number of protein copies on the biological activity of an albumin fusion protein (Ding et al. 2013). They produced three types of HSA fusion proteins with somatostatin fragment (SS14), namely, (SS14)2-HSA, (SS14)3-HSA, and HSA-(SS14)3 in an attempt to create long-acting SS14 derivatives. Among them, (SS14)2-HSA was found to be the most effective fusion protein from the standpoint of both expression yield and bioactivity. Same results were also observed for SS28 fused with HSA: (SS28)2-HSA, (SS28)3-HSA, and HSA-(SS28)2 (Ding et al. 2014a, b). They also conducted similar experiments using four different HSA-BNP fusion protein derivatives: BNP-HSA, (BNP)2-HSA, (BNP)4-HSA, and HSA-(BNP)2 (Ding et al. 2014a, b). Among them, HSA-(BNP)2 possessed the highest and most prolonged BNP activity in activating the natriuretic peptide receptor A. These results indicate that in the case of albumin fusion technology, more than three copies of protein fused to HSA may not be a suitable approach for enhancing the activity of the preparation. This strategy was applied to the HSA-thrombopoietin mimetic peptide (TMP) 2 fusion protein which exhibited a better stimulation of platelet production (Wang et al. 2016a, b).

4.3.1.3 Fusion Protein with Dual Functions by Two Different Partners

Bispecific antibodies such as tandem scFv molecules (taFv), diabodies (Db), or single-chain diabodies (scDb) are designed to target two different antigens so as to simultaneously retarget T lymphocytes to tumor cells, leading to tumor cells destruction. To extend their circulating half-lives, HSA-bispecific antibodies, directed against the tumor antigen carcinoembryonic antigen (CEA) and the T cell receptor complex molecule CD3, were constructed (scFv2-HSA, scDb-HSA, taFv-HSA) (Stork et al. 2009). These three constructs activated T cells but to a lesser extent than scFv2-HSA.

4.3.2 Modulation of Pharmacokinetics of Albumin Fusion Protein

To improve the pharmacokinetic properties of the partner protein more optimally for use as a therapeutic agent, three approaches have been examined: (1) using an HSA domain III mutant that enhances the binding affinity to FcRn to further extend the circulatory half-life of the fusion protein, (2) attachment of a homing peptide to the HSA fusion protein to achieve active targeting or to facilitate cell penetration, and (3) delivery of the HSA fusion protein gene and its expression in the body.

4.3.2.1 Engineered HSA Mutant as a Carrier in Albumin Fusion Technology

Serum albumin binds to the neonatal Fc receptor at low pH in endosomes after endocytosis and is transported back to the cellular surface, where serum albumin is released into the bloodstream, resulting in an extended circulation time in vivo. This implies that the development of HSA mutants with a higher FcRn affinity as a carrier for albumin fusion technology could be expected to further extend the blood circulation and pharmacodynamics of an albumin fusion protein. Actually, Andersen et al. found that K573P HSA mutant exhibited an increase in FcRn affinity by 12-fold, and the product showed an extended serum half-life in normal mice, transgenic mice for human FcRn, and cynomolgus monkeys (Andersen et al. 2014).

Novozyme Biopharma developed a unique platform, Veltis®, an innovated albumin fusion technology using engineered albumins in which the albumin-FcRn interaction that modulates the circulatory half-life of albumin enables biopharmaceutics to be manipulated, leading to superior pharmacokinetics tailored to the disease state and patient compliance. This system is able to provide once-weekly, once fortnightly, or once-monthly dosing of an albumin fusion protein.

4.3.2.2 Active Targeting of Albumin Fusion Protein to Specific Organ or Cell

In addition to improving pharmacokinetic profile for an impressive prolonged half-life, the specific delivery of a fusion protein to a targeted organ or cell has been attempted. In line with this concept, an HSA-tissue inhibitor of the metalloproteinase (TIMP) 2 fusion protein (HSA-TIMP2) containing a cyclic arginine-glycine-aspartate (cRGD) triad, which displayed a strong affinity and selectivity to the α(V)β(3), was developed to target tumor-associated cells expressing α (V)β(3) receptors (Lee et al. 2012). cRGD-HSA-TIMP2 improved uptake of preparation by tumors and the anticancer effect of HSA-TIMP2. Furthermore, when this fusion protein was labeled with ^{123}I and ^{68}Ga, it had biodistribution properties that made it feasible for use as a SPECT and PET imaging probe (Choi et al. 2011). Therefore, cRGD-HSA-TIMP2 has the potential for use, not only as an anticancer agent but also as a radioligand for the diagnosis of tumors.

Li et al. developed a novel tumor targeting carrier for theranosis using albumin fusion technology (Li et al. 2014a, b). They fused the N-terminal fragment (ATF) of urokinase that is able to bind a urokinase receptor to HSA because the urokinase receptor has been shown to have a high expression level in many tumors, but not in normal tissues. Thus, HSA-ATF has the potential as a versatile active targeting carrier for both cytotoxic and diagnosis agents to tumors that express urokinase receptors. In fact, HSA-ATF retained its ligand-binding ability as HSA, and monosubstituted beta-carboxy phthalocyanine zinc (CPZ), a hydrophobic photosensitizer, then stably formed a 1:1 complex with HSA-ATF (HSA-ATF/CPZ). Therefore, HSA-ATF/CPZ can function to both as a cytotoxic agent for photodynamic therapy and as a diagnostic agent for fluorescence molecular tomography in cancer treatment.

Inactivating hepatic stellate cells (HSCs) is a promising therapeutic strategy for the treatment of liver fibrosis. Lee et al. found a unique function of albumin that suppressed the activation of culture HSCs via the inhibition of retinoic acid signaling (Park et al. 2012). They further developed a novel antifibrotic drug, namely, the retinol-binding protein-HSA domain III that targeted albumin to stellate cells because of the marked accumulation of retinoic acid in HSCs (Lee et al. 2015). In vitro experiments clearly showed that this fusion protein downregulated retinoic acid signaling and, hence, inhibited the activation of stellate cells. In addition, the administration of retinol-binding protein-HSA domain III exhibited antifibrotic action against experimental liver fibrosis models.

α-Melanocyte-stimulating hormone (α-MSH) possesses potent anti-inflammatory action and protective effects with a wide therapeutic window in cases of brain damage. Thus, it would be expected that it might function as a novel therapeutics for various CNS disorders such as Parkinson's disease, Alzheimer's disease, ischemic stroke, and traumatic brain injury because neuroinflammation is significantly involved in the progression of CNS disorders. However, it is difficult to deliver MSH to the brain and its blood half-life is only a few minutes. To overcome these two limitations, Wang et al. attempted to link the N-terminus 11-amino acid transduction domain of the human immunodeficiency virus (TAT) and HSA to MSH (TAT-HSA-MSH) (Wang et al. 2016a, b). TAT-HSA-MSH showed a significant inhibitory effect for the NF-kB activation and TNF production in a cell line system. The intraperitoneal administration of TAT-HSA-MSH was detected in all regions of the brain including the hippocampus and cortex and hence exhibited a neuroprotective action against experimental disease model mice in vivo. Thus, TAT-HSA-MSH has the potential for use as an anti-neuroinflammation therapeutics for the treatment of CNS disorders.

4.3.2.3 Intracellular Delivery of Albumin Fusion Protein

Joshi et al. attempted to develop a novel fusion protein to facilitate the intracellular delivery of two anticancer agents, each with distinct but complimentary mechanisms, to produce synergistic efficacy. They constructed an HSA fusion protein with two antitumor peptides, a wild-type p53-derived peptide (p53i) or the high-affinity MDM2-binding peptide N8A-potent MDM2/MDMX peptide inhibitor (PMI) (Joshi et al. 2013). HSA-p53i and HSA-PMI are efficiently internalized by tumor cells, such as SJSA-1 cells. They also demonstrated that the fusion protein retained the ability to bind to long-chain fatty acids (FA). Therefore, the FA drug formed a stable complex with HSA-p53i or HSA-PMI. Actually, this hybrid fusion protein was the co-delivery carrier of p53i or PMI and FA-labeled drug to the same tumor cells.

4.3.2.4 Gene Delivery of Albumin Fusion Protein In Vivo

IFNγ has been shown to inhibit metastatic tumor growth and the onset of atopic dermatitis. Thus, its in vivo gene delivery represents a challenge as a new therapeutic strategy for the treatment of cancer. However, IFNγ rapidly disappears from the systemic circulation with a half-life of less than 3 min after an intravenous injection into mice, and its half-life in humans is about 4.5 h after an intramuscular injection. Miyakawa et al. attempted to increase the circulation half-life of mouse IFNγ after its gene delivery by designing a novel fusion protein of IFNγ with mouse serum albumin (MSA) (Miyakawa et al. 2011). The hydrodynamic injection of a plasmid expressing IFNγ-MSA resulted in a marked increase in the area under the concentration-time curve and sustained a mean residence time of IFNγ activity than those of IFNγ. Thus, the gene delivery of albumin is a promising approach for modulating the disposition and biological activities of partner proteins.

References

Andersen JT, Dalhus B, Viuff D, Ravn BT, Gunnarsen KS, Plumridge A, Bunting K, Antunes F, Williamson R, Athwal S, Allan E, Evans L, Bjørås M, Kjærulff S, Sleep D, Sandlie I, Cameron J (2014) Extending serum half-life of albumin by engineering neonatal Fc receptor (FcRn) binding. J Biol Chem 289:13492–13502

Arner ES, Holmgren A (2000) Physiological functions of thioredoxin and thioredoxin reductase. Eur J Biochem 267:6102–6109

Chapman AP (2002) PEGylated antibodies and antibody fragments for improved therapy: a review. Adv Drug Deliv Rev 54:531–545

Chen JH, Zhang XG, Jiang YT, Yan LY, Tang L, Yin YW, Cheng DS, Chen J, Wang M (2010) Bioactivity and pharmacokinetics of two human serum albumin-thymosin alpha1-fusion proteins, rHSA-Talpha1 and rHSA-L-Talpha1, expressed in recombinant Pichia pastoris. Cancer Immunol Immunother 59:1335–1345

Choi N, Kim SM, Hong KS, Cho G, Cho JH, Lee C, Ryu EK (2011) The use of the fusion protein RGD-HSA-TIMP2 as a tumor targeting imaging probe for SPECT and PET. Biomaterials 32:7151–7158

Choi S, Park S, Kim S, Lim C, Kim J, Cha DR, Oh J (2012) Recombinant fusion protein of albumin-retinol binding protein inactivates stellate cells. Biochem Biophys Res Commun 418:191–197

Chuang VT, Otagiri M (2007) Recombinant human serum albumin. Drugs Today 43:547–561

Chuang VT, Kragh-Hansen U, Otagiri M (2002) Pharmaceutical strategies utilizing recombinant human serum albumin. Pharm Res 19:569–577

de Bold MK, Sheffield WP, Martinuk A, Bhakta V, Eltringham-Smith L, de Bold AJ (2012) Characterization of a long-acting recombinant human serum albumin-atrial natriuretic factor (ANF) expressed in Pichia pastoris. Regul Pept 175:7–10

Ding Y, Fan J, Li W, Yang R, Peng Y, Deng L, Wu Y, Fu Q (2013) The effect of albumin fusion patterns on the production and bioactivity of the somatostatin-14 fusion protein in Pichia pastoris. Appl Biochem Biotechnol 170:1637–1648

Ding Y, Fan J, Li W, Peng Y, Yang R, Deng L, Fu Q (2014a) The effect of albumin fusion structure on the production and bioactivity of the somatostatin-28 fusion protein in Pichia pastoris. J Ind Microbiol Biotechnol 41:997–1006

Ding Y, Peng Y, Deng L, Wu Y, Fu Q, Jin J (2014b) The effects of fusion structure on the expression and bioactivity of human brain natriuretic peptide (BNP) albumin fusion proteins. Curr Pharm Biotechnol 15:856–863

Duttaroy A, Kanakaraj P, Osborn BL, Schneider H, Pickeral OK, Chen C, Zhang G, Kaithamana S, Singh M, Schulingkamp R, Crossan D, Bock J, Kaufman TE, Reavey P, Carey-Barber M, Krishnan SR, Garcia A, Murphy K, Siskind JK, McLean MA, Cheng S, Ruben S, Birse CE, Blondel O (2005) Development of a long-acting insulin analog using albumin fusion technology. Diabetes 54:251–258

Furukawa M, Tanaka R, Chuang VT, Ishima Y, Taguchi K, Watanabe H, Maruyama T, Otagiri M (2011) Human serum albumin-thioredoxin fusion protein with long blood retention property is effective in suppressing lung injury. J Control Release 154:189–195

Gao Y, LaFleur D, Shah R, Zhao Q, Singh M, Brimijoin S (2008) An albumin-butyrylcholinesterase for cocaine toxicity and addiction: catalytic and pharmacokinetic properties. Chem Biol Interact 175:83–87

Gerlza T, Winkler S, Atlic A, Zankl C, Konya V, Kitic N, Strutzmann E, Knebl K, Adage T, Heinemann A, Weis R, Kungl AJ (2015) Designing a mutant CCL2-HSA chimera with high glycosaminoglycan-binding affinity and selectivity. Protein Eng Des Sel 28:231–240

Golor G, Bensen-Kennedy D, Haffner S, Easton R, Jung K, Moises T, Lawo JP, Joch C, Veldman A (2013) Safety and pharmacokinetics of a recombinant fusion protein linking coagulation factor VIIa with albumin in healthy volunteers. J Thromb Haemost 11:1977–1985

Halpern W, Riccobene TA, Agostini H, Baker K, Stolow D, Gu ML, Hirsch J, Mahoney A, Carrell J, Boyd E, Grzegorzewski KJ (2002) Albugranin, a recombinant human granulocyte colony stimulating factor (G-CSF) genetically fused to recombinant human albumin induces prolonged myelopoietic effects in mice and monkeys. Pharm Res 19:1720–1729

Hou S, Li C, Huan Y, Liu S, Liu Q, Sun S, Jiang Q, Jia C, Shen Z (2015) Effects of E2HSA, a long-acting glucagon like peptide-1 receptor agonist, on glycemic control and beta cell function in spontaneous diabetic db/db mice. J Diabetes Res 2015:817839

Huang YS, Chen Z, Chen YQ, Ma GC, Shan JF, Liu W, Zhou LF (2008) Preparation and characterization of a novel exendin-4 human serum albumin fusion protein expressed in Pichia pastoris. J Pept Sci 14:588–595

Huang YS, Wen XF, Yang ZY, Wu YL, Lu Y, Zhou LF (2014) Development and characterization of a novel fusion protein of a mutated granulocyte colony-stimulating factor and human serum albumin in Pichia pastoris. PLoS One 9:e115840

Ikuta S, Chuang VT, Ishima Y, Nakajou K, Furukawa M, Watanabe H, Maruyama T, Otagiri M (2010) Albumin fusion of thioredoxin-the production and valuation of its biological activity for potential therapeutic applications. J Control Release 147:17–23

Joshi MR, Yao N, Myers KA, Li Z (2013) Human serum albumin and p53-activating peptide fusion protein is able to promote apoptosis and deliver fatty acid-modified molecules. PLoS One 8:e80926

Joung CH, Shin JY, Koo JK, Lim JJ, Wang JS, Lee SJ, Tan HK, Kim SL, Lim SM (2009) Production and characterization of long-acting recombinant human serum albumin-EPO fusion protein expressed in CHO cell. Protein Expr Purif 68:137–145

Kodama A, Watanabe H, Tanaka R, Tanaka H, Chuang VT, Miyamoto Y, Wu Q, Endo M, Hamasaki K, Ishima Y, Fukagawa M, Otagiri M, Maruyama T (2013) A human serum albumin-thioredoxin fusion protein prevents experimental contrast-induced nephropathy. Kidney Int 83:446–454

Kodama A, Watanabe H, Tanaka R, Kondo M, Chuang VT, Wu Q, Endo M, Ishima Y, Fukagawa M, Otagiri M, Maruyama T (2014) Albumin fusion renders thioredoxin an effective anti-oxidative and anti-inflammatory agent for preventing cisplatin-induced nephrotoxicity. Biochim Biophys Acta 1840:1152–1162

Kontermann RE, Brinkmann U (2015) Bispecific antibodies. Drug Discov Today 20:838–847

Lee MS, Kim YH, Kim YJ, Kwon SH, Bang JK, Lee SM, Song YS, Hahm DH, Shim I, Han D, Her S (2011) Pharmacokinetics and biodistribution of human serum albumin-TIMP-2 fusion protein using near-infrared optical imaging. J Pharm Sci 14:368–377

Lee MS, Jung JI, Kwon SH, Lee SM, Morita K, Her S (2012) TIMP-2 fusion protein with human serum albumin potentiates anti-angiogenesis-mediated inhibition of tumor growth by suppressing MMP-2 expression. PLoS One 7:e35710

Lee H, Jeong H, Park S, Yoo W, Choi S, Choi K, Lee MG, Lee M, Cha D, Kim YS, Han J, Kim W, Park SH, Oh J (2015) Fusion protein of retinol-binding protein and albumin domain III reduces liver fibrosis. EMBO Mol Med 7:819–830

Leung K (2008) ^{125}I-T84.66 scFv-human serum albumin. Molecular Imaging and Contrast Agent Database (MICAD) [Internet]

Li F, Meng F, Jin Q, Sun C, Li Y, Li H, Jin S (2014a) Fusion protein of single-chain variable domain fragments for treatment of myasthenia gravis. Neural Regen Res 9:851–856

Li R, Zheng K, Hu P, Chen Z, Zhou S, Chen J, Yuan C, Chen S, Zheng W, Ma E, Zhang F, Xue J, Chen X, Huang M (2014b) A novel tumor targeting drug carrier for optical imaging and therapy. Theranostics 4:642–659

Liu M, Huang Y, Hu L, Liu G, Hu X, Liu D, Yang X (2012) Selective delivery of interleukine-1 receptor antagonist to inflamed joint by albumin fusion. BMC Biotechnol 12:68

Marques JA, George JK, Smith IJ, Bhakta V, Sheffield WP (2001) A barbourin-albumin fusion protein that is slowly cleared in vivo retains the ability to inhibit platelet aggregation in vitro. Thromb Haemost 86:902–908

McDonagh CF, Huhalov A, Harms BD, Adams S, Paragas V, Oyama S, Zhang B, Luus L, Overland R, Nguyen S, Gu J, Kohli N, Wallace M, Feldhaus MJ, Kudla AJ, Schoeberl B, Nielsen UB (2012) Antitumor activity of a novel bispecific antibody that targets the ErbB2/ErbB3 oncogenic unit and inhibits heregulin-induced activation of ErbB3. Mol Cancer Ther 11:582–593

Melder RJ, Osborn BL, Riccobene T, Kanakaraj P, Wei P, Chen G, Stolow D, Halpern WG, Migone TS, Wang Q, Grzegorzewski KJ, Gallant G (2005) Pharmacokinetics and in vitro and in vivo anti-tumor response of an interleukin-2-human serum albumin fusion protein in mice. Cancer Immunol Immunother 54:535–547

Miyakawa N, Nishikawa M, Takahashi Y, Ando M, Misaka M, Watanabe Y, Takakura Y (2011) Prolonged circulation half-life of interferon γ activity by gene delivery of interferon γ-serum albumin fusion protein in mice. J Pharm Sci 100:2350–2357

Mu X, Hu K, Shen M, Kong N, Fu C, Yan W, Wei A (2016) Protection against influenza A virus by vaccination with a recombinant fusion protein linking influenza M2e to human serum albumin (HSA). J Virol Methods 228:84–90

Muller D, Karle A, Meissburger B, Hofig I, Stork R, Kontermann RE (2007) Improved pharmacokinetics of recombinant bispecific antibody molecules by fusion to human serum albumin. J Biol Chem 282:12650–12660

Müller N, Schneider B, Pfizenmaier K, Wajant H (2012) Superior serum half life of albumin tagged TNF ligands. Biochem Biophys Res Commun 396:793–799

Nakamura H, Hoshino Y, Okuyama H, Matsuo Y, Yodoi J (2009) Thioredoxin 1 delivery as new therapeutics. Adv Drug Deliv Rev 61:303–309

Nishida K, Watanabe H, Ogaki S, Kodama A, Tanaka R, Imafuku T, Ishima Y, Chuang VT, Toyoda M, Kondoh M, Wu Q, Fukagawa M, Otagiri M, Maruyama T (2015) Renoprotective effect of long acting thioredoxin by modulating oxidative stress and macrophage migration inhibitory factor against rhabdomyolysis-associated acute kidney injury. Sci Rep 5:14471

Nolte MW, Nichols TC, Mueller-Cohrs J, Merricks EP, Pragst I, Zollner S, Dickneite G (2012) Improved kinetics of rIX-FP, a recombinant fusion protein linking factor IX with albumin, in cynomolgus monkeys and hemophilia B dogs. J Thromb Haemost 10:1591–1599

Osborn BL, Sekut L, Corcoran M, Poortman C, Sturm B, Chen G, Mather D, Lin HL, Parry TJ (2002) Albutropin: a growth hormone-albumin fusion with improved pharmacokinetics and pharmacodynamics in rats and monkeys. Eur J Pharmacol 456:149–158

Park S, Choi S, Lee MG, Lim C, Oh J (2012) Retinol binding protein-albumin domain III fusion protein deactivates hepatic stellate cells. Mol Cells 34:17–22

Pratley RE, Nauck MA, Barnett AH, Feinglos MN, Ovalle F, Harman-Boehm I, Ye J, Scott R, Johnson S, Stewart M, Rosenstock J, HARMONY 7 study group (2014) Once-weekly albiglutide versus once-daily liraglutide in patients with type 2 diabetes inadequately controlled on oral drugs (HARMONY 7): a randomised, open-label, multicentre, non-inferiority phase 3 study. Lancet Diabetes Endocrinol 2:289–297

Richards DA, Braiteh FS, Garcia AA (2014) A phase 1 study of MM-111, a bispecific HER2/HER3 antibody fusion protein, combined with multiple treatment regimens in patients with advanced HER-positive solid tumors. J Clin Oncol 32:651

Rustgi VK (2009) Albinterferon alfa-2b, a novel fusion protein of human albumin and human interferon alfa-2b, for chronic hepatitis C. Curr Med Res Opin 25:991–1002

Santagostino E, Negrier C, Klamroth R, Tiede A, Pabinger-Fasching I, Voigt C, Jacobs I, Morfini M (2012) Safety and pharmacokinetics of a novel recombinant fusion protein linking coagulation factor IX with albumin (rIX-FP) in hemophilia B patients. Blood 120:240–211

Santagostino E, Martinowitz U, Lissitchkov T, Pan-Petesch B, Hanabusa H, Oldenburg J, Boggio L, Negrier C, Pabinger I, von Depka Prondzinski M, Altisent C, Castaman G, Yamamoto K, Álvarez-Roman MT, Voigt C, Blackman N, Jacobs I (2016) Long acting recombinant coagulation factor IX albumin fusion protein (rIX-FP) in hemophilia B: results of a phase 3 trial. Blood pii: blood-2015-09-669234

Sheffield WP, Mamdani A, Hortelano G, Gataiance S, Eltringham-Smith L, Begbie ME, Leyva RA, Liaw PS, Ofosu FA (2004) Effects of genetic fusion of factor IX to albumin on in vivo clearance in mice and rabbits. Br J Haematol 126:565–573

Sheffield WP, Gataiance S, Eltringham-Smith LJ (2007) Combined administration of barbourin-albumin and hirudin-albumin fusion proteins limits fibrin(ogen) deposition on the rabbit balloon-injured aorta. Thromb Res 119:195–207

Sinclair AM, Elliott S (2005) Glycoengineering: the effect of glycosylation on the properties of therapeutic proteins. J Pharm Sci 94:1626–1635

Stork R, Campigna E, Robert B, Müller D, Kontermann RE (2009) Biodistribution of a bispecific single-chain diabody and its half-life extended derivatives. J Biol Chem 284:25612–25619

Subramanian GM, Fiscella M, Lamousé-Smith A, Zeuzem S, McHutchison JG (2007) Albinterferon alpha-2b: a genetic fusion protein for the treatment of chronic hepatitis C. Nat Biotechnol 25:1411–1419

Tanaka R, Watanabe H, Kodama A, Chuang VT, Ishima Y, Hamasaki K, Tanaka K, Mizushima T, Otagiri M, Maruyama T (2013) Long-acting human serum albumin-thioredoxin fusion protein suppresses bleomycin-induced pulmonary fibrosis progression. J Pharmacol Exp Ther 345:271–283

Tanaka R, Ishima Y, Maeda H, Kodama A, Nagao S, Watanabe H, Chuang VT, Otagiri M, Maruyama T (2014a) Albumin fusion prolongs the antioxidant and anti-inflammatory activities of thioredoxin in mice with acetaminophen-induced hepatitis. Mol Pharm 11:1228–1238

Tanaka R, Ishima Y, Enoki Y, Kimachi K, Shirai T, Watanabe H, Chuang VT, Maruyama T, Otagiri M (2014b) Therapeutic impact of human serum albumin-thioredoxin fusion protein on influenza virus-induced lung injury mice. Front Immunol 5:561

Tomkin GH (2009) Albiglutide, an albumin-based fusion of glucagon-like peptide 1 for the potential treatment of type 2 diabetes. Curr Opin Mol Ther 11:579–588

Wang W, Ou Y, Shi Y (2004) AlbuBNP, a recombinant B-type natriuretic peptide and human serum albumin fusion hormone, as a long-term therapy of congestive heart failure. Pharm Res 21:2105–2111

Wang F, Wu M, Liu W, Shen Q, Sun H, Chen S (2013) Expression, purification, and lipolytic activity of recombinant human serum albumin fusion proteins with one domain of human growth hormone in Pichia pastoris. Biotechnol Appl Biochem 60:405–411

Wang M, Zhi D, Xian J, Ru Y, Wei S, Wang N, Liu Y, Wang H, Pei Y, Song M, Li Y, Li H (2016a) Functional expression of human serum albumin-tandem thrombopoietin mimetic peptide fusion protein as a novel thrombopoietin analog in Pichia pastoris. Biotechnol Lett 5:175–85

Wang M, Zhi D, Wang H, Ru Y, Ren H, Wang N, Liu Y, Li Y, Li H (2016b) TAT-HSA-α-MSH fusion protein with extended half-life inhibits tumor necrosis factor-α in brain inflammation of mice. Appl Microbiol Biotechno 100:5353–61

Weimer T, Wormsbacher W, Kronthaler U, Lang W, Liebing U, Schulte S (2008) Prolonged in-vivo half-life of factor VIIa by fusion to albumin. Thromb Haemost 99:659–667

Werle M, Bernkop-Schnurch A (2006) Strategies to improve plasma half life time of peptide and protein drugs. Amino Acids 30:351–367

Wu M, Shen Q, Yang Y, Zhang S, Qu W, Chen J, Sun H, Chen S (2013) Disruption of YPS1 and PEP4 genes reduces proteolytic degradation of secreted HSA/PTH in Pichia pastoris GS115. J Ind Microbiol Biotechnol 40:589–599

Xu X, Yang J, Liu Y, Shan C, Wang Q, Chen Z, Cheng Y (2015) The induction of prolonged myelopoietic effects in monkeys by GW003, a recombinant human granulocyte colony-stimulating factor genetically fused to recombinant human albumin. J Pharm Sci 104:760–767

Yang GG, Xu XY, Ding Y, Cui QQ, Wang Z, Zhang QY, Shi SH, Lv ZY, Wang XY, Zhang JH, Zhang RG, Xu CS (2015) Linker length affects expression and bioactivity of the onconase fusion protein in Pichia pastoris. Genet Mol Res 14:19360–19370

Yu X, Menard M, Prechl J, Bhakta V, Sheffield WP, Lazarus AH (2016) Monovalent Fc receptor blockade by an anti-Fcγ receptor/albumin fusion protein ameliorates murine ITP with abrogated toxicity. Blood 127:132–138

Yuan Q, Li L, Pian Y, Hao H, Zheng Y, Zang Y, Jiang H, Jiang Y (2016) Preliminary investigation of human serum albumin-Vβ inhibition on toxic shock syndrome induced by staphylococcus enterotoxin B in vitro and in vivo. Toxicon 113:55–59

Zeuzem S, Sulkowski MS, Lawitz EJ, Rustgi VK, Rodriguez-Torres M, Bacon BR, Grigorescu M, Tice AD, Lurie Y, Cianciara J, Muir AJ, Cronin PW, Pulkstenis E, Subramanian GM, McHutchison JG, ACHIEVE-1 Study Team (2010) Albinterferon Alfa-2b was not inferior to pegylated interferon-α in a randomized trial of patients with chronic hepatitis C virus genotype 1. Gastroenterology 139:1257–1266

Zhang Q, Lei J, Ding Y, Chen Y, Qu L, Chen S, Jin J (2009) Expression and purification of IFNbeta-HSA fusion protein in Pichia pastoris. Sheng Wu Gong Cheng Xue Bao 25:1746–1752

Zhang L, Wang L, Meng Z, Gan H, Gu R, Wu Z, Gao L, Zhu X, Sun W, Li J, Zheng Y, Dou G (2014) A novel exendin-4 human serum albumin fusion protein, E2HSA, with an extended half-life and good glucoregulatory effect in healthy rhesus monkeys. Biochem Biophys Res Commun 445:511–516

Zhao HL, Xue C, Wang Y, Sun B, Yao XQ, Liu ZM (2009) Elimination of the free sulfhydryl group in the human serum albumin (HSA) moiety of human interferon-alpha2b and HSA fusion protein increases its stability against mechanical and thermal stresses. Eur J Pharm Biopharm 72:405–411

Zhao HL, Xue C, Du JL, Ren M, Xia S, Liu ZM (2012a) Balancing the pharmacokinetics and pharmacodynamics of interferon-α2b and human serum albumin fusion protein by proteolytic or reductive cleavage increases its in vivo therapeutic efficacy. Mol Pharm 9:664–670

Zhao HL, Xue C, Du JL, Ren M, Xia S, Cheng YG, Liu ZM (2012b) Sustained and cancer cell targeted cytosolic delivery of Onconase results in potent antitumor effects. J Control Release 159:346–352

Zhao J, Si Y, Cheng M, Yang Y, Niu Y, Li X, Liu X, Yang W (2013a) Albumin fusion of interleukin-28B: production and characterization of its biological activities and protein stability. PLoS One 8:e64301

Zhao S, Zhang Y, Tian H, Chen X, Cai D, Yao W, Gao X (2013b) Extending the serum half-life of G-CSF via fusion with the domain III of human serum albumin. Biomed Res Int 2013:107238

Zhu RY, Xin X, Dai HY, Li Q, Lei JY, Chen Y, Jin J (2012) Expression and purification of recombinant human serum albumin fusion protein with VEGF165b in Pichia pastoris. Protein Expr Purif 85:32–37

Zollner S, Schuermann D, Raquet E, Mueller-Cohrs J, Weimer T, Pragst I, Dickneite G, Schulte S (2014) Pharmacological characteristics of a novel, recombinant fusion protein linking coagulation factor VIIa with albumin (rVIIa-FP). J Thromb Haemost 12:220–228

Chapter 5
Albumin Nanoparticles

Yasunori Iwao

Abstract Albumin is an attractive macromolecular carrier and is widely used in preparing nanoparticles due to its biodegradability, nontoxicity, and nonimmmunogenicity. Albumin nanoparticles themselves are biodegradable and can easily be prepared in defined sizes. Several protocols for preparing albumin nanoparticles that have been developed include: emulsification, desolvation, nab technology, thermal gelation, nano spray drying, and self-assembly techniques.

After preparing albumin nanoparticles, physicochemical properties such as particle size, particle size distribution, zeta potential, and morphology need to be carefully checked if they are intended to be used in a specific delivery system.

Albumin itself has several reactive amino acid groups on its surface, and, as a result, albumin nanoparticles also participate in the electrostatic adsorption of positively or negatively charged molecules. In addition, modifying the reactive amino acid groups on the surface can confer a variety of functionalities to albumin nanoparticles such as prolonged circulation half-life, enhanced nanosystem stability, sustained drug release, or targeting release.

Therefore, albumin nanoparticles represent one of the most important drug carriers for the delivery of therapeutic drugs.

Keywords Albumin nanoparticle • Emulsification • Desolvation • Nab technology • Physicochemical property • Surface modification

5.1 Introduction

Albumin is an attractive macromolecular carrier and has been widely used in preparing micro-/nanoparticles due to its biodegradability, nontoxicity, and nonimmunogenicity (Peters 1996; Kratz 2008). Both bovine serum albumin (BSA) and

Y. Iwao (✉)
Department of Pharmaceutical Engineering, School of Pharmaceutical Sciences, University of Shizuoka, 52-1 Yada, Suruga-ku, Shizuoka 422-8526, Japan
e-mail: yasuiwao@u-shizuoka-ken.ac.jp

M. Otagiri, V.T.G. Chuang (eds.), *Albumin in Medicine*,
DOI 10.1007/978-981-10-2116-9_5

human serum albumin (HSA) are frequently used for preparing nanoparticles in defined sizes. In addition, since albumin itself contains several reactive amino acid groups (thiol, amino, and carboxylic groups), albumin nanoparticles can participate in the electrostatic adsorption of positively or negatively charged molecules on its surface, without the assistance or need for any other compounds (Irache et al. 2005). In addition, these reactive amino acid groups can be used for surface modifications to impart various functionalities to albumin nanoparticles such as prolonged circulation half-life (eg, PEG) (Kouchakzadeh et al. 2010), enhanced nanosystem stability (eg, poly-l-lysine) (Singh et al. 2010), slow drug release (eg, cationic polymers) (Zhang et al. 2008), or targeted release (eg, folate or monoclonal antibodies) (Ulbrich et al. 2011; Wartlick et al. 2004).

There are numerous reviews regarding albumin nanoparticles. In this chapter, techniques in the preparation of such particles and potential applications of surface-modified albumin nanoparticles in drug delivery are in-depth reviewed.

5.2 Preparation Techniques of Albumin Nanoparticles

Several protocols for preparing albumin nanoparticles have been reported, namely, emulsification (Patil 2003; Sundar et al. 2010), desolvation (Langer et al. 2003; Weber et al. 2000), nab technology (Desai 2007), thermal gelation (Yu et al. 2006), nano spray drying (Lee et al. 2011), and self-assembly (Gong et al. 2009) techniques. In this section, emulsification, desolvation, and nab technology are discussed in detail, since these three methods are representative and are in widespread use.

5.2.1 Emulsification Method

The principle of nanoemulsion formation is based on the spontaneous emulsification that occurs when an aqueous phase is mixed with an organic phase. The organic phase is a homogeneous solution containing oil, a lipophilic surfactant, and a water miscible solvent, whereas the aqueous phase consists of a hydrophilic surfactant and water. Scheffel et al. (1972) initially reported on the preparation of albumin microspheres using this emulsification method, and, since then, several modified methods have also been reported (Patil 2003; Sundar et al. 2010). In this process, an aqueous solution containing albumin is converted into a homogeneous emulsion in plant oil (eg, cotton seed oil) at room temperature using a high-speed mechanical homogenizer. The above emulsion is added dropwise to a large volume of preheated oil with a temperature over 120 °C. This process results in the rapid evaporation of the existing water and irreversible albumin destruction. This process also induces the formation of nanoparticles (Fig. 5.1). The resulting suspension is then placed in an ice bath.

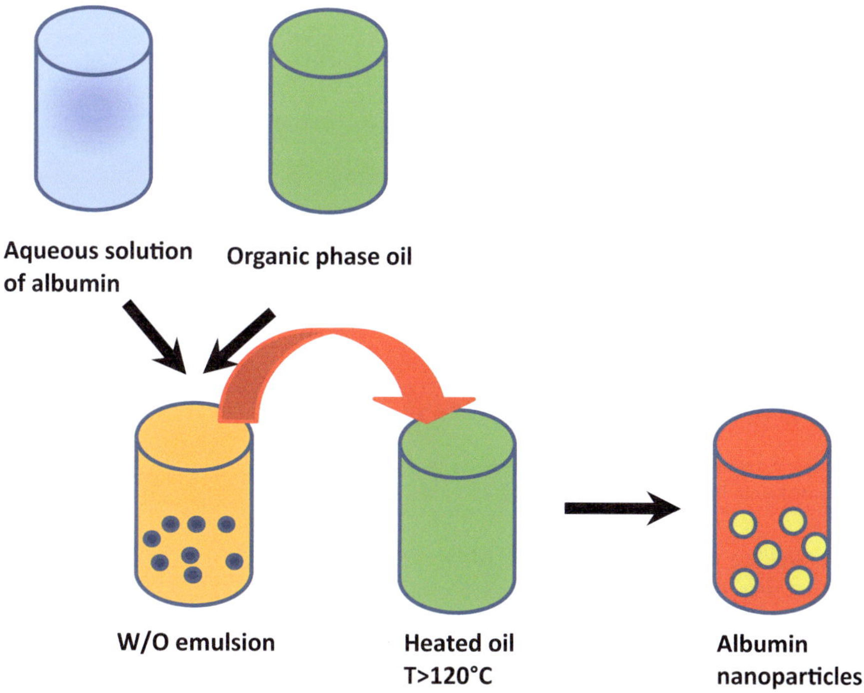

Fig. 5.1 Preparation of albumin nanoparticles by means of an emulsification method. The illustration is based on information found in Jahanshahi and Babaei (2008)

5.2.2 Desolvation Method

The main disadvantage of the above emulsion methods is the need for applying an organic solvent and the removal of the residual oil in the preparation process, and surfactants are required to produce a stabilized emulsion. Therefore, a desolvation process derived from the coacervation method used for microencapsulation was developed as an alternative method. Figure 5.2 outlines the method used to prepare albumin nanoparticles using a desolvating agent.

Nanoparticles are obtained by the continuous dropwise addition of an alcohol such as ethanol to an aqueous albumin solution under continuous stirring, until the solution becomes turbid. Ethanol changes the tertiary structure of the protein, and, during the addition of ethanol to the solution, albumin is phase separated due to its diminished water solubility (Langer et al. 2003). When a certain level of desolvation is reached, protein clumps are formed. The nanoparticles are then formed by the cross-linkage of the amino moieties in lysine residues and the guanidino side chains in arginine of albumin via a condensation reaction with the aldehyde group such as glutaraldehyde (Merodio et al. 2001). Weber et al. (2000) reported that the minimum required glutaraldehyde concentration for the preparation of stable nanoparticles was about 40 % with a reaction time of 24 h needed for the sufficient

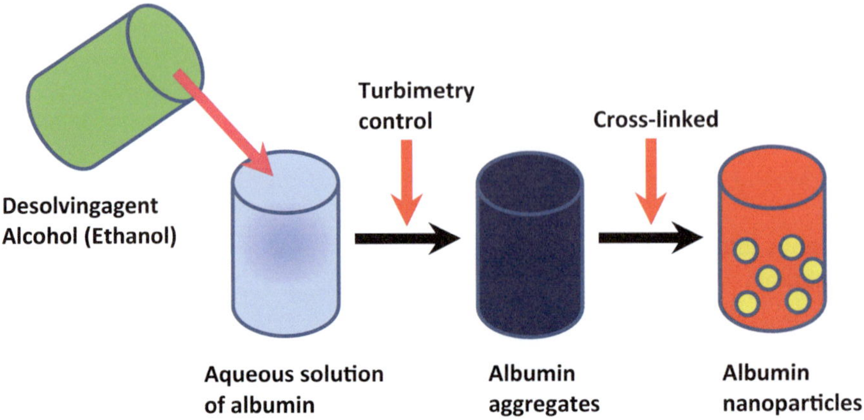

Fig. 5.2 Preparation of albumin nanoparticles by a desolvation method. The illustration is based on information found in Jahanshahi and Babaei (2008)

cross-linking of all amino groups. Methyl polyethylene glycol-modified oxidized dextran was reported as an alternative cross-linking agent to glutaraldehyde (Lin et al. 1994).

In order to obtain dispersed nanoparticles, the process must be stopped before particles start to aggregate and maintain nanodispersion conditions and a resolvating agent can be added. Alternatively, an aqueous solution containing lysine was added to cap the free aldehyde groups (Shen et al. 2011). After the elimination of ethanol by evaporation under reduced pressure, the resulting nanoparticles are purified by centrifugation to eliminate the free albumin and the excess cross-linking agent. Five percent mannitol was added as a cryoprotectant to the nanosuspension for subsequent freeze-drying to obtain a fine powder of the nanoparticles (Zhao et al. 2010).

Various parameters such as the initial protein concentration, temperature, pH, cross-linking agent concentration, agitation speed, and molar ratio of protein/alcohol affect the fabrication process to produce the desired properties of the nanoparticles. Langer et al. (2003) demonstrated that, when higher pH values are used, smaller nanoparticles result and the mean particle diameters of the nanoparticles could be adjusted to between 150 and 280 nm. They established a pump-controlled desolvation method which enabled the formation of nanoparticles with a narrow size distribution (Langer et al. 2003). In addition, Nguyen and Ko (2010) reported that the intermittent addition of a desolvating agent (ethanol) to albumin solution can improve the reproducibility of albumin nanoparticles with a narrow particle size distribution.

Queiroz et al. (2016) recently reported on a new technique using γ-irradiation to produce cross-linked BSA nanoparticles without the need for a cross-linking agent. Intermolecular dityrosine formation was mainly involved in this cross-linking of BSA nanoparticles by γ-irradiation (Fig. 5.3). Furthermore, Varca et al. (2016) evaluated the effect of the dose of γ-irradiation (2.5, 5, 7.5, and 10 kGy) over the devel-

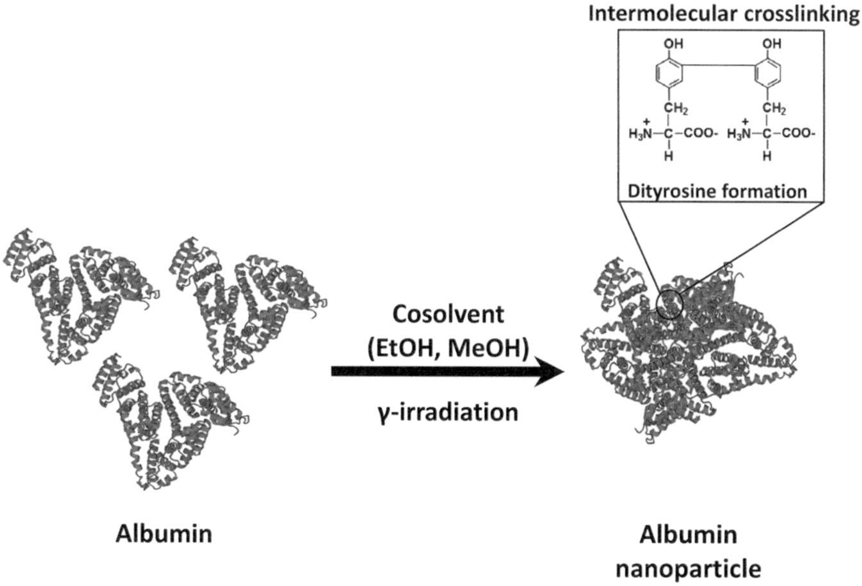

Fig. 5.3 Preparation of albumin nanoparticles by a new technique using γ-irradiation. The illustration is based on information found in Queiroz et al. (2016)

opment of protein-based nanoparticles with or without cosolvents. BSA solutions were irradiated using a γ-cell, and the optimized particle size and protein cross-linking degree occurred at 10 kG, demonstrating that γ-irradiation can be used for the preparation of BSA nanoparticles.

5.2.3 *Nab Technology*

A unique albumin-based nanoparticle technology (nanoparticle albumin-bound (nab) technology) to solubilize the lipophilic drugs in nanoparticles was developed by American Bioscience, Inc. The drug is mixed with HSA in an aqueous solvent and passed through a jet under high pressure to form albumin nanoparticles in the size range of 100–200 nm (Desai 2007). Abraxane® (nab-paclitaxel; paclitaxel–albumin nanoparticles) with an approximate diameter of 130 nm is the first FDA-approved nanotechnology-based chemotherapeutic product that has shown a significant benefit in the treatment of metastatic breast cancer. Several nab drugs are currently under development, including ABI-008 (nab-docetaxel) and ABI-009 (nab-rapamycin) (Desai 2007). Similarly, this nab technology could be applied for poorly water soluble drugs and curcumin-loaded HSA nanoparticles with a size range of 130–150 nm (Kim et al. 2011).

The reason why the albumin-based nanoparticle delivery systems shows enhanced therapeutic effects against solid tumors might be explained by changes in the pathophysiology of tumor tissue, characterized by angiogenesis, hypervasculature, defective vascular architecture, and impaired lymphatic drainage. In addition, the accumulation of nab-paclitaxel at the tumor site might be explained by the transcytosis initiated by the binding of albumin to a cell surface 60-kDa glycoprotein (gp60) receptor (albondin) as well as by the binding of albumin to secreted protein acidic and rich in cysteine (SPARC). Albumin binds to the gp60 receptor, which, in turn, results in the binding of gp60 with an intracellular protein (caveolin-1) and subsequent invagination of the cell membrane to form transcytotic vesicles, ie, caveolae (Arnedo et al. 2002). Furthermore, the tumor accumulation of nab-paclitaxel may be facilitated through the binding to SPARC, an extracellular matrix glycoprotein that is overexpressed and associated with poor prognosis in a variety of cancers including breast cancer (Cortes and Saura 2010).

5.3 Characterization of Albumin Nanoparticles

Physicochemical properties such as particle size, particle size distribution, and zeta potential (Jahanshahi et al. 2005) and the morphology of nanoparticles need to be carefully evaluated when designing nanoparticles for use as a drug delivery carrier. In addition, drug release needs to be also evaluated in order to achieve the site-specific action of the drug at the therapeutically optimal rate and dose regimen (Soppimath et al. 2001). Particle size and the size distribution of nanoparticles are particularly important physicochemical characteristics (Jahanshahi and Babaei 2008).

Nanoparticles have relatively higher intracellular uptake compared to microparticles and available to a wider range of biological targets due to their small size and relative mobility (Zauner et al. 2001). In addition, smaller nanoparticles have larger surface areas, and most of the drug would be released rapidly. However, smaller nanoparticles also have a greater risk of aggregation during storage. In this regard, the zeta potential which reflects the electrical potential of the nanoparticle should be evaluated. The zeta potential can be used to predict colloidal stability. In general, nanoparticles with a zeta potential above (+/−) 30 mV have been shown to be stable in suspension, as the surface charge prevents aggregation.

To evaluate particle size, particle size distribution, and zeta potential of nanoparticle with a range from a few nanometers to a few microns, dynamic light scattering (DLS) is often used. Furthermore, the morphology of nanoparticles was examined by two following techniques: atomic force microscopy (AFM) and scanning force microscope (SFM).

5.4 Potential Applications of Surface-Modified Albumin Nanoparticles in Drug Delivery

5.4.1 Albumin Nanoparticles with Prolonged Circulation Half-Life

When nanoparticles are administered intravenously, they are easily recognized by the immune systems and are cleared from the circulation by phagocytes (Muller and Wallis 1993). Apart from the size of nanoparticles, their surface hydrophobicity determines the amount of blood components, mainly proteins (opsonins) that are absorbed. The chemical modification of pharmaceutical nanocarriers with polyethylene glycol (PEG) is the most frequently used stabilizing agent for achieving in vivo long-term stability drug carriers, as first suggested for liposomes (Klibanov et al. 1990; Maruyama et al. 1991; Papadjopoulos et al. 1991; Suk et al. 2016). This polymer is nontoxic, non-immunogenic, nonantigenic, highly soluble in water, and FDA approved (Veronese and Pasut 2005).

Kouchakzadeh et al. (2010) reported that the PEG concentration used had the most significant effect on the amount of PEGylated amino groups and the pH had the least. They also reported that drug release from PEGylated nanoparticles was slower than that for non-PEGylated ones, probably due to the existence of a PEG layer around PEGylated particles which becomes an extra resistant layer to drug diffusion.

5.4.2 Albumin Nanoparticles with a Good Nanosystem Stability

The enzymatic degradation of albumin nanoparticles might be a major factor influencing the routes of administration, as well as site-specific drug delivery. Singh et al. (2010) prepared BSA NPs containing a short interfering ribonucleic acid (siRNA) via a coacervation technique and then stabilized the system by applying a cationic polymer, poly-l-lysine (PLL), coating with different molecular weights (MWs, 0.9–24 kDa) and concentrations (0.1–1.0 mg/ml). They reported that the stability of NPs in aqueous solution increased with increasing MW and PLL concentration. In addition, in the presence of trypsin, NPs coated with lower MW PLL were more stable than those with higher MW PLL, indicating that smaller molecules of PLL may be more suitable for particle coating for fabricating NPs with enhanced proteolytic resistance and better stability.

Furthermore, to reduce nonspecific protein adsorption, the same group conjugated PEG (1 kDa) units to PLL (4.2 and 24 kDa) and prepared BSA nanoparticles containing siRNA coated with these polymers (Yogasundaram et al. 2012). With smaller MW PLL, cellular uptake was not affected in the presence of PEG, but the PEG coating inhibited uptake in the case of higher MW PLL NPs. In addition, the higher MW PLL systems were cytotoxic, and this cytotoxicity was diminished when PEG was incorporated. Taken together, a PEG–PLL coating reduces enzymatic degradation, nonspecific protein adsorption, as well as cytotoxicity of albumin nanoparticles.

5.4.3 Albumin Nanoparticles with Suitable Drug Control Release Properties

Zhang et al. (2008) succeeded in preparing albumin nanoparticle with sustained drug release properties by applying a polyethylenimine (PEI) coating on the surface of nanoparticles. They reported that drug release from NPs could be controlled by controlling the PEI concentration. However, the PEI toxic effect against locally present cells needs to be given due consideration. Therefore, Zhang and coworkers attempted to synthesize PEI–PEG with different PEG substitutions, and albumin nanoparticles were coated with these polymers (Zhang et al. 2010). Cell viability assays showed that PEG substitution greatly reduced the cytotoxicity of the native PEI. In addition, an in vivo pharmacokinetics study after implanting these formulations showed that the drug retention in PEI–PEG-coated NPs in rats was significant. Taken together, these results indicate that PEGylated PEI-coated albumin nanoparticles would be useful to show favorable biocompatibility as well as sustained drug release properties.

5.4.4 Albumin Nanoparticles for Targeting Delivery

Nanoparticles represent useful drug delivery systems for the specific transport of drugs to tumor cells. Wartlick et al. (2004) reported that the use of albumin nanoparticles conjugated with a HER2 receptor-specific antibody could allow specific targeting delivery to HER2-overexpressing cells, indicating that albumin nanoparticles conjugated with an antibody against a specific tumor antigen hold promise as a selective drug delivery system for the treatment of tumors that are expressing a specific tumor antigen.

In addition, it was found that folate-conjugated HSA nanoparticles were successfully delivered to tumor cells and activated macrophages where folate receptor beta (FRβ) is specifically overexpressed (Ulbrich et al. 2011; Rollett et al. 2012).

5.5 Conclusion

Albumin nanoparticles have attracted considerable interest because they are biodegradable, biocompatible, and easy to prepare and can be engineered to provide adequate features to therapeutic applications using tunable surface modifications.

Up to now, numerous fundamental studies regarding their therapeutic applications have been reported against cancer, microbial infections, inflammation, and other diseases. In the near future, these albumin-based nanoparticle strategies promise to be promising for use in targeted drug delivery in clinical situations.

References

Arnedo A, Espuelas S, Irache JM (2002) Albumin nanoparticles as carriers for a phosphodiester oligonucleotide. Int J Pharm 244:59–72

Cortes J, Saura C (2010) Nanoparticle albumin-bound (nab™)-paclitaxel: improving efficacy and tolerability by targeted drug delivery in metastatic breast cancer. Eur J Cancer Suppl 8:1–10

Desai N (2007) Nanoparticle albumin bound (nab) technology: targeting tumors through the endothelial gp60 receptor and SPARC. Nanomedicine 3:337–346

Gong J, Huo M, Zhou J, Zhang Y, Peng X, Yu D, Zhang H, Li J (2009) Synthesis, characterization, drug-loading capacity and safety of novel octyl modified serum albumin micelles. Int J Pharm 376:161–168

Irache JM, Merodio M, Arnedo A, Camapanero MA, Mirshahi M, Espuelas S (2005) Albumin nanoparticles for the intravitreal delivery of anticytomegaloviral drugs. Mini Rev Med Chem 5:293–305

Jahanshahi M, Babaei Z (2008) Protein nanoparticle: a unique system as drug delivery vehicles. Afr J Biotechnol 7:4926–4934

Jahanshahi M, Zhang Z, Lyddiatt A (2005) Subtractive chromatography for purification and recovery of nano-bioproducts. J IET Nanobiotechnol 152:121–126

Kim TH, Jiang HH, Youn YS, Park CW, Tak KK, Lee S, Kim H, Jon S, Chen X, Lee KC (2011) Preparation and characterization of water-soluble albumin-bound curcumin nanoparticles with improved antitumor activity. Int J Pharm 403:285–291

Klibanov AL, Maruyama K, Torichilin VP, Huang L (1990) Amphipathic polyethyleneglycols effectively prolong the circulation time of liposomes. FEBS Lett 268:235–238

Kouchakzadeh H, Shojaosadati SA, Maghsoudi A, Farahani EV (2010) Optimization of PEGylation conditions for BSA nanoparticles using response surface methodology. AAPS PharmSciTech 11:1206–1211

Kratz F (2008) Albumin as a drug carrier: design of prodrugs, drug conjugates and nanoparticles. J Control Release 132:171–183

Langer K, Balthasar S, Vogel V, Dinauer N, von Briesen H, Schubert D (2003) Optimization of the preparation process for human serum albumin (HSA) nanoparticles. Int J Pharm 257:169–180

Lee SH, Heng D, Ng WK, Chan HK, Tan RB (2011) Nano spray drying: a novel method for preparing protein nanoparticles for protein therapy. Int J Pharm 403:192–200

Lin W, Coombes AG, Garnett MC, Davies MC, Schacht E, Davis SS, Illum L (1994) Preparation of sterically stabilized human serum albumin nanospheres using a novel Dextranox-MPEG crosslinking agent. Pharm Res 11:1588–1592

Maruyama K, Yuda T, Okamoto A, Ishikura C, Kojima S, Iwatsuru M (1991) Effect of molecular weight in amphipathic polyethyleneglycol on prolonging the circulation time of large unilamellar liposomes. Chem Pharm Bull 39:1620–1622

Merodio M, Arnedo A, Renedo MJ, Irache JM (2001) Ganciclovir-loaded albumin nanoparticles: characterization and in vitro release properties. Eur J Pharm Sci 12:251–259

Muller RH, Wallis KH (1993) Surface modification of i.v. injectable biodegradable nanoparticles with poloxamer polymers and poloxamine 908. Int J Pharm 89:25–31

Nguyen HH, Ko S (2010) Preparation of size-controlled BSA nanoparticles by intermittent addition of desolvating agent. IFMBE Proc 27:231–234

Papadjopoulos D, Allen TM, Gabizon A, Mayhew E, Matthay K, Huang SK, Lee KD, Woodle MC, Lasic DD, Redemon C (1991) Sterically stabilized liposomes: improvements in pharmacokinetics and antitumor therapeutic efficiency. Proc Natl Acad Sci U S A 88:11460–11464

Patil GV (2003) Biopolymer albumin for diagnosis and in drug delivery. Drug Dev Res 58:219–247

Peters T Jr (1996) All about albumin: biochemistry, genetics and medical application. Academic, San Diego

Queiroz RG, Varca GHC, Kadlubowski S, Ulanski P, Lugão AB (2016) Radiation-synthesized protein-based drug carriers: size-controlled BSA nanoparticles. Int J Biol Macromol 85:82–91

Rollett A, Reiter T, Nogueira P, Cardinale M, Loureiro A, Gomes A, Cavaco-Paulo A, Moreira A, Carmo AM, Guebitz GM (2012) Folic acid-functionalized human serum albumin nanocapsules for targeted drug delivery to chronically activated macrophages. Int J Pharm 427:460–466

Scheffel U, Rhodes BA, Natarajan TK, Wagner HN Jr (1972) Albumin microspheres for study of the reticuloendothelial system. J Nucl Med 13:498–503

Shen Z, Li Y, Kohama K, Oneill B, Bi J (2011) Improved drug targeting of cancer cells by utilizing actively targetable folic acid-conjugated albumin nanospheres. Pharmacol Res 63:51–58

Singh HD, Wang G, Uludağ H, Unsworth LD (2010) Poly-L-lysine-coated albumin nanoparticles: stability, mechanism for increasing. In vitro enzymatic resilience and siRNA release characteristics. Acta Biomater 6:4277–4284

Soppimath KS, Aminabhavi TM, Kulkarni AR, Rudzinski WE (2001) Biodegradable polymeric nanoparticles as drug delivery devices. J Control Release 7:1–20

Suk JS, Xu Q, Kim N, Hanes J, Ensign LM (2016) PEGylation as a strategy for improving nanoparticle-based drug and gene delivery. Adv Drug Deliv Rev 99:28–51

Sundar S, Kundu J, Kundu SC (2010) Biopolymeric nanoparticles. Sci Technol Adv Mater 11:1–13

Ulbrich K, Michaelis M, Rothweiler F, Knobloch T, Sithisarn P, Cinat J, Kreuter J (2011) Interaction of folate-conjugated human serum albumin (HSA) nano-particles with tumor cells. Int J Pharm 406:128–134

Varca GHC, Queiroz RG, Lugão AB (2016) Irradiation as an alternative route for protein cross-linking: cosolvent free BSA nanoparticles. Radia Phy Chem 124:111–115

Veronese FM, Pasut G (2005) PEGylation, successful approach to drug delivery. Drug Discov Today 10:1451–1458

Wartlick H, Michaelis K, Balthasar S, Strebhardt K, Kreuter J, Langer K (2004) Highly specific HER2-mediated cellular uptake of antibody-modified nanoparticles in tumor cells. J Drug Target 12:461–471

Weber C, Coester C, Kreuter J, Langer K (2000) Desolvation process and surface characteristics of protein nanoparticles. Int J Pharm 194:91–102

Yogasundaram H, Bahniuk MS, Singh HD, Aliabadi HM, Uludağ H, Unsworth LD (2012) BSA nanoparticles for siRNA delivery: coating effects on nanoparticle properties, plasma protein adsorption, and in vitro siRNA delivery. Int J Biomater 2012:584060

Yu S, Yao P, Jiang M, Zhang G (2006) Nanogels prepared by self-assembly of oppositely charged globular proteins. Biopolymers 83:148–158

Zauner W, Farrow NA, Haines AM (2001) In vitro uptake of polystyrene microspheres: effect of particles size, cell line and cell density. J Control Release 71:39–51

Zhang S, Wang G, Lin X, Chatzinikolaidou M, Jennissen H, Laub M (2008) Polyethylenimine-coated albumin nanoparticles for BMP-2 delivery. Biotechnol Prog 24:945–956

Zhang S, Kucharski C, Doschak MR, Sebald W, Uludag H (2010) Polyethylenimine–PEG coated albumin nanoparticles for BMP-2 delivery. Biomaterial 31:952–963

Zhao D, Zhao X, Zu Y, Li J, Zhang Y, Jiang R, Zhang Z (2010) Preparation, characterization, and in vitro targeted delivery of folate-decorated paclitaxel-loaded bovine serum albumin nanoparticles. Int J Nanomedicine 5:669–677

Chapter 6
Nanoparticle Albumin-Bound Paclitaxel (Abraxane®)

Neil Desai

Abstract Albumin has high binding affinity to hydrophobic molecules and is highly accumulated in tumors, making it an ideal carrier to transport water insoluble drugs to tumors. Nanoparticle albumin-bound (*nab*®) technology is an albumin-based nanoparticle drug delivery platform that preferentially delivers albumin-bound hydrophobic drugs to tumors without using toxic solvents.

Abraxane® (*nab*-paclitaxel) is the first approved product based on *nab* technology and the first protein nanotechnology-based chemotherapeutic. The conventional paclitaxel formulation utilizes Cremophor EL (CrEL) and ethanol as solvents, which lead to prolonged systemic exposure, slower tissue distribution, and increased drug toxicity. In contrast, preclinical and clinical studies have demonstrated that *nab*-paclitaxel displays distinct pharmacokinetics (PK) and biodistribution properties, increased antitumor efficacy, and improved safety profile compared with CrEL-paclitaxel. As a result, *nab*-paclitaxel has been approved for the treatment of multiple indications in oncology, including metastatic breast cancer, locally advanced or metastatic non-small cell lung cancer (NSCLC), metastatic adenocarcinoma of the pancreas, and advanced gastric cancer (in Japan). The clinical success of *nab*-paclitaxel demonstrates the great potential of *nab* technology and albumin-based drug delivery platforms in general through exploitation of the natural properties of albumin and tumor biology.

Keywords Paclitaxel • Caveolae-mediated transcytosis • Nanoparticle • Drug targeting

N. Desai (✉)
Strategic Platforms, Abraxis BioScience, A wholly owned subsidiary of Celgene Corporation, 11755 Wilshire Blvd., Suite 2300, Los Angeles, CA 90025, USA
e-mail: ndesai@celgene.com

M. Otagiri, V.T.G. Chuang (eds.), *Albumin in Medicine*,
DOI 10.1007/978-981-10-2116-9_6

6.1 Introduction

Albumin has unique properties that make it an ideal carrier for the delivery of hydrophobic drugs into tumors. Albumin is the most abundant plasmic protein in the blood with a relatively long half-life of approximately 19 days (Kratz 2008). It is a natural transporter for many physiologically important substances in the blood, such as hormones, fatty acids, bilirubin, calcium, and zinc (Kratz 2008). Albumin also has multiple specific and nonspecific binding sites for hydrophobic molecules and is known to bind a broad range of therapeutic agents, including sulfonamides, warfarin, penicillin, and paclitaxel (Kratz 2008; Purcell et al. 2000; Paal et al. 2001; Trynda-Lemiesz 2004). Furthermore, albumin is highly accumulated in tumor tissues, either due to the leaky capillary system and defective lymphatic drainage of tumors (Kratz 2008) or through an active caveolae-mediated transport process across tumor blood vessel endothelium (Desai et al. 2006; Minshall et al. 2000; Schnitzer 1992). Importantly, albumin is taken up by proliferating tumor cells via endocytosis and macropinocytosis, then catabolized by lysosomal degradation to support de novo protein synthesis, energy use, and tumor growth (Stehle et al. 1997; Commisso et al. 2013; Kremer et al. 2002; Kratz 2008). The accumulation of albumin in tumors provides potential rationale for albumin-based drug delivery systems to preferentially target tumors.

Traditionally, the effective delivery of hydrophobic drugs has been a significant hurdle and paclitaxel is a primary example of the challenge. Paclitaxel is a mitotic inhibitor isolated from the bark of the Pacific yew tree, *Taxus brevifolia*. Paclitaxel exerts its therapeutic effects as an anti-microtubule agent by stabilizing intracellular microtubules, thus blocking microtubule depolymerization (Schiff et al. 1979). This results in the disruption of normal disassembly and dynamic reorganization of the microtubule network necessary for vital interphase and mitotic cellular functions, which leads to cellular apoptosis and cell death (Schiff et al. 1979). Although paclitaxel is a potent anticancer agent with a broad spectrum of activity against solid tumors, its clinical use was delayed by the high hydrophobicity and great difficulties to formulate the drug. The conventional paclitaxel was first authorized as the proprietary product Taxol® (manufactured by Bristol-Myers Squibb, New York, New York), which consists of paclitaxel dissolved in a proprietary solvent, Cremophor® EL (CrEL, polyoxyethylated castor oil, BASF, Ludwigshafen, Germany) and ethanol (ten Tije et al. 2003; Taxol 2000).

The use of solvents in paclitaxel formulation is associated with their own toxicities and impairs paclitaxel drug distribution. Cremophor EL alone has been shown to cause toxicity including neuropathy and severe hypersensitivity reactions (sometimes lethal) in patients (ten Tije et al. 2003; Weiss et al. 1990; Dye and Watkins 1980; Irizarry et al. 2009; Mielke et al. 2006). As a result, CrEL-based paclitaxel requires slow infusion and premedications to prevent hypersensitivity reaction. In addition, CrEL can hinder the distribution and delivery of paclitaxel by forming micelles with highly hydrophobic interiors that entrap paclitaxel in circulation, thus reducing drug delivery and tumor exposure to paclitaxel while increasing drug toxicity such as risks of neutropenia by prolonging systemic exposure (van Zuylen et al. 2001; Gelderblom et al. 2001; Chen et al. 2014).

Nanoparticle albumin-bound (*nab*) technology is a proprietary nanotechnology-based drug delivery platform that utilizes the endogenous properties of albumin to achieve solvent-free and efficient delivery of hydrophobic drugs to target sites. As the first approved product based on the *nab* platform, *nab*-paclitaxel (ABRAXANE® for Injectable Suspension, ABI-007, manufactured by Abraxis BioScience, LLC, a wholly owned subsidiary of Celgene Corporation, Summit, New Jersey) was developed with intent to increase efficacy and reduce the toxicities associated with conventional solvent-based paclitaxel. *nab*-Paclitaxel is a solvent-free, albumin-stabilized nanoparticle formulation of paclitaxel. The use of human albumin via the *nab* technology allowed *nab*-paclitaxel to be dissolved in saline, thereby eliminating the need for premedications. Currently, *nab*-paclitaxel has been approved for the treatment of patients with metastatic breast cancer, locally advanced or metastatic non-small cell lung cancer (NSCLC), and metastatic adenocarcinoma of the pancreas in the United States, European Union, Japan, and multiple other countries around the world. In addition, *nab*-paclitaxel has been approved for treatment of advanced gastric cancer in Japan.

6.2 Nanoparticle Properties of *nab*-Paclitaxel

nab-Paclitaxel nanoparticles are complex three-dimensional structures comprised of paclitaxel in a noncrystalline amorphous state and human albumin with a mean particle size of approximately 130 nm (Fig. 6.1). Nanoparticles of *nab*-paclitaxel have a narrow size distribution as determined by dynamic laser light scattering (DLS), transmission electron microscopy (TEM), and Cryo-TEM (Gradishar 2006). Albumin has more than six specific and nonspecific sites that can bind paclitaxel

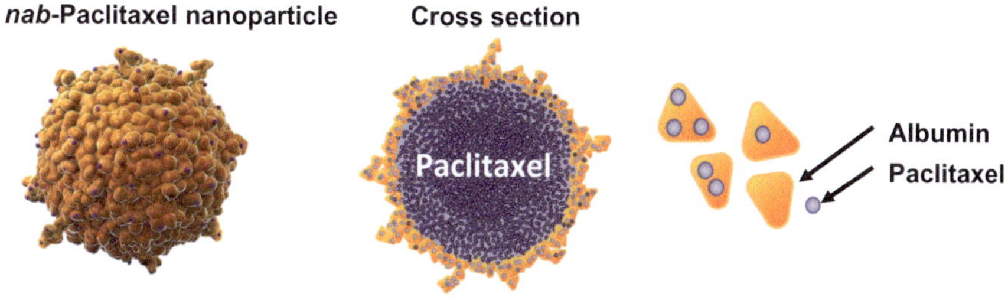

Fig. 6.1 Schematic representation of a *nab*-paclitaxel nanoparticle. Albumin-bound paclitaxel nanoparticle has a mean particle size of 130 nm. The interaction between albumin and paclitaxel is non-covalent. The cross-section view shows that a layer of albumin molecules with a certain level of cross-linking forms the nanoparticle surface, which surrounds a hydrophobic core containing amorphous and noncrystalline paclitaxel

with different affinities and with positive cooperativity (Paal et al. 2001). Using the natural affinity that exists between paclitaxel and albumin, *nab* technology creates *nab*-paclitaxel nanoparticles through non-covalent hydrophobic interaction. A layer of albumin molecules with a certain level of cross-linking forms the nanoparticle surface, surrounding a hydrophobic core containing amorphous and noncrystalline paclitaxel, as revealed by X-ray powder diffraction (Desai 2012a, b, 2013). The highly negative zeta potential of −31 mV and steric repulsion of the albumin surface prevent agglomeration and stabilize nanoparticles in aqueous suspension, which enables *nab*-paclitaxel nanoparticles to remain stable at room temperature for several days when reconstituted to a 5 mg paclitaxel/mL concentration solution with 0.9 % (w/v) saline (Desai 2012a, b; Desai 2013). The non-covalent binding nature and noncrystalline form of paclitaxel allow the drug to be readily available for rapid drug release and tissue distribution without the time lag and free energy needed to either cleave the covalent binding or dissolve crystalline paclitaxel as in the case of nanocrystals (Merisko-Liversidge et al. 1996).

6.3 Mechanism of Action of *nab*-Paclitaxel

Results from numerous preclinical and clinical studies have shed light on the mechanism of action of *nab*-paclitaxel, which uses albumin to enhance the delivery and bioavailability of paclitaxel and has features highly distinctive from conventional solvent-based paclitaxel.

6.3.1 Rapid Drug Release

Upon infusion into the circulation, *nab*-paclitaxel nanoparticles undergo a dynamic dissolution process into smaller nanoparticles and eventually to mostly albumin-bound paclitaxel complexes and a small fraction of unbound paclitaxel. In a randomized crossover pharmacokinetic study in patients with solid tumors, it was shown that protein-bound paclitaxel accounted for ~94 % of drug following *nab*-paclitaxel infusion. The mean fraction of unbound paclitaxel was 6.3 % with *nab*-paclitaxel, which was 2.6-fold higher than CrEL-paclitaxel (Gardner et al. 2008). Considering the higher dose and shorter infusion time, *nab*-paclitaxel allows ~tenfold increase in C_{max} and ~threefold higher AUC of free unbound paclitaxel compared with CrEL-paclitaxel, which may partially contribute to the increased delivery and improved efficacy of *nab*-paclitaxel (Fig. 6.2).

In contrast, the Cremophor EL/ethanol vehicle in CrEL-paclitaxel forms micelles that sequester paclitaxel (van Zuylen et al. 2001), preventing drug release and binding to plasmic proteins. It has been demonstrated that the presence of CrEL significantly inhibited paclitaxel binding to human serum albumin in a dose-dependent

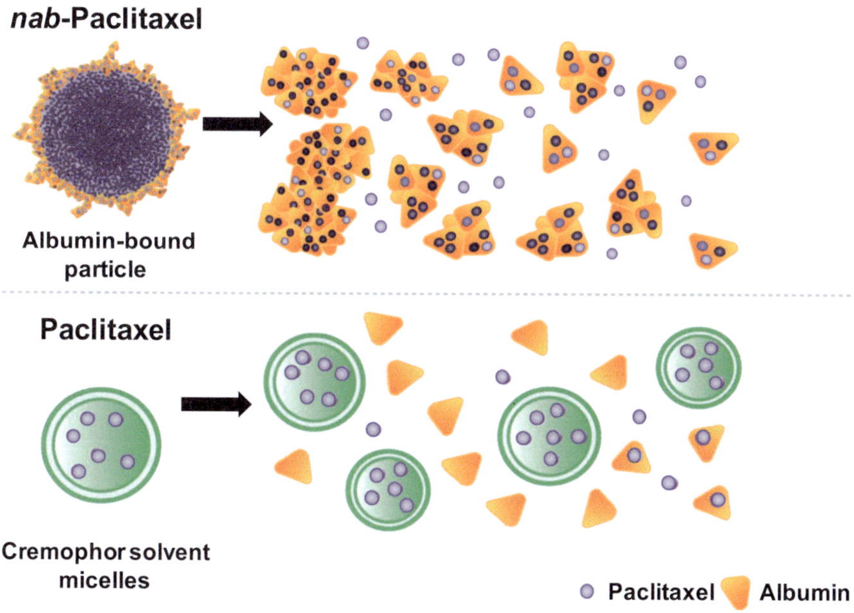

Fig. 6.2 Rapid drug release by *nab*-paclitaxel. In circulation, *nab*-paclitaxel nanoparticles undergo a dynamic dissolution process into smaller nanoparticles and eventually to mostly albumin-bound paclitaxel complexes and a small fraction of unbound paclitaxel. In contrast, the Cremophor EL/ethanol vehicle in CrEL-paclitaxel forms micelles that sequester paclitaxel, preventing drug release and binding to plasmic proteins

manner, with IC_{50} at 0.0017 % and almost complete inhibition at 0.3 % (Desai et al. 2006). These findings are clinically relevant, as the critical micellar concentration (CMC) of CrEL in aqueous solution is only 0.009 % (Kessel 1992), much lower than the peak plasma CrEL level of 0.3–0.5 % after intravenous (IV) administration of CrEL-paclitaxel (100–175 mg/m^2, over a 3-h period) (Sparreboom et al. 1998) and the plasma CrEL level of 0.1 % 24 h after infusion (Brouwer et al. 2000).

6.3.2 *Enhanced Transport Across Blood Vessel Endothelium*

Following the dissociation of *nab*-paclitaxel nanoparticles, albumin-paclitaxel complexes utilize the natural transport properties of albumin to distribute into tissues and preferentially tumors. The enhanced permeability and retention effect (EPR) has been proposed to enhance distribution of macromolecules and nanoparticles into tumors. The blood vessels of proliferating tumors have structural defects with fenestrations between 0.2 and 1.2 μm, which are highly permeable to macromolecules such as albumin and nanoparticles with sizes below 200 nm (Yuan et al. 1995;

Hobbs et al. 1998; Haley and Frenkel 2008), whereas the lack of proper lymphatic drainage in tumors decreased the clearance of albumin and other macromolecules with molecular weight greater than 40 kDa (Maeda et al. 2001). Theoretically, *nab*-paclitaxel nanoparticles and albumin-bound paclitaxel could potentially benefit from EPR to achieve higher accumulation in tumors. However, drug delivery to solid tumors in patients has been shown to be strongly hindered by major biological barriers including heterogeneous blood supply, elevated interstitial fluid pressure (IFP), and large transport distances in the tumor interstitium (Crommelin and Florence 2013; Nichols and Bae 2014). These factors seriously limit the potential clinical significance of EPR.

Based on current evidence, active transcytosis mediated by receptor and caveolae appears to be a more important transport pathway into tissues and tumors for albumin and *nab*-paclitaxel. Caveolae are a special type of lipid raft, 50–100 nm small invaginations of the plasma membrane in many cell types, especially in endothelial cells and adipocytes (Simionescu et al. 2002; Frank et al. 2009). Albumin binds to the albumin receptor glycoprotein gp60 (albondin) located on endothelial cell surface with nanomolar affinity (Schnitzer 1992), which triggers gp60 clustering and association with caveolar scaffolding protein caveolin-1, resulting in the activation of tyrosine kinase Src and the formation of caveolae (Tiruppathi et al. 1997). The plasmalemmal vesicles containing both gp60-bound and fluid phase albumin migrate from apical to basal membrane and release their contents by exocytosis into the subendothelial space, completing the transcytosis process (John et al. 2003; Tiruppathi et al. 2004).

Due to the absence of solvents, albumin-paclitaxel complexes from *nab*-paclitaxel can take full advantage of the natural albumin transcytosis pathway to enhance its tissue and tumor distribution. Albumin internalized by monolayer endothelial cells was found in endocytic plasmalemmal vesicles, some of which were early endosomes as indicated by the presence of the endosome biomarker early endosome antigen 1 (EEA1 protein). Very little albumin was found in lysosomes (Chen et al. 2015), indicating endocytic uptake of albumin and transendothelial trafficking of the molecule rather than breakdown of the protein in lysosomes. Live imaging demonstrated that albumin and fluorescent paclitaxel were present in punctae in endothelial cells and could be observed in very close proximity, suggesting cotransport (Chen et al. 2015). Consistent with the albumin vesicle trafficking pattern, no fluorescent paclitaxel was found in lysosomes as visualized by LysotrackerRed (Chen et al. 2015). As shown by in vitro drug uptake and permeability assays, when compared with CrEL-paclitaxel, *nab*-paclitaxel formulation increased the binding of paclitaxel to endothelial cells by 9.9-fold ($P<0.0001$) and the transport of paclitaxel across human umbilical vein endothelial cell (HUVEC) monolayers by 4.2-fold ($P<0.0001$) (Desai et al. 2006). The enhanced paclitaxel transcytosis was completely abolished by caveolar-disrupting agent methyl-β-cyclodextrin, demonstrating that the paclitaxel transport is mediated by caveolae (Desai et al. 2006).

On the other hand, the solvent CrEL severely hinders all the above albumin-facilitated paclitaxel transport processes, even at very low concentrations. Consistent with the inhibition of paclitaxel binding to albumin by CrEL discussed earlier, CrEL inhibited paclitaxel binding to endothelial cells (HUVECs) in a dose-dependent manner (IC_{50}: 0.010 %), with complete inhibition occurring at a concentration of 0.1 % (Desai et al. 2006). Live imaging showed a decrease in albumin uptake by HUVECs as measured by both anti-albumin immunofluorescent staining and HSA-TRITC with increasing concentrations of CrEL, and a fluorescence-activated cell sorting (FACS) assay demonstrated that the uptake of fluorescent-labeled paclitaxel in *nab*-paclitaxel formulation by HUVECs was also strongly inhibited by CrEL in a dose-dependent manner, with almost complete abolishment of cellular uptake at a concentration of 0.3 % (Chen et al. 2015). In the in vitro permeability assay, CrEL strongly inhibited paclitaxel transport across endothelial cells, with fit-determined concentrations required for 50 % inhibition of paclitaxel transcytosis (IC_{50}) of 0.19, 0.12, 0.16, and 0.22 % at 1, 2, 4, and 24 h, respectively (Chen et al. 2015). As discussed above, the CrEL concentrations in these in vitro assays are in the range of plasma CrEL levels in patients administered with CrEL-paclitaxel, suggesting that the presence of CrEL in formulation would have significant clinical implications. Taken together, these results clearly demonstrate that *nab*-paclitaxel, but not CrEL-paclitaxel, can utilize and leverage the active albumin transport mechanism for efficient paclitaxel distribution from the circulation.

6.3.3 Distinct Pharmacokinetic Profile

Due to its unique albumin-based nanoparticle formulation and mechanism of action, *nab*-paclitaxel displays distinct pharmacokinetic (PK) and biodistribution profiles compared with conventional CrEL-paclitaxel. In both preclinical and clinical studies, *nab*-paclitaxel exhibits a linear PK with rapid tissue distribution and increased distribution volume. In contrast, the concentration of CrEL increases with higher doses of CrEL-paclitaxel, leading to greater inhibitory effect on paclitaxel binding to albumin and tissue distribution; therefore, CrEL-paclitaxel displays slower tissue distribution and a nonlinear PK profile.

In rats and mice, CrEL-paclitaxel showed threefold higher plasma peak levels (C_{max}), higher plasma AUC (area under the concentration-time curve), and approximately a seven- to tenfold lower steady-state volume of distribution (Vdss) compared with *nab*-paclitaxel (Sparreboom et al. 2005). Clinically, the systemic drug exposure with intravenous *nab*-paclitaxel administration was approximately dose proportional from 80 to 300 mg/m^2 and was independent of the infusion duration (Ibrahim et al. 2002; Chen et al. 2014), whereas CrEL-paclitaxel displayed more than dose proportional increases in systemic exposure and infusion duration-dependent clearance (Gianni et al. 1995; van Tellingen et al. 1999). In a PK study comparing *na*b-paclitaxel (260 mg/m^2 IV over 30 min, q3w) and CrEL-paclitaxel

(175 mg/m^2 IV over 3 h, q3w) in patients with solid tumors, *nab*-paclitaxel displayed a significantly higher rate of clearance (21.13 vs 14.76 L/h/m^2, $P=0.048$) and a larger volume of distribution (663.8 vs 433.4 L/m^2, $P=0.040$) than CrEL-paclitaxel (Sparreboom et al. 2005).

The slow elimination of paclitaxel from circulation with CrEL causes prolonged systemic drug exposure and increased risk of the dose-limiting toxicity neutropenia. The exposure-neutropenia relationship has been described for CrEL-paclitaxel using a threshold model, with the duration of time that the plasma paclitaxel concentration was >0.05 μM being predictive of neutropenia in patients with ovarian cancer (Joerger et al. 2007). On the other hand, the faster tissue distribution by *nab*-paclitaxel causes a shorter duration of high plasmic drug concentration, reducing the risk of neutropenia. A population PK study with data from 150 patients in eight clinical trials revealed that for *nab*-paclitaxel, the time or AUC above the threshold concentration of 720 ng/mL (0.84 μM) correlated with the probability of experiencing a ≥50 % reduction in neutrophil count (Chen et al. 2014). The threshold concentration for *nab*-paclitaxel is nearly 17-fold higher than that observed for CrEL-paclitaxel. The difference in PK profiles may help explain the difference in clinical safety between *nab*-paclitaxel and CrEL-paclitaxel observed in large-scale randomized phase 3 studies. The incidence of grade 4 neutropenia was significantly lower in *nab*-paclitaxel arm vs CrEL-paclitaxel arm (phase 3 MBC trial, 9 % vs 22 %, $P<0.001$; phase 3 advanced NSCLC trial, 14 % vs 26 %, $P<0.001$), despite higher paclitaxel dose intensity delivered with *nab*-paclitaxel (Gradishar et al. 2005; Socinski et al. 2012).

In an analysis of population PK data with *nab*-paclitaxel and CrEL-paclitaxel using PK modeling, the plasma paclitaxel concentration versus time in solid tumor patients can best be described by a three-compartment pharmacokinetic model (Chen et al. 2014, 2015; Joerger et al. 2006): the central compartment (plasma and well-perfused organs), the first peripheral compartment (tissues/organ distribution through a saturable transporter-mediated mechanism), and the second peripheral compartment (tissue/organ distribution through a non-saturable passive diffusion). Consistent with observed results of gp60/caveolae-mediated albumin-paclitaxel transport and faster tissue distribution, the PK modeling also indicates that the distribution of *nab*-paclitaxel is more dependent upon transporter-mediated pathways, reflected as a more than twofold faster rate and a ninefold larger volume for saturable drug distribution to the first peripheral compartment compared to CrEL-paclitaxel (Chen et al. 2015). Conversely, drug delivery into tissue by CrEL-paclitaxel is more dependent upon passive diffusion. Further, the fraction of *nab*-paclitaxel dose delivered to tissues would remain relatively constant for either transporter-mediated or diffusion-related distribution over a broad clinical dose range, whereas transporter-mediated distribution decreases while diffusion-related distribution increases with higher dose of CrEL-paclitaxel (Chen et al. 2015). The PK study findings are consistent with the albumin receptor-facilitated transport mechanism of *nab*-paclitaxel, which allows faster and more efficient drug delivery to tumors compared with CrEL-paclitaxel (Fig. 6.3).

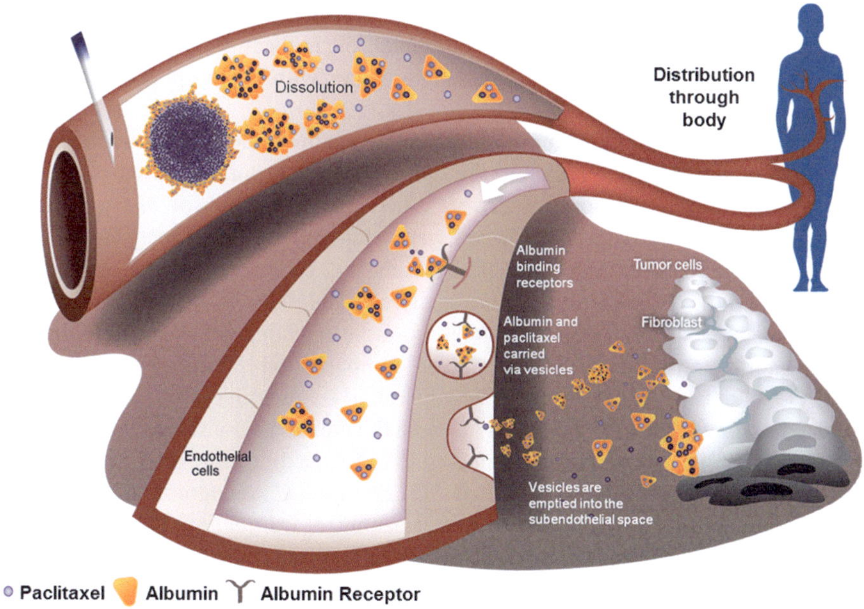

Fig. 6.3 Mechanisms for the transport of *nab*-paclitaxel into tumors. Following dissolution of *nab*-paclitaxel nanoparticles, the transcytosis of albumin-bound paclitaxel complexes across the endothelial barrier is facilitated by binding to the albumin receptor gp60 and caveolar transport. Upon entering the tumor, the natural demand for albumin by proliferating tumor cells enhances the accumulation, distribution, and penetration of albumin-bound paclitaxel

6.3.4 Efficient and Selective Tumor Accumulation and Penetration

There are also significant differences in tumor distribution of *nab*-paclitaxel and CrEL-paclitaxel. As discussed above, albumin is highly accumulated in tumors, as fast-growing tumor cells actively take up albumin through endocytosis and macropinocytosis (Stehle et al. 1997; Commisso et al. 2013). Catabolized by lysosomal degradation, albumin plays an essential role in supporting tumor cell proliferation by serving as a major energy and nutrient source and providing amino acids for protein synthesis (Stehle et al. 1997; Commisso et al. 2013).

The natural demand for albumin by solid tumors facilitates the delivery of active drug by *nab*-paclitaxel. In xenograft-bearing mice, radiolabeled paclitaxel from *nab*-paclitaxel distributed favorably into tumors versus normal tissues at the early time points compared with CrEL-paclitaxel, with a *nab*-paclitaxel/CrEL-paclitaxel ratio of 1.25 for tumor and 0.4–0.8 for normal tissue of different organs at 1 h post dose (Hawkins et al. 2003). In MX-1 human breast tumor xenografts, intravenous *nab*-paclitaxel achieved a 33 % higher intratumoral paclitaxel concentration at equal dose than CrEL-paclitaxel (Desai et al. 2006). In pediatric tumor models of rhabdomyosarcoma and neuroblastoma, a four- to sevenfold higher tumor/plasma paclitaxel drug ratio was observed for *nab*-paclitaxel compared with DMSO-paclitaxel (Zhang et al. 2013b).

Further, upon entering the tumor, *nab*-paclitaxel allows the active drug to distribute wider and penetrate deeper into the tumor tissue than solvent-based paclitaxel, resulting in greater antitumor efficacy. When equal amounts of *nab*-paclitaxel, CrEL-paclitaxel, and DMSO-paclitaxel were delivered through direct intratumoral microinjection into human pancreatic MIA PaCa-2 tumor xenografts, the area of response and the total fraction of mitotically arrested phospho-histone H3 (pHH3) positive cells at specific radial distances from the injection site were significantly greater for microinjected *nab*-paclitaxel compared with CrEL-paclitaxel and DMSO-paclitaxel (Chen et al. 2015). Similarly, microinjected *nab*-paclitaxel induced a significantly larger increase in both the area of response and total fraction of cells arrested in mitosis when compared to CrEL-paclitaxel-injected A2058 melanoma and DMSO-paclitaxel-injected H2122 NSCLC xenografts (Chen et al. 2015). The results clearly demonstrate that paclitaxel from *nab*-paclitaxel can distribute effectively and extensively within tumors, whereas other paclitaxel formulations such as a micellar formulation (CrEL-paclitaxel) or a solvent formulation (DMSO-paclitaxel) exhibit more limited intratumoral drug distribution and cellular uptake.

6.4 Nonclinical Studies of *nab*-Paclitaxel

Due to its distinct mechanisms of action, *nab*-paclitaxel demonstrated strong antitumor activity and improved therapeutic index in nonclinical studies both as a single agent and in combination with other therapeutic agents against a broad range of tumor models. In mice, the maximum tolerated dose (MTD) for *nab*-paclitaxel (30 mg/kg, qdx5) was substantially higher than for CrEL-paclitaxel (13.4 mg/kg, qdx5) (Desai et al. 2006). In nude mice bearing various human tumor xenografts treated with both agents at MTD (H522, lung; MX-1, breast; SK-OV-3, ovarian; PC3, prostate; and HT29, colon), *nab*-paclitaxel resulted in more complete regressions, longer time to recurrence and tumor doubling, and prolonged survival compared with CrEL-paclitaxel (Desai et al. 2006). The antitumor activity of *nab*-paclitaxel was also better or equal compared with polysorbate-based docetaxel at its MTD in various breast (MDA-MB-231 and MX-1), lung (LX-1), prostate (PC3), and colon (HT29) tumor xenograft models (Desai et al. 2008). In mice bearing MDA-MB-231 and MDA-MB-435 breast tumor xenografts, combined *nab*-paclitaxel and bevacizumab treatment significantly enhanced the antitumor activity and reduced both lymphatic and pulmonary metastasis compared with either drug administered as a single agent (Volk et al. 2008, 2011).

In nonclinical studies with pancreatic cancer models, the combination treatment of *nab*-paclitaxel and gemcitabine displayed strong antitumor activity over either agent alone and increased intratumoral gemcitabine levels by 2.8-fold (Von Hoff et al. 2011). The exact mechanism of enhanced gemcitabine tumor accumulation by *nab*-paclitaxel remains unclear. In separate studies using patient-derived and genetically engineered mouse model (GEMM) of pancreatic cancer, *nab*-paclitaxel disrupted tumor stroma as demonstrated by reduction in type I collagen and increased tumor vascularization (Alvarez et al. 2013; Von Hoff et al. 2011). In another study using KPC GEMM model, *nab*-paclitaxel increased intratumoral levels of the active gemcitabine metabolite gemcitabine triphosphate (dFdCTP), which was attributed to a marked decrease of cytidine deaminase, the primary gemcitabine metabolizing enzyme, following *nab*-paclitaxel treatment (Frese et al. 2012). In mice bearing subcutaneous AsPC-1 human pancreatic cancer xenografts, *nab*-paclitaxel demonstrated stronger antitumor activity and prolonged animal survival compared with polysorbate-based docetaxel (Awasthi et al. 2013).

In addition, *nab*-paclitaxel is highly active against gastric cancer xenograft models. In mice bearing subcutaneously or intraperitoneally (IP) implanted OCUM-2MD3 tumors, *nab*-paclitaxel (30 mg/kg/day, IV or IP) showed significantly greater antitumor activity than equitoxic dose of CrEL-paclitaxel (13.4 mg/kg/day, IP) (Kinoshita et al. 2014). In mice bearing SNU16 human gastric cancer xenografts, *nab*-paclitaxel treatment resulted in significantly stronger tumor growth suppression and longer animal survival compared with oxaliplatin or epirubicin treatment (Zhang et al. 2013a).

Recent preclinical studies also demonstrated dose-dependent antitumor activity of *nab*-paclitaxel against multiple types of pediatric solid tumors. *nab*-Paclitaxel displayed greater antitumor activity, better tolerability, and higher intratumor paclitaxel concentrations than DMSO-paclitaxel in several pediatric solid tumor models, including human neuroblastoma (SK-N-BE[2] and CHLA-20) and rhabdomyosarcoma (RH4 and RD) xenografts (Zhang et al. 2013b). In another study, *nab*-paclitaxel alone or in combination with gemcitabine displayed strong antitumor activity against both 143.98.2 osteosarcoma and A673 Ewing sarcoma xenografts (Wagner et al. 2014). Further, among 20 pediatric solid tumor xenograft models, single agent *nab*-paclitaxel was well tolerated and resulted in significant differences in event-free survival in 19 of 20 (95 %) solid tumors. Objective responses were observed in 12 of 20 (60 %) solid tumor xenografts. Complete responses (CR) or maintained CR were observed in five of eight Ewing sarcoma models and six of eight rhabdomyosarcomas (Houghton et al. 2015).

6.5 Key Clinical Results of *nab*-Paclitaxel

The unique properties of *nab*-paclitaxel confer it with distinct clinical efficacy and safety profiles from conventional CrEL-paclitaxel. Results from key clinical studies establish *nab*-paclitaxel as an important weapon in the chemotherapeutic arsenal with worldwide approvals in metastatic breast cancer (MBC), locally advanced or metastatic non-small cell lung cancer (NSCLC), metastatic adenocarcinoma of the pancreas, and advanced gastric cancer in Japan.

In phase 1 studies, it was established that *nab*-paclitaxel can be administered at a higher dose with shorter infusion duration than CrEL-paclitaxel (Taxol), without the need for premedication (Ibrahim et al. 2002; Hawkins et al. 2008). The MTD was 70% higher with *nab*-paclitaxel based on every 3-weeks dosing: 300 mg/m^2 (Ibrahim et al. 2002) versus 175 mg/m^2 for Taxol. The elimination of toxic solvents also enables *nab*-paclitaxel to be given in a shorter, more convenient infusion time of 30–40 min compared with 3–24 h with Taxol (Ibrahim et al. 2002). *nab*-Paclitaxel may be given without steroid and antihistamine premedication, which is required for Taxol to prevent solvent-related hypersensitivity reactions. Cremophor EL has been shown to leach plasticizers from polyvinyl chloride (PVC) bags and polyethylene-lined tubing (Gelderblom et al. 2001); therefore, Taxol needs to use glass, polypropylene, or polyolefin containers and non-PVC-containing infusion sets, whereas standard tubing and intravenous (IV) bags may be used for the IV administration of *nab*-paclitaxel (Ibrahim et al. 2002; Nyman et al. 2005).

For metastatic breast cancer, *nab*-paclitaxel was approved in the United States in 2005, in European Union in 2008, and in Japan in 2010. In a phase 2 study in patients with MBC whose disease progressed after weekly paclitaxel or docetaxel, *nab*-paclitaxel demonstrated promising disease control rate (31%) and >9 months of median overall survival (Blum et al. 2007). In a phase 3 study in 460 patients with metastatic breast cancer randomized to receive IV administration of either *nab*-paclitaxel at 260 mg/m^2 q3w or CrEL-paclitaxel at 175 mg/m^2 q3w (Gradishar et al. 2005), *nab*-paclitaxel showed statistically significantly higher response rates (33% versus 19%, $P=0.001$), longer time to tumor progression (5.3 versus 3.9 months, $P=0.006$), and increased survival in the subset of patients receiving second-line or greater treatment (12.9 versus 10.7 months, $P=0.024$). The incidence of grade 4 neutropenia was significantly lower with *nab*-paclitaxel than with CrEL-paclitaxel (9% versus 22%, $P<0.001$). No severe hypersensitivity reactions occurred with *nab*-paclitaxel despite the lack of premedication. Grade 3 neuropathy was higher for *nab*-paclitaxel (10% vs 2%, $P<0.001$) due to the approximately 50% higher dosage, but was easily manageable and improved quickly with a median recovery time of 22 days. Currently, *nab*-paclitaxel is indicated for the treatment of metastatic breast cancer, after failure of combination chemotherapy for metastatic disease or relapse within 6 months of adjuvant chemotherapy. Prior therapy should have included an anthracycline unless clinically contraindicated. The recommended dosage is 260 mg/m^2 IV over 30 min every 3 weeks.

For locally advanced or metastatic NSCLC, *nab*-paclitaxel was approved in the United States in 2012, in European Union in 2015, and in Japan in 2013. In a randomized phase 3 study in patients with advanced NSCLC (Socinski et al. 2010), patients received either the combination of *nab*-paclitaxel (100 mg/m^2, qw)/carboplatin (AUC 6, q3w) ($n=521$) or CrEL (200 mg/m^2, q3w)/carboplatin ($n=531$). The *nab*-paclitaxel arm demonstrated a significantly higher overall response rate (ORR) than the CrEL-paclitaxel arm (33% vs 25%; response rate ratio, 1.313; 95% confidence interval [CI], 1.082–1.593; $P=0.005$) with a favorable trend in progression-free survival (PFS; median, 6.3 vs 5.8 months; hazard ratio [HR], 0.902; 95 % CI, 0.767–1.060; $P=0.214$) and overall survival (OS; median, 12.1 vs 11.2 months; HR, 0.922; 95 % CI, 0.797–1.066; $P=0.271$). There were significantly less grade 3 and 4 neuropathy, neutropenia, arthralgia, and myalgia in the *nab*-paclitaxel arm, whereas less thrombocytopenia and anemia were observed in the CrEL-paclitaxel arm (Socinski et al. 2012). Currently, *nab*-paclitaxel is indicated for the treatment of locally advanced or metastatic NSCLC, as first-line treatment in combination with carboplatin, in patients who are not candidates for curative surgery or radiation therapy. The recommended dosage of *nab*-paclitaxel is 100 mg/m^2 intravenously over 30 min on days 1, 8, and 15 of each 21-day cycle, with carboplatin administered on day 1 of each 21-day cycle immediately after *nab*-paclitaxel.

For metastatic adenocarcinoma of the pancreas, *nab*-paclitaxel was approved in the United States in 2013, in European Union in 2013, and in Japan in 2014. While solvent-based taxanes failed to show clinically meaningful activity and adequate safety in several phase 2 studies of metastatic pancreatic cancer (Whitehead et al. 1997; Androulakis et al. 1999; Jacobs et al. 1999), *nab*-paclitaxel is highly active in this indication. In a randomized phase 3 study in 861 patients with metastatic pancreatic cancer, a combination of *nab*-paclitaxel (125 mg/m^2 weekly, 3 out of 4 weeks) and gemcitabine (1,000 mg/m^2 weekly, 3 out of 4 weeks) demonstrated significantly longer overall survival and improved clinical outcomes compared with the standard of care treatment of gemcitabine alone (Von Hoff et al. 2013). The median OS was 8.5 months for the *nab*-paclitaxel/gemcitabine arm vs 6.7 months for the gemcitabine arm (HR: 0.72; 95 % CI, 0.62–0.83; $P<0.001$). The median PFS was longer in the *nab*-paclitaxel/gemcitabine arm (5.5 vs 3.7 months HR: 0.69; 95 % CI, 0.58–0.82; $P<0.001$); the ORR was 23 % vs 7 % in the two groups ($P<0.001$). The survival rate was higher with the *nab*-paclitaxel/gemcitabine arm (35 % vs 22 % at 1 year and 9 % vs 4 % at 2 years). The most common adverse events of grade 3 or higher were neutropenia (38 % in the *nab*-paclitaxel/gemcitabine arm vs 27 % in the gemcitabine arm), fatigue (17 % vs 7 %), and neuropathy (17 % vs 1 %). Febrile neutropenia occurred in 3 % vs 1 % of the patients in the two arms, respectively. In the *nab*-paclitaxel/gemcitabine arm, neuropathy of grade 3 improved to grade ≤1 with a median duration of 29 days. Currently, *nab*-paclitaxel is indicated for the treatment of metastatic adenocarcinoma of the pancreas as first-line treatment, in combination with gemcitabine. The recommended dosage of *nab*-paclitaxel is 125 mg/m^2 IV over 30–40 min on days 1, 8, and 15 of each 28-day cycle, with gemcitabine administered on days 1, 8, and 15 of each 28-day cycle immediately after *nab*-paclitaxel.

For advanced gastric cancer, *nab*-paclitaxel was approved in Japan in 2013. In a multicenter phase 2 study in 56 Japanese patients with unresectable or recurrent gastric cancer who had received a prior round of fluoropyrimidine-containing chemotherapy (Sasaki et al. 2014), *nab*-paclitaxel administered IV at 260 mg/m^2 q3w resulted in an ORR of 27.8 % (15/54 evaluable patients; 95 % confidence interval [CI], 16.5–41.6) with one complete response and a disease control rate of 59.3 % (32/54; 95 % CI, 45.0–72.4). The median PFS and OS were 2.9 months (95 % CI, 2.4–3.6) and 9.2 months (95 % CI, 6.9–11.4), respectively. The most common grade

3/4 toxicities were neutropenia (49.1 %), leucopenia (20.0 %), lymphopenia (10.9 %), and peripheral sensory neuropathy (23.6 %).

Overall, in head-to-head clinical comparison with CrEL-paclitaxel, *nab*-paclitaxel consistently demonstrates greater clinical benefits and better safety profile in multiple cancer types. Despite premedications, 171 unique cases of hypersensitivity following CrEL-paclitaxel infusion were identified in a review of adverse event reports submitted to regulatory agencies in the United States, Europe, and Japan between 1997 and 2007, of which 34 % were fatal (Irizarry et al. 2009). In phase 3 studies in patients with MBC and NSCLC, no severe hypersensitivity reaction occurred in the *nab*-paclitaxel arm even in the absence of premedication requirement, while three occurred in the CrEL-paclitaxel arm in spite of premedications (two patients with MBC and one patient with advanced NSCLC) (Gradishar et al. 2005; Socinski et al. 2012). Because of the rapid tissue distribution of paclitaxel from circulation by *nab*-paclitaxel, the incidence of grade 4 neutropenia was significantly lower in *nab*-paclitaxel arm vs CrEL-paclitaxel arm (phase 3 MBC trial, 9 % vs 22 %, $P<0.001$; phase 3 advanced NSCLC trial, 14 % vs 26 %, $P<0.001$), despite higher paclitaxel dose intensity delivered with *nab*-paclitaxel (Gradishar et al. 2005; Socinski et al. 2012). For neuropathy, the incidence of grade 3 neuropathy was more common in the *nab*-paclitaxel arm compared with CrEL-paclitaxel arm based on an every 3-week dosing schedule in the randomized phase 3 study in patients with MBC (10 % vs 2 %, $P<0.001$) (Gradishar et al. 2005), whereas significantly lower neuropathy rates were observed in the weekly *nab*-paclitaxel vs q3w CrEL-paclitaxel arm (grade 3: 3 % vs 11 %) in the phase 3 study in patients with advanced NSCLC (Socinski et al. 2012), suggesting that the incidence of neuropathy is dose and schedule dependent. Importantly, neuropathy caused by *nab*-paclitaxel treatment was reversible, and patients in general recovered quickly with treatment interruption or dose reduction.

6.6 Conclusions and Outlook

Currently, there are numerous phase 3 clinical trials ongoing for *nab*-paclitaxel in different solid tumor indications, including triple negative MBC, squamous cell NSCLC, adjuvant pancreatic adenocarcinoma, and gastric cancer. Studies are also exploring the combination of *nab*-paclitaxel with novel therapeutic agents, such as stroma modulating agent PEGylated recombinant human hyaluronidase (PEGPH20, Halozyme Therapeutics, Inc.) and immunotherapy agent anti-PD-L1 antibody atezolizumab (Genentech/Roche).

Several other drugs based on the *nab* technology platform are also undergoing preclinical and clinical development for oncology and vascular disease indications. The mammalian target of rapamycin, mTOR, is a key regulator of cell proliferation and an important target in cancer and proliferative vascular diseases (Dancey 2010; Goncharova 2013). *nab*-Rapamycin (ABI-009) is an albumin-bound injectable form of rapamycin. In a phase 1 study in 26 heavily pretreated patients with advanced solid tumors, *nab*-rapamycin was well tolerated with MTD established at 100 mg/m^2 IV weekly and showed evidence of responses and stable disease with various solid tumors including renal cell carcinoma and bladder cancer, both known for mTOR overexpression (Gonzalez-Angulo et al. 2013).

In conclusion, *nab* technology-based drugs such as *nab*-paclitaxel are highly complex nanoparticle products that utilize the natural transport pathways and tumor accumulation properties of albumin to achieve improved distribution, tumor targeting, efficacy, and safety. The *nab* technology represents a major development in the delivery of hydrophobic drugs, with *nab*-paclitaxel serving as a strong testament for the potential of albumin-based drug delivery system.

Acknowledgements Dr. Shihe Hou's expert editorial and writing assistance of this manuscript is greatly appreciated.

References

Alvarez R, Musteanu M, Garcia-Garcia E, Lopez-Casas PP, Megias D, Guerra C, Munoz M, Quijano Y, Cubillo A, Rodriguez-Pascual J, Plaza C, de Vicente E, Prados S, Tabernero S, Barbacid M, Lopez-Rios F, Hidalgo M (2013) Stromal disrupting effects of nab-paclitaxel in pancreatic cancer. Br J Cancer 109:926–933

Androulakis N, Kourousis C, Dimopoulos MA, Samelis G, Kakolyris S, Tsavaris N, Genatas K, Aravantinos G, Papadimitriou C, Karabekios S, Stathopoulos GP, Georgoulias V (1999) Treatment of pancreatic cancer with docetaxel and granulocyte colony-stimulating factor: a multicenter phase II study. J Clin Oncol 17:1779–1785

Awasthi N, Zhang C, Schwarz AM, Hinz S, Wang C, Williams NS, Schwarz MA, Schwarz RE (2013) Comparative benefits of Nab-paclitaxel over gemcitabine or polysorbate-based docetaxel in experimental pancreatic cancer. Carcinogenesis 34:2361–2369

Blum JL, Savin MA, Edelman G, Pippen JE, Robert NJ, Geister BV, Kirby RL, Clawson A, O'Shaughnessy JA (2007) Phase II study of weekly albumin-bound paclitaxel for patients with metastatic breast cancer heavily pretreated with taxanes. Clin Breast Cancer 7:850–856

Brouwer E, Verweij J, De Bruijn P, Loos WJ, Pillay M, Buijs D, Sparreboom A (2000) Measurement of fraction unbound paclitaxel in human plasma. Drug Metab Dispos 28:1141–1145

Chen N, Li Y, Ye Y, Palmisano M, Chopra R, Zhou S (2014) Pharmacokinetics and pharmacodynamics of nab-paclitaxel in patients with solid tumors: disposition kinetics and pharmacology distinct from solvent-based paclitaxel. J Clin Pharmacol 54:1097–1107

Chen N, Brachmann C, Liu X, Pierce DW, Dey J, Kerwin WS, Li Y, Zhou S, Hou S, Carleton M, Klinghoffer RA, Palmisano M, Chopra R (2015) Albumin-bound nanoparticle (nab) paclitaxel exhibits enhanced paclitaxel tissue distribution and tumor penetration. Cancer Chemother Pharmacol 76:699–712

Commisso C, Davidson SM, Soydaner-Azeloglu RG, Parker SJ, Kamphorst JJ, Hackett S, Grabocka E, Nofal M, Drebin JA, Thompson CB, Rabinowitz JD, Metallo CM, Vander Heiden MG, Bar-Sagi D (2013) Macropinocytosis of protein is an amino acid supply route in Ras-transformed cells. Nature 497:633–637

Crommelin DJ, Florence AT (2013) Towards more effective advanced drug delivery systems. Int J Pharm 454:496–511

Dancey J (2010) mTOR signaling and drug development in cancer. Nat Rev Clin Oncol 7:209–219

Desai N (2012a) Albumin drug nanoparticles. In: Kratz F, Senter P, Steinhagen H (eds) Drug delivery in oncology: from basic research to cancer therapy, 1st edn. Wiley-VCH Verlag GmbH & Co. KGaA, Weinheim, pp 1133–1161
Desai N (2012b) Challenges in development of nanoparticle-based therapeutics. AAPS J 14:282–295
Desai N (2013) Integration of nab-technology into clinical drug development. In: Bischoff J (ed) Nanotechnologie beim Mammakarzinom- Grundlagen und aktuelle Perspektiven. UNI-MED Verlag AG, Bremen, pp 22–31
Desai N, Trieu V, Yao Z, Louie L, Ci S, Yang A, Tao C, De T, Beals B, Dykes D, Noker P, Yao R, Labao E, Hawkins M, Soon-Shiong P (2006) Increased antitumor activity, intratumor paclitaxel concentrations, and endothelial cell transport of cremophor-free, albumin-bound paclitaxel, ABI-007, compared with cremophor-based paclitaxel. Clin Cancer Res 12:1317–1324
Desai NP, Trieu V, Hwang LY, Wu R, Soon-Shiong P, Gradishar WJ (2008) Improved effectiveness of nanoparticle albumin-bound (nab) paclitaxel versus polysorbate-based docetaxel in multiple xenografts as a function of HER2 and SPARC status. Anticancer Drugs 19:899–909
Dye D, Watkins J (1980) Suspected anaphylactic reaction to Cremophor EL. Br Med J 280:1353
Frank PG, Pavlides S, Lisanti MP (2009) Caveolae and transcytosis in endothelial cells: role in atherosclerosis. Cell Tissue Res 335:41–47
Frese KK, Neesse A, Cook N, Bapiro TE, Lolkema MP, Jodrell DI, Tuveson DA (2012) nab-Paclitaxel potentiates gemcitabine activity by reducing cytidine deaminase levels in a mouse model of pancreatic cancer. Cancer Discov 2:260–269
Gardner ER, Dahut WL, Scripture CD, Jones J, Aragon-Ching JB, Desai N, Hawkins MJ, Sparreboom A, Figg WD (2008) Randomized crossover pharmacokinetic study of solvent-based paclitaxel and nab-paclitaxel. Clin Cancer Res 14:4200–4205
Gelderblom H, Verweij J, Nooter K, Sparreboom A (2001) Cremophor EL: the drawbacks and advantages of vehicle selection for drug formulation. Eur J Cancer 37:1590–1598
Gianni L, Kearns CM, Giani A, Capri G, Vigano L, Locatelli A, Bonadonna G, Egorin MJ (1995) Nonlinear pharmacokinetics and metabolism of paclitaxel and its pharmacokinetic/pharmacodynamic relationships in humans. J Clin Oncol 13:180–190
Goncharova EA (2013) mTOR and vascular remodeling in lung diseases: current challenges and therapeutic prospects. FASEB J 27:1796–1807
Gonzalez-Angulo AM, Meric-Bernstam F, Chawla S, Falchook G, Hong D, Akcakanat A, Chen H, Naing A, Fu S, Wheler J, Moulder S, Helgason T, Li S, Elias I, Desai N, Kurzrock R (2013) Weekly nab-rapamycin in patients with advanced nonhematologic malignancies: final results of a phase i trial. Clin Cancer Res 19:5474–5484
Gradishar WJ (2006) Albumin-bound paclitaxel: a next-generation taxane. Expert Opin Pharmacother 7:1041–1053
Gradishar WJ, Tjulandin S, Davidson N, Shaw H, Desai N, Bhar P, Hawkins M, O'Shaughnessy J (2005) Phase III trial of nanoparticle albumin-bound paclitaxel compared with polyethylated castor oil-based paclitaxel in women with breast cancer. J Clin Oncol 23:7794–7803
Haley B, Frenkel E (2008) Nanoparticles for drug delivery in cancer treatment. Urol Oncol 26:57–64
Hawkins MJ, Desai N, Soon-Shiong P (eds) (2003) Rationale, preclinical support, and clinical proof-of-concept for formulating waterinsoluble therapeutics as albumin-stabilized nanoparticles: experience with Paclitaxel. [abstr 442]. American Association for Cancer Research (AACR) Annual Meeting, Anaheim
Hawkins MJ, Soon-Shiong P, Desai N (2008) Protein nanoparticles as drug carriers in clinical medicine. Adv Drug Deliv Rev 60:876–885
Hobbs SK, Monsky WL, Yuan F, Roberts WG, Griffith L, Torchilin VP, Jain RK (1998) Regulation of transport pathways in tumor vessels: role of tumor type and microenvironment. Proc Natl Acad Sci U S A 95:4607–4612
Houghton PJ, Kurmasheva RT, Kolb EA, Gorlick R, Maris JM, Wu J, Tong Z, Arnold MA, Chatterjee M, Williams TM, Smith MA (2015) Initial testing (stage 1) of the tubulin binding

agent nanoparticle albumin-bound (nab) paclitaxel (Abraxane((R))) by the Pediatric Preclinical Testing Program (PPTP). Pediatr Blood Cancer 62:1214–1221

Ibrahim NK, Desai N, Legha S, Soon-Shiong P, Theriault RL, Rivera E, Esmaeli B, Ring SE, Bedikian A, Hortobagyi GN, Ellerhorst JA (2002) Phase I and pharmacokinetic study of ABI-007, a cremophor-free, protein-stabilized, nanoparticle formulation of paclitaxel. Clin Cancer Res 8:1038–1044

Irizarry L, Luu T, McKoy J, Samaras A, Fisher M, Carias E, Raisch D, Calhoun E, Bennett C (2009) Cremophor EL-containing paclitaxel-induced anaphylaxis: a call to action. Community Oncol 6:132–134

Jacobs AD, Otero H, Picozzi V (1999) Gemcitabine (G) and Taxotere® (T) in patients with unresectable pancreatic carcinoma. Am Soc Clin Oncol 18:1103A

Joerger M, Huitema AD, van den Bongard DH, Schellens JH, Beijnen JH (2006) Quantitative effect of gender, age, liver function, and body size on the population pharmacokinetics of Paclitaxel in patients with solid tumors. Clin Cancer Res 12:2150–2157

Joerger M, Huitema AD, Richel DJ, Dittrich C, Pavlidis N, Briasoulis E, Vermorken JB, Strocchi E, Martoni A, Sorio R, Sleeboom HP, Izquierdo MA, Jodrell DI, Calvert H, Boddy AV, Hollema H, Fety R, Van der Vijgh WJ, Hempel G, Chatelut E, Karlsson M, Wilkins J, Tranchand B, Schrijvers AH, Twelves C, Beijnen JH, Schellens JH (2007) Population pharmacokinetics and pharmacodynamics of paclitaxel and carboplatin in ovarian cancer patients: a study by the European organization for research and treatment of cancer-pharmacology and molecular mechanisms group and new drug development group. Clin Cancer Res 13:6410–6418

John TA, Vogel SM, Tiruppathi C, Malik AB, Minshall RD (2003) Quantitative analysis of albumin uptake and transport in the rat microvessel endothelial monolayer. Am J Physiol Lung Cell Mol Physiol 284:L187–L196

Kessel D (1992) Properties of cremophor EL micelles probed by fluorescence. Photochem Photobiol 56:447–451

Kinoshita J, Fushida S, Tsukada T, Oyama K, Watanabe T, Shoji M, Okamoto K, Nakanuma S, Sakai S, Makino I, Furukawa H, Hayashi H, Nakamura K, Inokuchi M, Nakagawara H, Miyashita T, Tajima H, Takamura H, Ninomiya I, Fujimura T, Masakazu Y, Hirakawa K, Ohta T (2014) Comparative study of the antitumor activity of Nab-paclitaxel and intraperitoneal solvent-based paclitaxel regarding peritoneal metastasis in gastric cancer. Oncol Rep 32:89–96

Kratz F (2008) Albumin as a drug carrier: design of prodrugs, drug conjugates and nanoparticles. J Control Release 132:171–183

Kremer P, Hartung G, Bauder-Wust U, Schrenk HH, Wunder A, Heckl S, Zillmann U, Sinn H (2002) Efficacy and tolerability of an aminopterin-albumin conjugate in tumor-bearing rats. Anticancer Drugs 13:615–623

Maeda H, Sawa T, Konno T (2001) Mechanism of tumor-targeted delivery of macromolecular drugs, including the EPR effect in solid tumor and clinical overview of the prototype polymeric drug SMANCS. J Control Release 74:47–61

Merisko-Liversidge E, Sarpotdar P, Bruno J, Hajj S, Wei L, Peltier N, Rake J, Shaw JM, Pugh S, Polin L, Jones J, Corbett T, Cooper E, Liversidge GG (1996) Formulation and antitumor activity evaluation of nanocrystalline suspensions of poorly soluble anticancer drugs. Pharm Res 13:272–278

Mielke S, Sparreboom A, Mross K (2006) Peripheral neuropathy: a persisting challenge in paclitaxel-based regimes. Eur J Cancer 42:24–30

Minshall RD, Tiruppathi C, Vogel SM, Niles WD, Gilchrist A, Hamm HE, Malik AB (2000) Endothelial cell-surface gp60 activates vesicle formation and trafficking via G(i)-coupled Src kinase signaling pathway. J Cell Biol 150:1057–1070

Nichols JW, Bae YH (2014) EPR: evidence and fallacy. J Control Release 190:451–464

Nyman DW, Campbell KJ, Hersh E, Long K, Richardson K, Trieu V, Desai N, Hawkins MJ, Von Hoff DD (2005) Phase I and pharmacokinetics trial of ABI-007, a novel nanoparticle formulation of paclitaxel in patients with advanced nonhematologic malignancies. J Clin Oncol 23:7785–7793

Paal K, Muller J, Hegedus L (2001) High affinity binding of paclitaxel to human serum albumin. Eur J Biochem 268:2187–2191
Purcell M, Neault JF, Tajmir-Riahi HA (2000) Interaction of taxol with human serum albumin. Biochim Biophys Acta 1478:61–68
Sasaki Y, Nishina T, Yasui H, Goto M, Muro K, Tsuji A, Koizumi W, Toh Y, Hara T, Miyata Y (2014) Phase II trial of nanoparticle albumin-bound paclitaxel as second-line chemotherapy for unresectable or recurrent gastric cancer. Cancer Sci 105:812–817
Schiff PB, Fant J, Horwitz SB (1979) Promotion of microtubule assembly in vitro by taxol. Nature 277:665–667
Schnitzer JE (1992) gp60 is an albumin-binding glycoprotein expressed by continuous endothelium involved in albumin transcytosis. Am J Physiol 262:H246–H254
Simionescu M, Gafencu A, Antohe F (2002) Transcytosis of plasma macromolecules in endothelial cells: a cell biological survey. Microsc Res Tech 57:269–288
Socinski MA, Vinnichenko I, Okamoto I, Hon JK, Hirsh V (eds) (2010) Results of a randomized, phase 3 trial of nab-Paclitaxel (nab-P) and Carboplatin (C) compared with Cremophor-based Paclitaxel (P) and carboplatin as first-line therapy in advanced non-small cell lung cancer (NSCLC). In: Proceedings of the 46th American Society of Clinical Oncology Annual Meeting (ASCO); 2010 Jun 4–8; Chicago
Socinski MA, Bondarenko I, Karaseva NA, Makhson AM, Vynnychenko I, Okamoto I, Hon JK, Hirsh V, Bhar P, Zhang H, Iglesias JL, Renschler MF (2012) Weekly nab-paclitaxel in combination with carboplatin versus solvent-based paclitaxel plus carboplatin as first-line therapy in patients with advanced non-small-cell lung cancer: final results of a phase III trial. J Clin Oncol 30:2055–2062
Sparreboom A, Verweij J, van der Burg ME, Loos WJ, Brouwer E, Vigano L, Locatelli A, de Vos AI, Nooter K, Stoter G, Gianni L (1998) Disposition of Cremophor EL in humans limits the potential for modulation of the multidrug resistance phenotype in vivo. Clin Cancer Res 4:1937–1942
Sparreboom A, Scripture CD, Trieu V, Williams PJ, De T, Yang A, Beals B, Figg WD, Hawkins M, Desai N (2005) Comparative preclinical and clinical pharmacokinetics of a cremophor-free, nanoparticle albumin-bound paclitaxel (ABI-007) and paclitaxel formulated in Cremophor (Taxol). Clin Cancer Res 11:4136–4143
Stehle G, Sinn H, Wunder A, Schrenk HH, Stewart JC, Hartung G, Maier-Borst W, Heene DL (1997) Plasma protein (albumin) catabolism by the tumor itself – implications for tumor metabolism and the genesis of cachexia. Crit Rev Oncol Hematol 26:77–100
Taxol P (2000) Taxol® (paclitaxel) for injection, BMS insert
ten Tije AJ, Verweij J, Loos WJ, Sparreboom A (2003) Pharmacological effects of formulation vehicles: implications for cancer chemotherapy. Clin Pharmacokinet 42:665–685
Tiruppathi C, Song W, Bergenfeldt M, Sass P, Malik AB (1997) Gp60 activation mediates albumin transcytosis in endothelial cells by tyrosine kinase-dependent pathway. J Biol Chem 272:25968–25975
Tiruppathi C, Naqvi T, Wu Y, Vogel SM, Minshall RD, Malik AB (2004) Albumin mediates the transcytosis of myeloperoxidase by means of caveolae in endothelial cells. Proc Natl Acad Sci U S A 101:7699–7704
Trynda-Lemiesz L (2004) Paclitaxel-HSA interaction. Binding sites on HSA molecule. Bioorg Med Chem 12:3269–3275
van Tellingen O, Huizing MT, Panday VR, Schellens JH, Nooijen WJ, Beijnen JH (1999) Cremophor EL causes (pseudo-) non-linear pharmacokinetics of paclitaxel in patients. Br J Cancer 81:330–335
van Zuylen L, Karlsson MO, Verweij J, Brouwer E, de Bruijn P, Nooter K, Stoter G, Sparreboom A (2001) Pharmacokinetic modeling of paclitaxel encapsulation in Cremophor EL micelles. Cancer Chemother Pharmacol 47:309–318

Volk LD, Flister MJ, Bivens CM, Stutzman A, Desai N, Trieu V, Ran S (2008) Nab-paclitaxel efficacy in the orthotopic model of human breast cancer is significantly enhanced by concurrent anti–vascular Endothelial Growth Factor A therapy. Neoplasia 10:613–623
Volk LD, Flister MJ, Chihade D, Desai N, Trieu V, Ran S (2011) Synergy of nab-paclitaxel and bevacizumab in eradicating large orthotopic breast tumors and preexisting metastases. Neoplasia 13:327–338
Von Hoff DD, Ramanathan RK, Borad MJ, Laheru DA, Smith LS, Wood TE, Korn RL, Desai N, Trieu V, Iglesias JL, Zhang H, Soon-Shiong P, Shi T, Rajeshkumar NV, Maitra A, Hidalgo M (2011) Gemcitabine plus nab-paclitaxel is an active regimen in patients with advanced pancreatic cancer: a phase I/II trial. J Clin Oncol 29:4548–4554
Von Hoff DD, Ervin T, Arena FP, Chiorean EG, Infante J, Moore M, Seay T, Tjulandin SA, Ma WW, Saleh MN, Harris M, Reni M, Dowden S, Laheru D, Bahary N, Ramanathan RK, Tabernero J, Hidalgo M, Goldstein D, Van Cutsem E, Wei X, Iglesias J, Renschler MF (2013) Increased survival in pancreatic cancer with nab-paclitaxel plus gemcitabine. N Engl J Med 369:1691–1703
Wagner LM, Yin H, Eaves D, Currier M, Cripe TP (2014) Preclinical evaluation of nanoparticle albumin-bound paclitaxel for treatment of pediatric bone sarcoma. Pediatr Blood Cancer 61:2096–2098
Weiss RB, Donehower RC, Wiernik PH, Ohnuma T, Gralla RJ, Trump DL, Baker JR Jr, Van Echo DA, Von Hoff DD, Leyland-Jones B (1990) Hypersensitivity reactions from taxol. J Clin Oncol 8:1263–1268
Whitehead RP, Jacobson J, Brown TD, Taylor SA, Weiss GR, Macdonald JS (1997) Phase II trial of paclitaxel and granulocyte colony-stimulating factor in patients with pancreatic carcinoma: a Southwest Oncology Group study. J Clin Oncol 15:2414–2419
Yuan F, Dellian M, Fukumura D, Leunig M, Berk DA, Torchilin VP, Jain RK (1995) Vascular permeability in a human tumor xenograft: molecular size dependence and cutoff size. Cancer Res 55:3752–3756
Zhang C, Awasthi N, Schwarz MA, Hinz S, Schwarz RE (2013a) Superior antitumor activity of nanoparticle albumin-bound paclitaxel in experimental gastric cancer. PLoS One 8:e58037
Zhang L, Marrano P, Kumar S, Leadley M, Elias E, Thorner P, Baruchel S (2013b) Nab-Paclitaxel is an active drug in preclinical model of pediatric solid tumors. Clin Cancer Res 19:5972–5983

Chapter 7
Optison™ Albumin Microspheres in Ultrasound-Assisted Gene Therapy and Drug Delivery

Alex Jackson, Jason W. Castle, Adrian Smith, and Christina K. Kalli

Abstract Optison™ (Perflutren Protein-Type A Microspheres Injectable Suspension, USP) is a sterile non-pyrogenic suspension of microspheres of human serum albumin with perflutren (also known as perfluoropropane). Optison™ microspheres are micrometre-sized gas-filled bubbles that have a shell consisting of human albumin. The size range is 2–4 μm in diameter and 95 % are less than 10 μm. This means that intravenously injected Optison™ may pass through the pulmonary capillary bed and access all parts of the systemic vasculature. The size distribution, shell properties and gas core provide a bubble that oscillates in response to and scatters ultrasound at frequencies useful for clinical imaging. The established use of Optison™ is in the field of echocardiography, where it provides echogenic contrast enhancement for suboptimal echocardiograms.

Optison™ is currently marketed in North America and Europe where researchers may gain access to the product for research and experimental use under their own institutional processes. One such experimental use that has shown promise is the use of ultrasound combined with a microbubble agent to induce transient changes to biological tissue with the aim of increasing delivery and penetration of therapeutic molecules. It has been known for decades that microbubbles can act as nucleation sites for a range of ultrasound-induced physical effects such as stable cavitation, inertial cavitation and jetting. These phenomena have been shown to have direct physical effects on biological membranes in the vicinity of the microbubble such as the creation of pores (sonoporation) which can persist for seconds up to several minutes depending on their size and the level of impact to the host cell. This approach has been applied as an alternative to viral vectors to address the significant challenge of delivering genetic material for anticancer and cardiovascular gene therapy. There are approximately 50 research papers on the use of Optison™ to enhance the transfection of oligonucleotides and plasmid DNA. This chapter will introduce

A. Jackson (✉) • A. Smith • C.K. Kalli
GE Healthcare, Life Sciences, The Grove Centre,
White Lion Road, Amersham, Buckinghamshire HP7 9LL, UK
e-mail: Alex.Jackson@ge.com

J.W. Castle
General Electric Global Research, Niskayuna, NY, USA

M. Otagiri, V.T.G. Chuang (eds.), *Albumin in Medicine*,
DOI 10.1007/978-981-10-2116-9_7

Optison™ and its characteristics in the context of established use in diagnostic clinical imaging and experimental use in the delivery of therapeutic molecules. The methodology and results in the field of drug delivery and gene therapy will be reviewed and summarised in the context of effectiveness, potential for clinical translation and potential impact in medicine.

Keywords Optison™ • Albumin • Microbubble • Microsphere • Ultrasound • Sonoporation • Drug delivery • Gene delivery

7.1 Introduction

Contrast-enhanced ultrasound (CEUS) is a technique that improves the visualisation and assessment of cardiac function and tissue perfusion by using gas-filled microspheres to increase the echogenicity of the blood following intravenous injection. The safety and efficacy of ultrasound contrast agents for diagnostic use in humans has been proven in clinical trials. Non-invasive, real-time nature of CEUS provided a technique that can be used repeatedly during various radiological applications such as diagnostic imaging, monitoring for tumour progression and recurrence as well as treatment planning, and for echocardiography.

Even before the 1970s, micrometre-sized bubbles producing acoustic interference had been used to aid echocardiography. These studies used first-generation ultrasound contrast agents (USCAs) which were little more than air-filled microbubbles of agitated saline. However, microbubbles produced in this manner were of variable size and short duration and when injected intravenously enabled ultrasound imaging of the right side of the heart but were unable to traverse the pulmonary circulation to aid imaging of the left chambers of the heart and other parts of the body. Further work led to stabilisation of microbubbles of a consistently smaller size by using shells of albumin and lipids. Increased echogenicity was also achieved by replacing the air with heavy gases such as perfluorocarbons (Goldberg et al. 1994; Stride and Saffari 2003).

7.2 Optison™ Product Information and Development

Microbubble contrast agents are small gas-filled microspheres with a typical diameter between 1 and 10 μm (Sirsi and Borden 2009), similar to normal human red blood cells (6–8 μm). First-generation air-filled microbubbles (e.g. Albunex®, Levovist®, Echovist®) (Kim et al. 1996; Bao et al. 1997) encapsulated within a thin (10–15 nm) protein shell were characterised by low stability and limited utility due to high solubility of air in blood limiting their persistence within the vasculature (Hernot and Klibanov 2008). Second-generation microbubbles such as Optison™,

Sonazoid™ (perfluorobutane microbubbles, GE Healthcare, Oslo, Norway), SonoVue® (sulphur hexafluoride microbubbles, Bracco Imaging S.p.A., Milan, Italy) and Definity® (perflutren lipid microspheres, Lantheus Medical Imaging Inc., North Billerica, MA) are used today in clinical practice as tracers for imaging of the vasculature in organs such as the heart, liver and breast. In these agents, hydrophobic gases such as perfluoropropane, perfluorobutane and sulphur hexafluoride, rather than air, are encapsulated in thin, stable, biocompatible shells, phospholipids and proteins (between 10 and 200 nm thick). The nature of the gas and shell increases the microbubble resistance to arterial pressure preventing them from rapidly dissolving in the blood stream (Alter et al. 2009; Wei et al. 1998; Averkiou et al. 2004).

Optison™ was launched as the first of the second-generation of USCAs in the North America market in January 1998 and in Europe in May of the same year. The second-generation USCAs had markedly improved echogenicity and stability sufficient to traverse the pulmonary circulation and provide contrast in the systemic circulation following intravenous injection. Consisting of a simple albumin shell (which is readily metabolised) and inert perfluoropropane[1] gas (which is readily expired), Optison™ provides up to 5 min of left ventricular ultrasound contrast (Cohen et al. 1998).

7.3 Contrast-Enhanced Echocardiography

Contrast-enhanced echocardiography is used to enhance colour and spectral Doppler flow signals from the cavities and vessels; delineate the endocardium by left ventricular cavity opacification, which is particularly useful during stress echocardiography; and image perfusion of the myocardium in patients with myocardial infarction and chronic ischaemic heart disease. Additional clinical applications such as assessment of cardiac shunts are evaluated based on detection of contrast flowing through a "hole" in the right atrium or ventricle allowing passage of microbubbles into the left atrium or ventricle before traversing the pulmonary circulation. The visualisation of contrast in the aorta and peripheral circulation without left heart opacification indicates an extracardiac shunt, whilst the appearance of contrast in the left atrium via the pulmonary veins indicates pulmonary arteriovenous fistula. An enhanced colour and spectral Doppler signal allows for the evaluation of tricuspid regurgitation, whilst contrast in the left atrium immediately after appearance in the right atrium reflects patent foramen ovale. Further, diagnosis of a persistent left superior vena cava is identified from visualisation of contrast in an enlarged coronary sinus and right atrium following injection into the left antecubital vein (Stewart 2003; Becher and Burns 2000). Optison™ is approved in the US and Europe for echocardiography: Optison™ is indicated for use in "patients with suboptimal

[1] Perfluoropropane is also known as perflutren and octafluoropropane, the chemical formula of which is C_3F_8.

echocardiograms to opacify the left ventricle (LV) and to improve the delineation of the left ventricular endocardial borders". Therefore, any other clinical applications are strictly "off-label" and may only be carried out in compliance with the appropriate legal/regulatory framework for clinical research.

7.4 Ultrasound Contrast Agents in Gene Therapy and Drug Delivery

The goal of localised gene therapy and drug delivery is to increase efficacy and safety by reducing systemic toxicity, eliminating potential immunogenicity and reducing non-specific gene/drug delivery or viral chromosomal insertion, whilst preserving or enhancing therapeutic effects. The need for safe and efficient transfection delivery systems in vitro and subsequently in vivo has led researchers to investigate non-viral techniques. One of the most important emerging non-viral techniques to assist gene therapy and drug delivery is the use of ultrasound with microbubble contrast agents such as Optison™.

Microbubbles can act as vehicles incorporating the drug within the polymer, protein or lipid shell, inside the gas core or in liposomes associated with the microbubble, whilst genes can be loaded onto lipid or protein microbubbles. When microbubbles act as carrier vehicles for drugs and genes, they can release their content to the targeted area when activated by the applied ultrasound acoustic pressure or ultrasound-induced hyperthermia resulting in microbubble disruption or complete destruction. Additionally, studies have shown that enhanced delivery can be achieved when the drug or gene is not directly bound to the microbubble, but when they are introduced separately. This occurs by a mechanism known as "sonoporation", whereby the drug or gene enters the cell via pores created by ultrasound applied in the presence of microbubbles. Using these approaches, the delivery of the therapeutic agent may be targeted to a particular organ or tissue by the shape of the applied ultrasound beam. In some circumstances this can be an advantage over the targeting achieved, for example, by innate tissue tropism of viral vectors or lack of targeting in drugs.

7.5 Sonoporation: Definition

Sonoporation is defined as the interaction of ultrasound with microbubble contrast agents resulting in transient or permanent cell membrane permeability. Sonoporation and consequently drug delivery and gene transfection efficacy are affected by the applied ultrasound wave parameters. These parameters include frequency, acoustic pressure, ultrasound exposure time, pulse repetition frequency (pulse interval and length), duty cycle[2] (DC) and the mechanical index[3] (MI) (Escoffre et al. 2013; Mehier-Humbert and Guy 2005). During transient pore formation, uptake of drugs, gene complexes and therapeutic agents from an extracellular environment is enhanced. In such cases, membrane alteration is brief following ultrasound exposure; the pore resealing allows for drug/gene compounds to be retained within the cytoplasm. This process is defined as reversible sonoporation. When sonoporation results in permanent cell membrane permeability, subsequently leading to cell lysis and apoptosis, the phenomenon is defined as irreversible sonoporation (Endoh et al. 2002; Taniyama et al. 2002a; Sheikh et al. 2011) and is generally avoided for drug delivery.

7.6 Ultrasound and Microbubble Interactions: Inducing Sonoporation

Although the precise physical and biochemical mechanisms of sonoporation are not yet fully understood, the necessity of microbubbles indicates that they induce the process (Bao et al. 1997; Greenleaf et al. 1998). As the microbubbles undergo oscillations within an ultrasound field, they interact with the membranes of nearby cells or tissue. As the microbubbles are exposed to the acoustic waves going through phases of negative and positive pressure, the compressible microbubbles will volumetrically expand and contract (Qin et al. 2009).

During low pressure amplitudes, whilst the microbubble undergoes linear oscillations (stable linear cavitation), their radius successively decreases and increases. During microbubble expansion, the surface of the cell membrane is pushed away (Fig. 7.1a), whilst during contraction, the cell membrane is pulled towards the microbubble (Fig. 7.1b). This expansion and compression perturbs the cell membrane, resulting in pore formation. In addition, an imparting motion and a translational displacement upon the oscillating microbubble occur with ultrasonic radiation (Lum et al. 2006; Urban et al. 2010). The microbubble may lose part of its shell material, and thus any bound drug or gene may be delivered into the intracellular region (Fig. 7.1c) (Delalande et al. 2013).

As high pressure amplitudes are approached, the microbubble undergoes asymmetric nonlinear oscillation thus growing in successive cycles. Once an unstable size is reached, the microbubble collapses violently and asymmetrically (inertial cavitation) giving rise to highly energetic liquid jets. The liquid jets move with sonic speeds and are capable of penetrating the cellular membrane (Fig. 7.1d). As the microbubble collapses, micro-streaming (local steady flow) (Fig. 7.1e), shear

---

[2] Duty cycle, DC, is the percentage time that ultrasound transmission takes place during treatment. Therefore, DC of 100 % is continuous wave.

[3] Mechanical index is a measure of ultrasound intensity. MI = peak negative pressure/square root of ultrasound centre frequency.

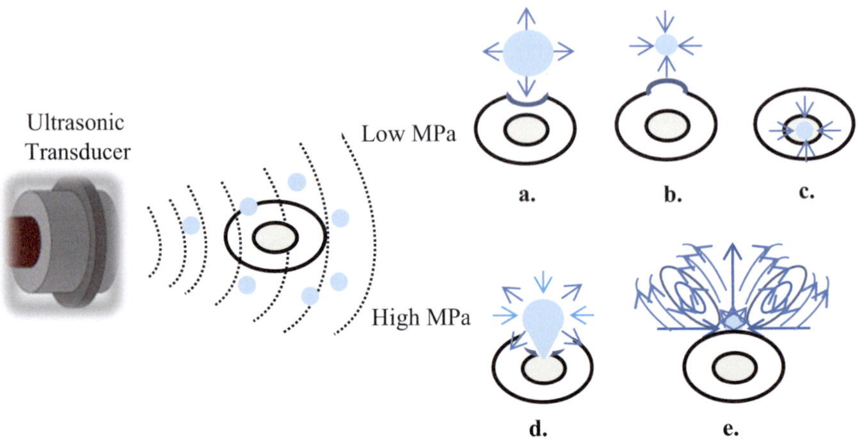

Fig. 7.1 Microbubble and cell interaction during ultrasound exposure (Kalli et al. 2014)

stresses, free radicals and shock waves are produced (Doinikov and Bouakaz 2010; Kuliszewski et al. 2011; Wu et al. 2002). During inertial cavitation, extremely high pressures and temperatures can be reached within the microbubble (Liang et al. 2010). These phenomena change the microbubble-to-cell environment, increasing heat transfer at the microbubble-to-cell interface. The friction along the interface results in pore formation on the cell membrane thus increasing permeability to extracellular molecules and substances. Further, an imparting motion and a translational displacement upon the oscillating microbubble occur with ultrasonic radiation (Lum et al. 2006; Urban et al. 2010). As with stable cavitation, the microbubble can pass through the cell membrane, lose part of its shell and deliver drugs into the intracellular compartment (Delalande et al. 2013).

7.7 Optison™ In Vitro Gene Therapy and Drug Delivery

Optison™ in vitro studies for gene therapy and drug delivery investigate the interaction and bioeffects of therapeutic agents and microbubble-enhanced ultrasound on cell cultures, suspensions and monolayers. The aim is to define the optimal acoustic conditions for sonoporation that will result in maximum transfection or delivery efficacy whilst maintaining high cell viability. Depending on the application, the acoustic conditions may be optimised to achieve a balance between viability and, e.g., gene transfection. Current applications are mainly in gene transfection for research or to serve as preliminary experiments prior to in vivo studies. In the future, application could expand to include modification of cells for therapeutic purposes. Researchers must choose between viral vectors, which may give near quantitative transfection but with some drawbacks, and non-viral methods, which have different attributes.

Published studies include work with cell lines such as blood cells, MCF7 breast cancer cells, human gingival squamous carcinoma cells (Ca9-22) and epidermoid cell monolayers (Jelenc et al. 2012; Cochran and Wheatley 2013; Qiu et al. 2010; Iwanaga et al. 2007). Experimental setups involve transducer placement either in direct contact with the cell culture or with the cell culture in a water bath, whilst the transducer is placed at a distance. For in vitro studies, the reported frequencies range from 1 to 10 MHz, whilst acoustic pressure is set at 0.06 MPa/MHz (Miller et al. 2008).

Iwanaga et al. (2007) observed the local delivery of chemotherapeutic cytotoxic drugs to human gingival squamous carcinoma cells (Ca9-22). The cells were exposed to sonoporation at a frequency of 1 MHz, an intensity of 11 W/cm^2, a DC of 10 % and exposure duration of 20 s. Sonoporation was performed in the presence of Optison™ (600 μL), to deliver bleomycin (BLM) and transfect gene pVIVO1-cdtB. Flow cytometry determined the percentage of apoptotic Ca9-22 cells based on the presence of hypodiploid DNA post-sonoporation; 17.7 % of the cells exposed to a combination of sonoporation and BLM were apoptotic compared to 8.4 % and 8.5 % when exposed to sonoporation or BLM alone. The transfection of pVIVO1-cdtB was evaluated from the number of cells expressing β-galactosidase. Following sonoporation by cavitation of Optison™, the cell count was remarkably high (8×10^2/cm^2) in comparison with the control and ultrasound alone groups where cell count was null. Results support the potential use of cytotoxic chemotherapeutic drugs with sonoporation for improved cancer therapy.

In vitro experiments by Eshet at al. (Duvshani-Eshet et al. 2007) on human (LNCaP) and murine (PC2) prostate cancer cells and endothelial cells (EC) showed transfection of cDNA-PEX using therapeutic ultrasound (TUS: 1 MHz, 2 W/cm^2, DC of 30 %, 30 min) and Optison™ (10 % v/v) microbubbles. The biological activity of haemopexin-like domain fragment (PEX) expression was assessed based on significant inhibition of proliferation (<65 %), migration (<50 %) and increase in apoptosis relative to control (-TUS, -PEX). Incubation of human umbilical vein endothelial cells (HUVEC) with conditioned media taken from LNCaP and PC2 resulted in an increase in apoptosis of 18 ± 4 % ($P<0.001$) and 17 ± 4 % ($P<0.001$), respectively, from control 7 ± 2 %. Results signify the efficacy of therapeutic ultrasound in delivering antiangiogenic drugs for prostate cancer therapy.

Optimal drug and gene delivery in human breast cancer cells (MCF7) by ultrasound-induced cavitation (3 MHz, 3 W/cm^2, DC of 20 %, 1 min) with Optison™ (200 μL) were studied by Larina et al. (2005). Delivery of macromolecular anti-drugs (fluorescein isothiocyanate-dextrans) simulating antisense oligonucleotides (10 kDa), antibodies (70 kDa) and genes (2000 kDa) in MCF7 cells were 73.5 ± 3.3 %, 72.7 ± 0.9 % and 62.7 ± 2.1 %, respectively. Optimal parameters provided 36.7 ± 4.9 % pEGFP plasmid DNA transfection of cells that survived, whilst apoptotic cells were determined by flow cytometry to be 13.5 ± 1.6 %. This study indicates the potential of applying optimised therapeutic parameters to obtain efficient drug and gene delivery in cancer chemo- and biotherapy.

Miller et al. (2003) investigated DNA plasmid transfer in epidermoid cell monolayers (A434) with 2 % v/v Optison™, using a diagnostic ultrasound scanner

(1.5 MHz, 2.3 MPa). Induced gene transfer following ultrasound exposure (90 s) was assessed based on green fluorescent protein (GFP) levels. GFP expression of 3.7 % (1.2 % SD) was reported for the exposed group compared to 0.4 % (0.7 % SD) for the nonexposed group ($P<0.01$). The percentage of apoptotic cells was greater at 28.6 % (6.3 % SD) for the exposed group compared to the control group at 3.4 % (1.7 % SD). These results indicate the potential for the future use of diagnostic ultrasound scanners and ultrasound contrast agents for gene transfer and clinical therapeutic applications.

Guzman et al. (2003) studied the bioeffects caused by changes in acoustic cavitation and bubble density on human DUI145 prostate cancer cells at varying concentrations (2.5×10^5–4.0×10^7 cells/mL). Cells were exposed to a range of ultrasound energy (500 kHz, 2–817 J/cm^2, 0.64–2.96 MPa, 120–2000 ms) over a range of Optison™ concentrations (3.6×10^4–9.3×10^7 bubbles/mL). Results based on flow cytometry indicated the increase of calcein uptake (subgroups by calcein uptake level: nominal (0.013 ± 0.007 μM), low (1.49 ± 0.26 μM), high (11.2 ± 1.7 μM)) and cell viability based on cell concentration increase, whilst increase in bubble density did not affect calcein uptake and decreased cell viability. By correlating results with the cell-to-bubble ratio, bubble-mediated bioeffects were shown to have increased with cell concentration increase and decreased with bubble density increase. An estimate over which concentration of bubbles destroyed or permeabilised cells was established with maximum "blast radius" (3–90 times the bubble radius). The results of these experiments show that both high cell viability and molecular uptake can be achieved with low energy ultrasound exposure.

Lawrie et al. (2000) investigated the effect of cavitation for vascular gene delivery in porcine vascular smooth muscle cells (VSMC). The transfection of luciferase reporter DNA plasmids (pGL3 and pRSVLUC), liposomal plasmid (Promega Tfx-50-complexed plasmid DNA) and non-lipid polyamine plasmid (TransIT-LT1)-mediated cell transfection was investigated. The cells were exposed to ultrasound settings at a frequency of 956 kHz, MI of 2 and DC of 6 % for 60 s in the presence of Albunex® or Optison™ (10 % v/v). Luciferase activity enhancement for pGL3 in the presence of either microbubble contrast agent was similar at more than 200-fold. Enhancements observed for Albunex® and Optison™ were 137.2 ± 89.7 LU/mg (range 1.6–487.9) and 128.8 ± 72.0 LU/mg (range 8.7–327.9), respectively, whilst representing equivalence to liposome-mediated transfection (82.9 ± 42.9 LU/mg) using Tfx-50. Luciferase activity enhancement in the presence of Optison™ was also similar in parallel experiments for pRSVLUC and pGL3 (approximately 300-fold) with representative enhancements of 112.7 ± 33.4 LU/mg (range 23–290) for pRSVLUC and 110.7 ± 27.9 LU/mg (range 20–254) for pGL3. In addition, cavitation effects in the presence of Optison™ on luciferase activity following TransIT-LT1 indicated an increase in transgene expression of approximately 3000-fold. Results suggest the essential role of acoustic cavitation in the field of targeted cardiovascular gene therapy.

An overview of a number of available in vitro studies with Optison™ for gene therapy and drug delivery is displayed in Table 7.1.

Table 7.1 Optison™ in vitro studies for gene therapy and drug delivery

Year [reference]	Target tissue/cell type	Ultrasound settings	Contrast agent and result
Bao et al. (1997)	Chinese hamster ovary cells	2.25 MHz, 0.8 MPa	Albunex (10 %); luciferase plasmid; 0.33 ng luciferase /10^6 cells
Greenleaf et al. (1998)	Human chondrocytes	1 MHz, 0.32–0.41 MPa	Albunex, transfection of GFP plasmid: 43 %
Lawrie et al. (2000)	Vascular smooth muscle cells	956 kHz, 39–128 W/cm^2	pGL3: 110.7–112.7 LU/mg, pRSVLUC: 112.7 ± 33.4 LU/mg
			300-fold
Frenkel et al. (2002)	293 cell line	1.3 MHz, 1.6 MI	Albumin microbubble, luciferase plasmid: 41 ± 3 % transfection
Eshet et al. (Duvshani-Eshet et al. 2007)	Human prostate cancer cell lines (LNCaP), murine prostate cancer cells (PC2)	1 MHz, 2 W/cm^2, DC: 30 %, 30 min	Optison™ (10 %). Proliferation (<65 %), migration (<50 %), apoptosis: LNCaP: 18 ± 4 %, PC2: 17 ± 4 %
Larina et al. (2005)	MCF7 human breast adenocarcinoma,	3 MHz, 0.17 MPa, 180 J/cm^2, 3 W/cm^2, DC: 20 %, 1 min	Optison™ (200 µl/1 ml sample), FITC dextrans: 73.5 ± 3.3 %, 72.7 ± 0.9 %, 62.7 ± 2.1 %; pEGFP plasmid DNA: 36.7 ± 4.9 %; apoptosis: 13.5 ± 1.6 %
Kinoshita and Hynynen (2005)	HeLa and BJAB cells	1.696 MHz, 0.5 MPa, 1.8–73 J/cm^2, DC: 100 %, 2–20 s and PW (50 ms pulse length, 2 Hz pulse frequency, 100 s)	Optison™ (2 %), apoptosis: 35 %
Eshet et al. (Duvshani-Eshet et al. 2006)	Baby hamster kidney cells	1 MHz, 2 W/cm^2 (0.159 MI), DC: 30 %, 30 min	Optison™ (10 %), DNA in cytoplasm: 16 % of cells, DNA in nucleus: 14 % of cells
Kinoshita and Hynynen (2006)	BJAB and HeLa cells, C166 cells	1.696 MHz, 2.75 W, DC: 10 %	Optison™ (2 and 5 %), apoptosis, 35 %; suppressed expression of EGFP in C166 cells, 15 %; and cell ratio, 25 %
Lawrie et al. (2003)	Porcine vascular smooth muscle cells (VSMC)	956 kHz, MI: 2, DC: 6 %, 60 s, 120 W/cm^2	Optison™ (10 %), pRSV-LUC: 95 LU/µg. Maxl hydrogen peroxide production: 1.2 µM
Miller et al. (2003)	Epidermoid cell (A434) monolayers	1.5 MHz, 2.3 MPa	Optison™ (2 %), apoptosis: 28.6 %. GFP: 3.7 %
		Pulse duration: 1.5 µs, exposure duration for entire chamber 90 s	

(continued)

Table 7.1 (continued)

Year [reference]	Target tissue/cell type	Ultrasound settings	Contrast agent and result
Lawrie et al. (2000)	Porcine vascular smooth muscle cells (VSMC)	956 kHz, MI: 2, 120 W/cm^2, DC: 6 %, 60 s	pGL3 DNA delivery: Optison™, >200-fold increase, 128.8 ± 72.0 LU/mg; liposome-DNA delivery, 82.9 ± 42.9 LU/mg. pRSVLUC: Optison™, 300-fold increase, 112.7 ± 33.4 LU/mg; TransIT-LT1, 927 ± 359 LU/mg, 3000-fold. H_2O_2: 34 μM
Iwanaga et al. (2007)	Human gingival squamous carcinoma cell line, Ca9-22	1 MHz, 1 W/cm^2, DC: 10 %, 20 s	Optison™ (600 μl), pEF1a-b-GAL: 800/cm^2 (cell count/cm^2)
Forbes et al. (2008)	Chinese hamster	Centre frequency: 3.15 MHz	Optison™ (1.55 × 10^7/ml), viable cells, 0.63–10.2 %, non-viable cells, 0.74–3.9 %
	Ovary cells	PRF: 10 Hz, 120 kPa–3.5 MPa five cycles, 30 s	
Ward et al. (2000)	Jurkat lymphocytes cells	2 MHz, PRF: 10 kHz, 0.2 MPa	Reparable sonoporation observed at a larger bubble-cell spacing vs lethal sonoporation (50 mm vs. 39.7 mm)
		DC: 10 %	
Hallow et al. (2006)	DU145 human prostate cancer cells, human aortic smooth muscle cells (AoSMC)	11 and 3.1 MHz, 0.5–2 MPa	Higher frequency yielded lower levels of cavitation activity. Cavitation activity decreased with decreasing pressure. Increasing Optison™ concentration increased amount and duration of cavitation activity
		0.03–405 J/cm^2, DC: 1 %, 3–3000 ms	
Guzman et al. (2003)	DU145 human prostate cancer cells	500 kHz, 0.9–4.2 MPa, 2–817 J/cm^2, DC: 6 %, 20–2000 ms	Calcein uptake: nominal uptake 0.013 ± 0.007 μM, low uptake 1.49 ± 0.26 μM, high uptake 11.2 ± 1.7 μM; cell death rate at 2Hz = 3× cell death rate at 1 Hz (95 % nonviable cells)
Karshafian et al. (2005)	KHT-C cells	500 kHz, 0.18–0.8 MPa, 6–125 J/cm^2	Optison™ (7 %), 80 % permeabilisation
Guzman et al.	DU145 prostate cancer and aortic smooth muscle cells	500 kHz, 0.6–3.0 MPa, DC: 6 %	With increasing pressure and exposure time, molecular uptake of calcein increased and approached equilibrium with the extracellular solution. Cell viability decreased
		Pulse lengths: 0.02–60 ms, 120–2000 ms	
Guzman et al. (2001)	DU145 human prostate cancer cells, aortic smooth muscle cells	500 kHz, 0.6–3.0 MPa, DC: 6 %	Nominal uptake 10^4–10^5, molecules/ cell, low uptake 10^6 molecules/cell, high uptake 10^7 molecules/cell
		Pulse lengths: 0.02–60 ms, 120–2000 ms	

7.8 Optison™ In Vivo Gene Therapy and Drug Delivery

Since the discovery of ultrasound-enhanced delivery mechanisms, numerous studies reporting biological mechanistic and ultrasound parameter optimisation have been published (see Tables 7.1 and 7.2). Many of the published studies include control groups that either omit Optison™ or ultrasound enhancement. These experiments confirm the specific combined effect of both the microbubble agent and ultrasound. This is important for both the optimisation of delivery and, in the long term, for assessing the safety of therapies that may be developed using this approach, since transfection or delivery should be minimal in other nontarget tissues. Some of the published studies confirm this by sampling nontarget tissue, e.g., the liver for gene or protein products (Azuma et al. 2008).

Several ultrasound parameter optimisation studies have used skeletal muscle delivery as a model system (Chen et al. 2011; Sakai et al. 2009). However, skeletal muscle is also a target tissue with clinical relevance in treatment of peripheral artery disease (PAD). Morishita et al. (Jaroslav Turánek et al. 2015) studied the effectiveness of intramuscular injection of hepatocyte growth factor plasmid with Optison™ followed by fairly intense ultrasound pulses (1 min at 1 MHz, 2.5 W/cm^2). Efficacy was demonstrated by increased angiographic score and capillary density. Research aiming to demonstrate the potential of non-viral gene therapy and targeted drug delivery techniques extends to organ systems such as the liver, kidney, lungs, brain, spinal cord, knee joints and glands, as well as to numerous tumour types and disease models (see Table 7.2) (Hashiya et al. 2004; Kondo et al. 2004; Lan et al. 2003; Lu et al. 2003; Pislaru et al. 2003; Shimamura et al. 2004, 2005 Song et al. 2002; Takeuchi et al. 2003; Taniyama et al. 2002a, b; Choi et al. 2007a, b; Duvshani-Eshet et al. 2007, Duvshani-Eshet and Machluf 2007; Howard et al. 2006; Ka et al. 2007; Koike et al. 2005; Miao et al. 2005; Nakaya et al. 2005; Ng et al. 2005; Alter et al. 2009; Aoi et al. 2008; Azuma et al. 2008; Hart et al. 2008; McDannold et al. 2008; Sheikov et al. 2008; Shen et al. 2008; Treat et al. 2007; Xenariou et al. 2007; Castle et al. 2015; Ozaki et al. 2012; Wang 2011). Depending on the biological target of interest and the desired therapeutic effect, there is a balance between achieving high delivery/transfection efficiency whilst limiting undesired bioeffects such as inflammation and vascular damage. It may be the case that unique tailored conditions will be needed for each biological target and therapy. The example above for PAD may be one such application where some unwanted bioeffects may be acceptable in order to achieve a high level of transfection. Most of the gene/drug delivery studies report the use of a 1 MHz ultrasound, although a range of 0.3–14 MHz has also been utilised. Acoustic pressure varies from 1.2 to 4.6 MPa under vigorous conditions to 0.1–0.5 MPa under mild conditions (Qiu et al. 2012; Hu et al. 2012).

Management of restenosis post-angioplasty is a therapeutic area with a clear medical need. Following an in vitro study, Morishita et al. (Taniyama et al. 2002b) showed in model studies that delivery of luciferase plasmid to the carotid artery in a rat balloon injury model is dramatically increased 1000-fold compared to naked plasmid alone, by simultaneous treatment with Optison™ and moderate energy

Table 7.2 Summary of in vivo studies

Year [reference]	Type of study	Target tissue/organ (animal)	Target disease/model	Therapeutic agent or probe/ultrasound settings
2002 (Taniyama et al. 2002a)	Therapy feasibility	Skeletal muscle (mouse)	PAD	Hepatocyte growth factor plasmid/1 MHz, 1 min at 2.5 W/cm^2
2002 (Song et al. 2002)	Optimisation	Skeletal muscle (mouse)	N/A	Fluorescent polymeric microparticle/1 MHz, 0.75 MI
2002 (Taniyama et al. 2002b)	Therapy feasibility	Carotid (rat)	Cardiovascular restenosis (balloon injury model)	p53 plasmid/1 MHz, 2 min at 2.5 W/cm^2
2003 (Lan et al. 2003)	Therapy feasibility	Kidney (rat)	End-stage renal disease: UUO model	Smad7 plasmid/1 MHz at 5 % power (full detail not reported)
2003 (Lu et al. 2003)	Optimisation	Skeletal muscle (mouse)	N/A	GFP plasmid/1 MHz, 3 W/cm^2, 60 s exposure, DC 20 %
2003 (Pislaru et al. 2003)	Optimisation	Skeletal muscle (rat)	N/A	Tissue factor pathway inhibitor or luciferase plasmid/1.7 MHz, MI 0.9 or 1.7
2003 (Takeuchi et al. 2003)	Therapy feasibility	Carotid (rat)	Cardiovascular restenosis (balloon injury model)	C-type natriuretic peptide (CNP)/1.8 MHz, MI 1.0
2004 (Shimamura et al. 2004)	Feasibility	Central nervous system (CNS) (rat)	N/A	Luciferase plasmid/1 MHz, 26 % DC, 5 W/cm^2
2004 (Hashiya et al. 2004)	Therapy feasibility	Carotid (rat)	Cardiovascular restenosis (balloon injury model)	Transcription factor EF2 decoy DNA/1 MHz, 2.5 W/cm^2 for 2 min
2004 (Kondo et al. 2004)	Therapy feasibility	Myocardium (rat)	Acute myocardial infarction	Hepatocyte growth factor (Vax1) plasmid/1.3 MHz, 2.2 Pa for 2 min
2005 (Koike et al. 2005)	Feasibility	Kidney (rat)	N/A	GFP or luciferase plasmid/1 MHz at 5 % power (full detail not reported)
2005 (Ng et al. 2005)	Therapy feasibility	Kidney (rat)	Renal inflammation in remnant kidney model	Smad7 plasmid/1 MHz for 2 min (full details not reported)
2005 (Shimamura et al. 2005)	Feasibility	Spinal cord (rat)	N/A	Luciferase plasmid/1 MHz, DC 20 %, 0.5 W/cm^2

2005 (Nakaya et al. 2005)	Therapy feasibility	Knee joint synovium (rabbit)	Arthritis (antigen-induced arthritis model)	Methotrexate/1 MHz, 2.0 W/cm^2, DC 50 %, 2 min
2005 (Miao et al. 2005)	Therapy feasibility	Liver (rat)	Haemophilia	Human factor IX plasmid/1.18 MHz, 4 MPa
2006 (Howard et al. 2006)	Optimisation/mechanistic	Tumour xenograft (mouse)	Cancer	GFP plasmid/4.0 MHz, duration 4 min, 657, 506 and 304 kPa
2007 (Duvshani-Eshet and Machluf 2007)	Therapy feasibility	Tumour xenograft (mouse)	Cancer	GFP or luciferase plasmid/1 MHz, 30 % DC, 2 W/cm^2
2007 (Choi et al. 2007b)	Feasibility	Brain (mouse)	N/A	Gadolinium MR contrast agent/1.525 MHz, DC 20 %, 2.0, 2.5 and 2.7 MPa
2007 (Ka et al. 2007)	Therapy feasibility	Kidney (mouse)	Renal inflammation and fibrosis	Smad7 plasmid/1 MHz for 30 s (full details not reported)
2007 (Choi et al. 2007a)	Feasibility/optimisation	Brain (mouse)	N/A	Gadolinium MR contrast agent/1.525 MHz, 0.5–1.1 MPa, DC: 20 %
2007 (Takahashi et al. 2007)	Feasibility	Spinal cord (mouse)	N/A	Luciferase plasmid/950 kHz, 1.3 W/cm^2, 0.6 MPa, DC, 20 %
2007 (Treat et al. 2007)	Therapy feasibility	Brain (rat)	CNS tumour	Doxorubicin/1.5 or 1.7 MHz, 0.6 W/cm^2, 1.1 MPa, DC 1 %
2007 (Duvshani-Eshet et al. 2007)	Therapy feasibility	Tumour xenograft (mouse)	Prostate cancer	pPEX plasmid/1 MHz, 2 W/cm^2, 30 % DC
2007 (Xenariou et al. 2007)	Feasibility	Lung (mouse)	Cystic fibrosis (proposed)	Luciferase plasmid/1 MHz, 3 W/cm^2, 0.74 MPa, 20 % DC
2008 (Hart et al. 2008)	Therapy feasibility	Myocardium (rat)	Myocardial infarction model	GFP or C-terminal domain of PYK2 (CRNK) plasmid/10 MHz, 1.5 MI
2008 (Sheikov et al. 2008)	Mechanistic	Brain (rat)	N/A	Horse radish peroxidase/1.5 MHz, 1.1 MPa, 1 % DC

(continued)

Table 7.2 (continued)

Year [reference]	Type of study	Target tissue/organ (animal)	Target disease/model	Therapeutic agent or probe/ultrasound settings
2008 (McDannold et al. 2008)	Optimisation	Brain (rat)	N/A	Gadolinium MR contrast agent/690 kHz, 0.1–1.5 MPa
2008 (Aoi et al. 2008)	Therapy feasibility	Tumour xenograft (mouse)	Cancer (ganciclovir therapy)	Herpes simplex thymidine kinase plasmid/1 MHz; 3.0 W/cm^2, 60 s
2008 (Azuma et al. 2008)	Therapy feasibility	Liver (mouse)	Liver metastasis	NFkB decoy gene/2 MHz, 2.5 W/cm^2
2008 (Shen et al. 2008)	Optimisation	Liver (mouse)	Haemophilia (proposed)	Luciferase plasmid/1 MHz, 0–4.3 MPa
2009 (Alter et al. 2009)	Mechanistic/optimisation	Heart (mouse)	N/A	Luciferase plasmid/14 MHz, MI 1.8
2011 (Wang 2011)	Mechanistic	Skeletal muscle	N/A	GFP plasmid/1 MHz, 2 W/cm^2
2012 (Ozaki et al. 2012)	Therapy feasibility	Colon (rat)	Irritable bowel disease/colitis	Ribbon-type NFkB decoy oligonucleotides/2.5 W/cm^2 (frequency not stated)
2015 (Castle et al. 2015)	Therapy feasibility	Liver (rat)	High-density lipoprotein modification therapy	ApoA-I plasmid/1.54 MHz, MI 1.3, 2 s duration, 10 s. RF, ten cycles

ultrasound (2 min at 1 MHz and 2.5 W/cm^2). In the same study, therapeutic potential was demonstrated by transfection of p53, since this protein is known to suppress neointimal formation. An approximate fivefold increase in p53 expression was observed, and neointimal formation was significantly inhibited in the injured section. Importantly, no vascular damage was observed in control uninjured carotid segments subjected to ultrasound and Optison™ treatment. A related study from the same group reported local delivery of E2F decoy oligodeoxynucleotides under similar experimental conditions (Takeuchi et al. 2003). The Optison™ and ultrasound group showed significant reduction in intimal/medial ratio (I/M ratio) vs controls, and transfection was confirmed by immunohistochemistry (IHC). A further study examined enhanced delivery of C-type natriuretic peptide (CNP) with Optison™ and a long ultrasound exposure (1.8 MHz, MI 1.0, 30 Hz frame rate, applied for 10 min, repeated five times at 3 min intervals). Treatment with CNP has been shown to prevent neointimal formation when administered in high doses. The treatment group showed 17 % of the intimal/medial ratio (I/M ratio) vs controls, and the effect persisted up to 28 days post injury. A high dose of CNP alone was also effective (I/M ratio was 18 % of control) demonstrating that the dose of a therapeutic peptide can be reduced whilst maintaining efficacy.

Related to the above studies in the carotid artery injury model, there is interest in novel approaches to reducing LV remodelling following acute myocardial infarction. One approach is to introduce or promote the production of proangiogenesis signalling molecules which have the ability to promote new blood vessel growth thus reducing ischaemia. In vivo studies show that increased levels of hepatocyte growth factor (HGF) have both a protective and pro-recovery role in models of myocardial ischaemia. Kondo et al. (2004) reported the transfection of HGF pVax1 gene plasmid in an acute myocardial infarction rat model. Plasmid was administered systemically with a 20 % v/v solution of Optison™, and ultrasound settings were 1.3 MHz for 2 min at a peak negative pressure of 2.2 MPa. There was a 60 % reduction in infarct size compared with infarcts in controls, which was a greater effect than observed in previous in vivo HGF studies. The treated group also displayed increased angiogenesis and a low level of LV remodelling as determined by myocardial contrast echocardiography and histopathology. In a related study, Samarel et al. (Hart et al. 2008) studied the effect of Ca^{2+}-dependent, nonreceptor protein tyrosine kinase (PYK2) inhibition by Cell Adhesion Kinase-β-Related Non-Kinase (CRNK) gene therapy in an acute myocardial infarction rat model. PYK2 has been implicated in LV remodelling and heart failure. In this case, an adenovirus CRNK construct rather than a plasmid was used, and it was delivered to the myocardium using a novel catheter-based technique and a high ultrasound frequency of 10 MHz and 1.5 MI delivered transoesophageally. The therapy increased survival and ventricular shortening (measured by echocardiography) compared to control groups, and CRNK overexpression was confirmed with IHC.

The ability to target genes or small molecular agents to the brain or spinal cord is an attractive target because the blood-brain barrier (BBB) presents an obstacle that limits intravascular therapeutic approaches. It has been shown in several studies that ultrasound with a microbubble can selectively open the BBB allowing the delivery of therapeutic molecules that are usually blocked from transport into the brain. In ground breaking work, this principle has recently been demonstrated in the clinic through the intact skull using MR-guided ultrasound (Lipsman et al. 2014). Studies carried out in preclinical models have demonstrated the feasibility of this approach and have allowed the optimisation of, e.g., ultrasound settings for a plasmid and Optison™ delivery method using model probes such as gadolinium-based MR contrast agents (Choi et al. 2007a, b; Takahashi et al. 2007) or plasmid DNA of luciferase or horseradish peroxidase (Sheikov et al. 2008; Takahashi et al. 2007). Most studies use an ultrasound frequency of 0.69–1.5 MHz and moderate pressure amplitudes of 0.55–1.1 MPa. Generally, greater than a tenfold increase in gene transfection is observed, but up to a 1000-fold increase is possible. It has also been observed that the spatial distribution of deposited probe is determined by both the ultrasound beam shape and the vasculature of the brain region (Howard et al. 2006). These factors will have to be considered carefully, depending on the particular region of interest in the brain. In one study aimed at developing a novel antinociceptive therapy, delivery of pCMV-luciferase-GL3 plasmid DNA to mouse spinal cord was investigated (Hart et al. 2008). After transdural injection of plasmid and Optison™, ex vivo IHC showed a large increase in transfection compared to controls. However, no significant effects were observed in behavioural models.

Ultrasound-enhanced delivery of therapeutic agents seems ideally suited to use in highly perfused parenchymal organs such as the liver and kidney since these organs are already routinely examined by ultrasound in the diagnostic field (Piscaglia et al. 2011). Potentially, the liver could be a relevant target for transfection of many genes to achieve a therapeutic effect (Sakai et al. 2009; Pislaru et al. 2003; Miao et al. 2005; Nakaya et al. 2005; Ng et al. 2005; Aoi et al. 2008; Xenariou et al. 2007; Wang 2011). The therapeutic potential of Smad7 overexpression has been evaluated in several animal models such as the end-stage renal disease (unilateral ureteral obstruction, UUO) rat model (Pislaru et al. 2003), renal inflammation remnant kidney rat model (Aoi et al. 2008) and renal inflammation and fibrosis in mice (Miao et al. 2005). Smad7 is a mediator for TGF-β, which is a key mediator in renal fibrosis. Ultrasound settings were 1 MHz with continuous wave in 30 s intervals for 30–120 s total. Throughout these studies, very high transfection rates were observed in glomerular cells giving up to a fivefold increase in Smad7 expression. In one study where plasmid and Optison™ were injected into the renal artery prior to ultrasound, tubulointerstitial myofibroblast accumulation was reduced by 85 %, and collagen I and III mRNA and protein expression were reduced by 60–70 %. Some novel applications for gene transfection in the liver have been investigated. Miao et al. (2005) propose a treatment for haemophilia and demonstrate feasibility of transfection of human factor IX plasmid in a mouse model using relatively high-intensity ultrasound (1.18 MHz and peak negative pressure of 4 MPa). Gene delivery was enhanced (25-fold) in the best ultrasound group giving up to 63 ng/mL factor IX, which could be clinically therapeutic. Castle et al. (Jaroslav Turánek et al. 2015) demonstrate hepatic transfection of the apoA-I gene to increase the production of HDL-C in rats. Tail-vein infusion and hepatic ultrasound resulted in elevated serum HDL-C (increased by up to 61 %) which could be clinically relevant in modifying risk factors post-myocardial infarction if translated to humans.

Targeted delivery of anticancer agents is an area that has potential benefits in terms of reduced systemic toxicity for drugs, or enhanced delivery of genes. With this approach, it may be possible to use fairly intense ultrasound in order to achieve high transfection of, e.g., antiangiogenic genes or a high concentration of cytotoxic drugs, since other potential ultrasound-induced bioeffects may be tolerable. In a mouse model of prostate cancer, plasmid antiangiogenic PEX gene delivery with a fairly long ultrasound exposure (1 MHz, 2 W/cm^2 for 20 min 30 % DC) gave a reduction in tumour volume with repeat treatments (Larina et al. 2005). The effect was modest, but significant in the Optison™ groups. In another study, subcutaneous tumours in mice were transfected (2 MHz, 2.5 W/cm^2) with herpes simplex thymidine kinase gene for ganciclovir anticancer therapy (Aoi et al. 2008). Recurrent treatment gave fourfold reductions in tumour volume.

Further studies have demonstrated the feasibility of gene delivery to other tissues and organs such as the joint, bowel and lung (Ozaki et al. 2012; Xenariou et al. 2007; Nakaya et al. 2005). There is scope for many further in vivo studies to develop optimal ultrasound settings for particular therapies, for both of these and additional targets.

7.9 Early Clinical Results

7.9.1 Sonothrombolysis

One of the most clinically advanced applications of ultrasound and microbubble-mediated therapy is its use for the dissolution of clots following a heart attack. When a patient presents with acute ST segment elevation myocardial infarction (STEMI), current clinical intervention includes the use of lytic agents in addition to percutaneous angioplasty and stenting. Aside from patient discomfort post procedure, a major concern following the use of thrombolytic agents is the risk of bleeding. The advent of microbubble-assisted sonothrombolysis may offer an effective alternative to alleviate both of these side effects (Roos et al. 2014; Unger et al. 2014; Wu et al. 2014). In this context, USCAs are not only used diagnostically to identify location of thrombus but also used as a tool to impart mechanical energy at the site of blockage. This is accomplished by increasing the acoustic output, namely, the MI, precisely at the point of obstruction causing inertial cavitation and bubble destruction. As recently reported, several clinical trials are underway using sonothrombolysis in the acute care setting (Mathias et al. 2015; Kamp 2015; Kamp n.d.; Tavares et al. 2015; Porter n.d.). These studies aim to demonstrate the safety and feasibility of diagnostic ultrasound in conjunction with microbubbles to restore epicardial flow in acute STEMI. Early results are promising as the treatment has been well tolerated and efficacious in patients whilst being readily incorporated into patient workflow in the emergency setting.

7.9.2 *Pancreatic Cancer*

In a first of its kind study, a group of researchers have shown promising result in a pilot human study by increasing the overall survival in patients with pancreatic adenocarcinoma. In addition to the standard chemotherapeutic regimen, following injection of USCA, sonoporation was induced to increase efficacy of the circulating chemotherapeutic agent. Using modified acoustics from a commercial ultrasound system, patients were on average able to tolerate 75 % more cycles of gemcitabine (Kotopoulis et al. 2013; Dimcevski n.d.; Gilje 2015). This improved tolerance leads to a reduction of tumour size or, at minimum, a reduced rate of tumour growth. With improved quality of life, the combination of chemotherapy with sonoporation has provided hope for patients with inoperable tumours who are otherwise out of treatment options.

7.10 Hardware and Software Requirements

In the case of sonothrombolysis once the region of interest – the location of thrombus – is selected, the ultrasound transducer should only transmit a high MI acoustic pulse for microbubble cavitation when appropriate, i.e., when USCA is present. This would enable complete microbubble destruction whilst allowing a sufficient period of low MI imaging to ensure adequate reperfusion, thereby minimising the total amount of acoustic energy delivered and optimising appropriate amount of therapy.

Likewise to optimise sonoporation for enhanced efficacy of chemotherapeutics, researchers must work with industry to develop application-specific improvements in probe design. Unlike for sonothrombolysis, where the target treatment area is of a very limited size, other drug delivery systems must be capable of affecting much larger diseased or target regions. Whether this is in the pancreas, liver or other organ, the therapeutic acoustics will need to be able to trigger sonoporation in three dimensions to be fully efficacious.

As this technology continues to advance from the preclinical setting to early clinical trials, the requisite components will also need to progress. The translation of this technology into clinical practice will be greatly facilitated in large part by improved hardware and software. Such things as automation of contrast detection for impulse triggering and improved transducer performance for depicting human anatomy accurately will be critical. Generally, a strategy of using commercially available ultrasound systems as well as USCAs could offer several distinct advantages. Using well-established and regulated apparatus may provide a lower barrier to entry of clinical acceptance; this includes the use of USCA dosing and acoustic energy within existing diagnostic guidelines.

7.11 Conclusions

Ongoing in vitro and in vivo investigations, together with early clinical study results represent strong evidence that the use of microbubbles offers a novel and promising approach for ultrasound-mediated targeted gene therapy and drug delivery. The central concept is to restrict molecular uptake and thus treatment to the region of interest, minimising toxic effects on healthy tissue. Although progress has been made in both technological and biological research areas, there is still the need of an increased understanding around delivery mechanisms before translation into routine clinical practice can be established to treat cancer, cardiovascular diseases and genetic disorders.

One of the most imminent needs in gene therapy and drug delivery is the development of ideal vectors that reflect characteristics of sustained high transfection efficacy and expression whilst safely delivering multiple genes/drugs to specific target regions through a simple non-viral transfer procedure (Nozaki et al. 2006).

Focus should also be turned to the design components of the microbubble contrast agents to ensure their safety profile along with their stability for longer circulation in the vasculature. Since microbubbles can act as potential vehicles to deliver agents, components that directly affect transfection efficacy during ultrasound exposure need to be reviewed, such as shell composition and optimal microbubble size, loading capacity, circulation life, drug release mechanisms and biodistribution (Mehier-Humbert et al. 2007; Kotopoulis et al. 2013). Target-specific gene/drug delivery could also be improved with the design of tumour or organ-specific microbubbles.

Further insight into defining optimised ultrasound parameters for a homogeneous acoustic field with increased target specificity and efficiency for mediating therapeutic effects is an investigational priority. Great effort is required to devise technological advancements involving both design and parameter optimisation in both in vitro and in vivo systems regarding ultrasound hardware and software. Direct strategic partnerships amongst academia, industry and pharmaceutical companies are of high importance in order to yield new technologies that share the same development goals and concerns of cellular physiology, genetics, physical chemistry and acoustic physics affecting microbubble ultrasound-mediated delivery and therapy (Castle et al. 2013, Castle and Feinstein 2014).

References

Alter J, Sennoga CA, Lopes DM, Eckersley RJ, Wells DJ (2009) Microbubble stability is a major determinant of the efficiency of ultrasound and microbubble mediated in vivo gene transfer. Ultrasound Med Biol 35:976–984

Aoi A, Watanabe Y, Mori S, Takahashi M, Vassaux G, Kodama T (2008) Herpes simplex virus thymidine kinase-mediated suicide gene therapy using nano/microbubbles and ultrasound. Ultrasound Med Biol 34:425–434

Averkiou M, Bruce M, Jensen S, Rafter P, Brock-Fisher T, Powers J (eds) (2004) Pulsing schemes for the detection of nonlinear echoes from contrast microbubbles. In: 9th European symposium on ultrasound contrast imaging, Rotterdam

Azuma H, Tomita N, Sakamoto T, Kiyama S, Inamoto T, Takahara K, Kotake Y, Segawa N, Morishita R, Takahara S, Hayasaki H, Otsuki Y, Horie S, Tanigawa N, Katsuoka Y (2008) Marked regression of liver metastasis by combined therapy of ultrasound-mediated NF kappaB-decoy transfer and transportal injection of paclitaxel, in mouse. Int J Cancer J Int Cancer 122:1645–1656

Bao S, Thrall BD, Miller DL (1997) Transfection of a reporter plasmid into cultured cells by sonoporation in vitro. Ultrasound Med Biol 23:953–959

Becher H, Burns PN (2000) Handbook of contrast echocardiography: left ventricular function and myocardial perfusion. Springer, Berlin

Castle J, Feinstein SB (2014) Ultrasound-directed, site-specific gene delivery. Methods Mol Biol (Clifton, NJ) 1141:67–76

Castle J, Butts M, Healey A, Kent K, Marino M, Feinstein SB (2013) Ultrasound-mediated targeted drug delivery: recent success and remaining challenges. Am J Physiol Heart Circ Physiol 304:H350–H357

Castle JW, Kent KP, Fan Y, Wallace KD, Davis CE, Roberts JC, Marino ME, Thomenius KE, Lim HW, Coles E, Davidson MH, Feinstein SB, DeMaria A (2015) Therapeutic ultrasound: increased HDL-cholesterol following infusions of acoustic microspheres and apolipoprotein A-I plasmids. Atherosclerosis 241:92–99

Chen Y-C, Jiang L-P, Liu N-X, Ding L, Liu X-L, Wang Z-H, Hong K, Zhang Q-P (2011) Enhanced gene transduction into skeletal muscle of mice in vivo with pluronic block copolymers and ultrasound exposure. Cell Biochem Biophys 60:267–273

Choi JJ, Pernot M, Brown TR, Small SA, Konofagou EE (2007a) Spatio-temporal analysis of molecular delivery through the blood-brain barrier using focused ultrasound. Phys Med Biol 52:5509–5530

Choi JJ, Pernot M, Small SA, Konofagou EE (2007b) Noninvasive, transcranial and localized opening of the blood-brain barrier using focused ultrasound in mice. Ultrasound Med Biol 33:95–104

Cochran M, Wheatley MA (2013) In vitro gene delivery with ultrasound-triggered polymer microbubbles. Ultrasound Med Biol 39:1102–1119

Cohen JL, Cheirif J, Segar DS, Gillam LD, Gottdiener JS, Hausnerova E, Bruns DE (1998) Improved left ventricular endocardial border delineation and opacification with OPTISON (FS069), a new echocardiographic contrast agent. Results of a phase III Multicenter Trial. J Am Coll Cardiol 32:746–752

Delalande A, Kotopoulis S, Postema M, Midoux P, Pichon C (2013) Sonoporation: mechanistic insights and ongoing challenges for gene transfer. Gene 6:00364–00368

Dimcevski G (n.d.) Ultrasound-assisted treatment of inoperable pancreatic cancer. ClinicalTrials.gov. https://clinicaltrials.gov/show/NCT01674556

Doinikov AA, Bouakaz A (2010) Acoustic microstreaming around an encapsulated particle. J Acoust Soc Am 127:1218–1227

Duvshani-Eshet M, Machluf M (2007) Efficient transfection of tumors facilitated by long-term therapeutic ultrasound in combination with contrast agent: from in vitro to in vivo setting. Cancer Gene Ther 14:306–315

Duvshani-Eshet M, Baruch L, Kesselman E, Shimoni E, Machluf M (2006) Therapeutic ultrasound-mediated DNA to cell and nucleus: bioeffects revealed by confocal and atomic force microscopy. Gene Ther 13:163–172

Duvshani-Eshet M, Benny O, Morgenstern A, Machluf M (2007) Therapeutic ultrasound facilitates antiangiogenic gene delivery and inhibits prostate tumor growth. Mol Cancer Ther 6:2371–2382

Endoh M, Koibuchi N, Sato M, Morishita R, Kanzaki T, Murata Y, Kaneda Y (2002) Fetal gene transfer by intrauterine injection with microbubble-enhanced ultrasound. Mol Ther: J Am Soc Gene Ther 5:501–508

Escoffre JM, Zeghimi A, Novell A, Bouakaz A (2013) In-vivo gene delivery by sonoporation: recent progress and prospects. Curr Gene Ther 13:2–14

Forbes MM, Steinberg RL, O'Brien WD Jr (2008) Examination of inertial cavitation of Optison in producing sonoporation of chinese hamster ovary cells. Ultrasound Med Biol 34:2009–2018

Frenkel PA, Chen S, Thai T, Shohet RV, Grayburn PA (2002) DNA-loaded albumin microbubbles enhance ultrasound-mediated transfection in vitro. Ultrasound Med Biol 28:817–822

Gilje OH (ed) (2015) Recent Norwegian experiences with Sonazoid. The 30th anniversary bubble conference, Chicago, 10–12 Sept 2015

Goldberg BB, Liu J-B, Forsberg F (1994) Ultrasound contrast agents: a review. Ultrasound Med Biol 20:319–333

Greenleaf WJ, Bolander ME, Sarkar G, Goldring MB, Greenleaf JF (1998) Artificial cavitation nuclei significantly enhance acoustically induced cell transfection. Ultrasound Med Biol 24:587–595

Guzman HR, Nguyen DX, Khan S, Prausnitz MR (2001) Ultrasound-mediated disruption of cell membranes. II. Heterogeneous effects on cells. J Acoust Soc Am 110:597–606

Guzman HR, McNamara AJ, Nguyen DX, Prausnitz MR (2003) Bioeffects caused by changes in acoustic cavitation bubble density and cell concentration: a unified explanation based on cell-to-bubble ratio and blast radius. Ultrasound Med Biol 29:1211–1222

Hallow DM, Mahajan AD, McCutchen TE, Prausnitz MR (2006) Measurement and correlation of acoustic cavitation with cellular bioeffects. Ultrasound Med Biol 32:1111–1122

Hart DL, Heidkamp MC, Iyengar R, Vijayan K, Szotek EL, Barakat JA, Leya M, Henze M, Scrogin K, Henderson KK, Samarel AM (2008) CRNK gene transfer improves function and reverses the myosin heavy chain isoenzyme switch during post-myocardial infarction left ventricular remodeling. J Mol Cell Cardiol 45:93–105

Hashiya N, Aoki M, Tachibana K, Taniyama Y, Yamasaki K, Hiraoka K, Makino H, Yasufumi K, Ogihara T, Morishita R (2004) Local delivery of E2F decoy oligodeoxynucleotides using ultrasound with microbubble agent (Optison) inhibits intimal hyperplasia after balloon injury in rat carotid artery model. Biochem Biophys Res Commun 317:508–514

Hernot S, Klibanov AL (2008) Microbubbles in ultrasound-triggered drug and gene delivery. Adv Drug Deliv Rev 60:1153–1166

Howard CM, Forsberg F, Minimo C, Liu JB, Merton DA, Claudio PP (2006) Ultrasound guided site specific gene delivery system using adenoviral vectors and commercial ultrasound contrast agents. J Cell Physiol 209:413–421

Hu X, Kheirolomoom A, Mahakian LM, Beegle JR, Kruse DE, Lam KS, Ferrara KW (2012) Insonation of targeted microbubbles produces regions of reduced blood flow within tumor vasculature. Invest Radiol 47:398

Iwanaga K, Tominaga K, Yamamoto K, Habu M, Maeda H, Akifusa S, Tsujisawa T, Okinaga T, Fukuda J, Nishihara T (2007) Local delivery system of cytotoxic agents to tumors by focused sonoporation. Cancer Gene Ther 14:354–363

Turánek J, Miller AD, Kauerová Z, Lukáč R, Mašek J, Koudelka S, Raška M (2015) Advances in bioengineering, Prof. Pier Andrea Serra (ed), ISBN: 978-953-51-2141-1, InTech, doi:10.5772/59870. Available from: http://www.intechopen.com/books/advances-in-bioengineering/lipid-based-nanoparticles-and-microbubbles-multifunctional-lipid-based-biocompatible-particles-for-i. Lipid-based nanoparticles and microbubbles – multifunctional lipid-based biocompatible particles for in vivo imaging and theranostics, advances in bioengineering

Jelenc J, Jelenc J, Miklavčič D, Lebar AM (2012) Low-frequency sonoporation in vitro: experimental system evaluation. Strojniški Vestn-J Mech Eng 58:319–326

Ka SM, Huang XR, Lan HY, Tsai PY, Yang SM, Shui HA, Chen A (2007) Smad7 gene therapy ameliorates an autoimmune crescentic glomerulonephritis in mice. J Am Soc Nephrol: JASN 18:1777–1788

Kalli C, Teoh WC, Leen E (2014) Introduction of genes via sonoporation and electroporation. Adv Exp Med Biol 818:231–254

Kamp O (ed) (2015) Reduction of microvasculature injury using a theranostic imaging strategy, an unexpected finding. In: The 30th anniversary bubble conference, Chicago, 10–12 Sept 2015

Kamp O, Dilkmas PA (n.d.) Ultrasound contrast agents to facilitate sonothrombolysis in patients with acute myocardial infarction. http://www.trialregister.nl/trialreg/admin/rctview.asp?TC=161

Karshafian R, Bevan PD, Burns PN, Karshafian R, Samac S, Banerjee M, Bevan PD (eds) (2005) Ultrasound-induced uptake of different size markers in mammalian cells. Ultrasonics symposium, 2005 IEEE; 18–21 Sept 2005

Kim HJ, Greenleaf JF, Kinnick RR, Bronk JT, Bolander ME (1996) Ultrasound-mediated transfection of mammalian cells. Hum Gene Ther 7:1339–1346

Kinoshita M, Hynynen K (2005) Intracellular delivery of Bak BH3 peptide by microbubble-enhanced ultrasound. Pharm Res 22:716–720

Kinoshita M, Hynynen K (2006) Intracellular delivery of peptides and siRNAs using microbubble enhanced focused ultrasound. AIP Conf Proc 829:538–542

Koike H, Tomita N, Azuma H, Taniyama Y, Yamasaki K, Kunugiza Y, Tachibana K, Ogihara T, Morishita R (2005) An efficient gene transfer method mediated by ultrasound and microbubbles into the kidney. J Gene Med 7:108–116

Kondo I, Ohmori K, Oshita A, Takeuchi H, Fuke S, Shinomiya K, Noma T, Namba T, Kohno M (2004) Treatment of acute myocardial infarction by hepatocyte growth factor gene transfer: the first demonstration of myocardial transfer of a "functional" gene using ultrasonic microbubble destruction. J Am Coll Cardiol 44:644–653

Kotopoulis S, Dimcevski G, Gilja OH, Hoem D, Postema M (2013) Treatment of human pancreatic cancer using combined ultrasound, microbubbles, and gemcitabine: a clinical case study. Med Phys 40:072902

Kuliszewski MA, Kobulnik J, Lindner JR, Stewart DJ, Leong-Poi H (2011) Vascular gene transfer of SDF-1 promotes endothelial progenitor cell engraftment and enhances angiogenesis in ischemic muscle. Mol Ther 19:895–902. doi:10.1038/mt.2011.1018. Epub 2011 Mar 1031

Lan HY, Mu W, Tomita N, Huang XR, Li JH, Zhu HJ, Morishita R, Johnson RJ (2003) Inhibition of renal fibrosis by gene transfer of inducible Smad7 using ultrasound-microbubble system in rat UUO model. J Am Soc Nephrol: JASN 14:1535–1548

Larina IV, Evers BM, Esenaliev RO (2005) Optimal drug and gene delivery in cancer cells by ultrasound-induced cavitation. Anticancer Res 25:149–156

Lawrie A, Brisken AF, Francis SE, Cumberland DC, Crossman DC, Newman CM (2000) Microbubble-enhanced ultrasound for vascular gene delivery. Gene Ther 7:2023–2027

Lawrie A, Brisken AF, Francis SE, Wyllie D, Kiss-Toth E, Qwarnstrom EE, Dower SK, Crossman DC, Newman CM (2003) Ultrasound-enhanced transgene expression in vascular cells is not dependent upon cavitation-induced free radicals. Ultrasound Med Biol 29:1453–1461

Liang HD, Tang J, Halliwell M (2010) Sonoporation, drug delivery, and gene therapy. Proc Inst Mech Eng Part H J Eng Med 224:343–361

Lipsman N, Mainprize TG, Schwartz ML, Hynynen K, Lozano AM (2014) Intracranial applications of magnetic resonance-guided focused ultrasound. Neurother: J Am Soc Exp Neurother 11:593–605

Lu QL, Liang HD, Partridge T, Blomley MJ (2003) Microbubble ultrasound improves the efficiency of gene transduction in skeletal muscle in vivo with reduced tissue damage. Gene Ther 10:396–405

Lum AFH, Borden MA, Dayton PA, Kruse DE, Simon SI, Ferrara KW (2006) Ultrasound radiation force enables targeted deposition of model drug carriers loaded on microbubbles. J Control Release 111:128–134

Mathias W, Kamp O, Porter T (eds) Safety and feasibility of diagnostic ultrasound high mechanical Index impulses in restoring epicardial flow in acute ST segment elevation myocardial infarction in humans. In: The 30th anniversary bubble conference, Chicago, 10–12 Sept 2015

McDannold N, Vykhodtseva N, Hynynen K (2008) Effects of acoustic parameters and ultrasound contrast agent dose on focused-ultrasound induced blood-brain barrier disruption. Ultrasound Med Biol 34:930–937

Mehier-Humbert S, Guy RH (2005) Physical methods for gene transfer: improving the kinetics of gene delivery into cells. Adv Drug Deliv Rev 57:733–753

Mehier-Humbert S, Yan F, Frinking P, Schneider M, Guy RH, Bettinger T (2007) Ultrasound-mediated gene delivery: influence of contrast agent on transfection. Bioconjug Chem 18:652–662

Miao CH, Brayman AA, Loeb KR, Ye P, Zhou L, Mourad P, Crum LA (2005) Ultrasound enhances gene delivery of human factor IX plasmid. Hum Gene Ther 16:893–905

Miller DL, Dou C, Song J (2003) DNA transfer and cell killing in epidermoid cells by diagnostic ultrasound activation of contrast agent gas bodies in vitro. Ultrasound Med Biol 29:601–607

Miller DL, Averkiou MA, Brayman AA, Everbach EC, Holland CK, Wible JH Jr, Wu J (2008) Bioeffects considerations for diagnostic ultrasound contrast agents. J Ultrasound Med: Off J Am Inst Ultrasound Med 27:611–632; quiz 633–616

Nakaya H, Shimizu T, Isobe K, Tensho K, Okabe T, Nakamura Y, Nawata M, Yoshikawa H, Takaoka K, Wakitani S (2005) Microbubble-enhanced ultrasound exposure promotes uptake of methotrexate into synovial cells and enhanced antiinflammatory effects in the knees of rabbits with antigen-induced arthritis. Arthritis Rheum 52:2559–2566

Ng YY, Hou CC, Wang W, Huang XR, Lan HY (2005) Blockade of NFkappaB activation and renal inflammation by ultrasound-mediated gene transfer of Smad7 in rat remnant kidney. Kidney Int Suppl 67(94):S83–S91

Nozaki T, Ogawa R, Watanabe A, Nishio R, Fuse H, Kondo T (2006) Ultrasound-mediated gene transfection: problems to be solved and future possibilities. J Med Ultrason 33:135–142

Ozaki K, Makino H, Aoki M, Miyake T, Yasumasa N, Osako MK, Nakagami H, Rakugi H, Morishita R (2012) Therapeutic effect of ribbon-type nuclear factor-kappaB decoy oligonucleotides in a rat model of inflammatory bowel disease. Curr Gene Ther 12:484–492

Piscaglia F, Nolsoe C, Fau–Dietrich CF, Dietrich Cf Fau–Cosgrove DO, Cosgrove Do Fau–Gilja OH, Gilja Oh Fau–Bachmann Nielsen M, Bachmann Nielsen M Fau–Albrecht T, Albrecht T Fau–Barozzi L, Barozzi L Fau–Bertolotto M, Bertolotto M Fau–Catalano O, Catalano O Fau–Claudon M, Claudon M Fau–Clevert DA, Clevert Da Fau–Correas JM, Correas Jm Fau–D'Onofrio M, D'Onofrio M Fau–Drudi FM, Drudi Fm Fau–Eyding J, Eyding J Fau–Giovannini M, Giovannini M Fau–Hocke M, Hocke M Fau–Ignee A, Ignee A Fau–Jung EM, Jung Em Fau–Klauser AS, Klauser As Fau–Lassau N, Lassau N Fau–Leen E, Leen E Fau–Mathis G, Mathis G Fau–Saftoiu A, Saftoiu A Fau–Seidel G, Seidel G Fau–Sidhu PS, Sidhu Ps Fau–ter Haar G, ter Haar G Fau–Timmerman D, Timmerman D Fau–Weskott HP, Weskott HP (2011) The EFSUMB guidelines and recommendations on the clinical practice of contrast enhanced ultrasound (CEUS): update 2011 on non-hepatic applications

Pislaru SV, Pislaru C, Kinnick RR, Singh R, Gulati R, Greenleaf JF, Simari RD (2003) Optimization of ultrasound-mediated gene transfer: comparison of contrast agents and ultrasound modalities. Eur Heart J 24:1690–1698

Porter T (n.d.) Microvascular reperfusion utilizing sonothrombolysis in acute myocardial infarction (MRUSMI TRIAL). ClinicalTrials.gov. https://clinicaltrials.gov/ct2/show/NCT02170103

Qin S, Caskey CF, Ferrara KW (2009) Ultrasound contrast microbubbles in imaging and therapy: physical principles and engineering. Phys Med Biol 54:R27–R57. doi:10.1088/0031-9155/1054/1086/R1001, Epub 2009 Feb 1019

Qiu Y, Luo Y, Zhang Y, Cui W, Zhang D, Wu J, Zhang J, Tu J (2010) The correlation between acoustic cavitation and sonoporation involved in ultrasound-mediated DNA transfection with polyethylenimine (PEI) in vitro. J Control Release 145:40–48

Qiu Y, Zhang C, Tu J, Zhang D (2012) Microbubble-induced sonoporation involved in ultrasound-mediated DNA transfection in vitro at low acoustic pressures. J Biomech 45:1339–1345

Roos ST, Juffermans LJM, Slikkerveer J, Unger EC, Porter TR, Kamp O (2014) Sonothrombolysis in acute stroke and myocardial infarction: a systematic review. IJC Heart Vessel 4:1–6

Sakai T, Kawaguchi M, Kosuge Y (2009) siRNA – mediated gene silencing in the salivary gland using in vivo microbubble-enhanced sonoporation. Oral Dis 15:505–511

Sheikh S, Pallagatti S, Singh B, Puri N, Singh R, Kalucha A (2011) Sonoporation, a redefined ultrasound modality as therapeutic aid: a review. J Clin Exp Dent 3:e228–e234

Sheikov N, McDannold N, Sharma S, Hynynen K (2008) Effect of focused ultrasound applied with an ultrasound contrast agent on the tight junctional integrity of the brain microvascular endothelium. Ultrasound Med Biol 34:1093–1104

Shen ZP, Brayman AA, Chen L, Miao CH (2008) Ultrasound with microbubbles enhances gene expression of plasmid DNA in the liver via intraportal delivery. Gene Ther 15:1147–1155

Shimamura M, Sato N, Taniyama Y, Yamamoto S, Endoh M, Kurinami H, Aoki M, Ogihara T, Kaneda Y, Morishita R (2004) Development of efficient plasmid DNA transfer into adult rat central nervous system using microbubble-enhanced ultrasound. Gene Ther 11:1532–1539

Shimamura M, Sato N, Taniyama Y, Kurinami H, Tanaka H, Takami T, Ogihara T, Tohyama M, Kaneda Y, Morishita R (2005) Gene transfer into adult rat spinal cord using naked plasmid DNA and ultrasound microbubbles. J Gene Med 7:1468–1474

Sirsi S, Borden M (2009) Microbubble compositions, properties and biomedical applications. Bubble Sci Eng Technol 1:3–17

Song J, Chappell JC, Qi M, VanGieson EJ, Kaul S, Price RJ (2002) Influence of injection site, microvascular pressure and ultrasound variables on microbubble-mediated delivery of microspheres to muscle. J Am Coll Cardiol 39:726–731

Stewart MJ (2003) Contrast echocardiography. Heart 89:342–348

Stride E, Saffari N (2003) Microbubble ultrasound contrast agents: a review. Proc Inst Mech Eng Part H J Eng Med 217:429–447

Takahashi M, Kido K, Aoi A, Furukawa H, Ono M, Kodama T (2007) Spinal gene transfer using ultrasound and microbubbles. J Control Release: Off J Control Release Soc 117:267–272

Takeuchi H, Ohmori K, Kondo I, Oshita A, Shinomiya K, Yu Y, Takagi Y, Mizushige K, Kangawa K, Kohno M (2003) Potentiation of C-type natriuretic peptide with ultrasound and microbubbles to prevent neointimal formation after vascular injury in rats. Cardiovasc Res 58:231–238

Taniyama Y, Tachibana K, Hiraoka K, Aoki M, Yamamoto S, Matsumoto K, Nakamura T, Ogihara T, Kaneda Y, Morishita R (2002a) Development of safe and efficient novel nonviral gene transfer using ultrasound: enhancement of transfection efficiency of naked plasmid DNA in skeletal muscle. Gene Ther 9:372–380

Taniyama Y, Tachibana K, Hiraoka K, Namba T, Yamasaki K, Hashiya N, Aoki M, Ogihara T, Yasufumi K, Morishita R (2002b) Local delivery of plasmid DNA into rat carotid artery using ultrasound. Circulation 105:1233–1239

Tavares BG, Aguiar MO, Tsutsui JM, Garcia D, De Oliveira AG, Ramires JAF, Filho RK, Porter T, Wilson M (eds) (2015) Safety and feasibility of diagnostic ultrasound high mechanical index impulses in restoring epicardial flow in acute st segment elevation myocardial infarction in humans. The American Society of Echocardiography, Seattle, 10–14 Jun 2015; 10–12 Sept 2015

Treat LH, McDannold N, Vykhodtseva N, Zhang Y, Tam K, Hynynen K (2007) Targeted delivery of doxorubicin to the rat brain at therapeutic levels using MRI-guided focused ultrasound. Int J Cancer J Int Cancer 121:901–907

Unger E, Porter T, Lindner J, Grayburn P (2014) Cardiovascular drug delivery with ultrasound and microbubbles. Adv Drug Deliv Rev 72:110–126

Urban MW, Fatemi M, Greenleaf JF (2010) Modulation of ultrasound to produce multifrequency radiation force. J Acoust Soc Am 127:1228–1238

Wang XH (2011) Role of constituents of Optison in Optison-mediated gene transfection enhancement in skeletal muscle in vivo. J Ultrasound Med: Off J Am Inst Ultrasound Med 30:325–332

Ward M, Wu J, Chiu JF (2000) Experimental study of the effects of Optison concentration on sonoporation in vitro. Ultrasound Med Biol 26:1169–1175

Wei K, Jayaweera AR, Firoozan S, Linka A, Skyba DM, Kaul S (1998) Basis for detection of stenosis using venous administration of microbubbles during myocardial contrast echocardiography: bolus or continuous infusion? J Am Coll Cardiol 32:252–260

Wu J, Ross JP, Chiu JF (2002) Reparable sonoporation generated by microstreaming. J Acoust Soc Am 111:1460–1464

Wu J, Xie F, Kumar T, Liu J, Lof J, Shi W, Everbach EC, Porter TR (2014) Improved sonothrombolysis from a modified diagnostic transducer delivering impulses containing a longer pulse duration. Ultrasound Med Biol 40:1545–1553

Xenariou S, Griesenbach U, Liang HD, Zhu J, Farley R, Somerton L, Singh C, Jeffery PK, Ferrari S, Scheule RK, Cheng SH, Geddes DM, Blomley M, Alton EW (2007) Use of ultrasound to enhance nonviral lung gene transfer in vivo. Gene Ther 14:768–774

Chapter 8
S-Nitroso Adducts of Albumin Analogs: Characterization, Categorization, and Possible Future Therapeutic Applications

Yu Ishima, Ulrich Kragh-Hansen, and Masaki Otagiri

Abstract Nitric oxide (NO) is a ubiquitous messenger molecule that is involved in multiple cellular functions. In particular, NO has multiple important actions that contribute to the maintenance of vascular homeostasis. Thus, an inappropriate production of NO leads to a disease state. To date, pharmacologically active compounds that release NO within the body, such as organic nitrates, have been used as therapeutic agents, but their efficacy is significantly limited by undesirable side effects. Therefore, novel NO donors with better pharmacological and pharmacokinetic properties would be highly desirable. The *S*-nitrosothiol fraction in plasma is largely composed of endogenous *S*-nitrosated human serum albumin (Mono-SNO-HSA), and that is why we are testing whether this albumin form has the potential to be therapeutically useful. Recently, we developed SNO-HSA analogs such as SNO-HSA with many conjugated SNO groups (Poly-SNO-HSA) which were prepared using chemical modification. Unexpectedly, we found striking inverse effects between Poly-SNO-HSA and Mono-SNO-HSA. Despite the fact that Mono-SNO-HSA inhibits apoptosis, Poly-SNO-HSA exerts very strong proapoptotic effects against tumor cells. Furthermore, Poly-SNO-HSA can reduce or perhaps completely eliminate the multidrug resistance often developed by cancer cells. In addition, we recently developed an *S*-nitrosated HSA dimer (SNO-HSA dimer) as a novel enhanced permeability and retention (EPR) effect enhancer. The SNO-HSA dimer increases the tumor accumulation of macromolecular antitumor drugs by a factor 3-4 and thereby their antitumor effects. In this review, we first summarize the effective factors on the *S*-nitrosation and *S*-denitrosation of HSA both in vitro and

Y. Ishima (✉)
Department of Pharmacokinetics and Biopharmaceutics, Institute of Health Biosciences, The University of Tokushima, 1-78-1, Sho-machi, Tokushima 770-8505, Japan
e-mail: ishima.yuu@tokushima-u.ac.jp

U. Kragh-Hansen
Department of Biomedicine, University of Aarhus,
Ole Worms Allé 6, Building 1180, DK-8000 Aarhus C, Denmark

M. Otagiri
Faculty of Pharmaceutical Sciences, Sojo University,
4-22-1 Ikeda, Kumamoto 860-0082, Japan

M. Otagiri, V.T.G. Chuang (eds.), *Albumin in Medicine*,
DOI 10.1007/978-981-10-2116-9_8

in biological systems. Finally, we propose the possibility that Poly-SNO-HSA and SNO-HSA dimer can be used as a safe and effective multifunctional antitumor agent and as a nano-EPR enhancer.

Keywords *S*-nitrosation • *S*-transnitrosation • Ligand binding • Cytoprotection • Cytotoxicity • Antibacterial effect • Cancer therapy • Enhanced permeability and retention effect

8.1 Introduction

Nitric oxide (NO), a free radical, possesses various modulatory effects on biological systems (Moncada et al. 1991; Mizutani and Layon 1996; Stamler et al. 1992a, b, 1993; Nathan 1992; Iyengar et al. 1987; Hibbs et al. 1988; Clancy and Abramson 1995; Clancy et al. 1994; Lancaster and Hibbs 1990; Furchgott and Zawadzki 1980; Palmer et al. 1987; Myers et al. 1990; Gaston et al. 1993; Lancaster 1994; Drapier et al. 1991; Stadler et al. 1993; Mülsch et al. 1993; Geng et al. 1994). For example, low concentrations of NO produced by constitutive NO synthase (cNOS) isoforms have several cytoprotective effects, such as the regulation of local blood flow as an endothelium-derived relaxing factor (Palmer et al. 1987), removal of superoxides (Gryglewski et al. 1986), and inhibition of superoxide anion production by neutrophils (Ródenas et al. 1998). Defects in NO production can lead to many cardiovascular abnormalities such as essential hypertension, stroke, atherosclerosis, and ischemia/reperfusion injury (Ignarro and Wei 2002). Therefore, replacement of or supplementing endogenous NO production by exogenously administered NO is an important and effective treatment for cardiovascular diseases. However, NO therapy still has some problems that need to be overcome, such as the following. First, the administration of NO gas has limited utility, partly because of its short half-life in vivo (~0.1 s) (Ignarro 2000). Secondly, it is important to be able to control reaction selectivity and the dose of the NO donor against reactive oxygen species such as in NO therapy in inflammatory diseases. Thirdly, a local application of NO may be a very effective and safe form of NO therapy. Thus, it is essential to develop reliable NO donors with better pharmacological and pharmacokinetic parameters.

In our search for a reliable and safe NO donor, we pursued a novel approach, namely, to examine the possibility of using an "NO-traffic protein." An NO-traffic protein is meant to be a protein with (i) a high efficiency of *S*-nitrosation, (ii) a high stability of the *S*-nitroso form in the circulation, and (iii) a high efficiency of *S*-transnitrosation into cells. As a candidate in this respect, we focused on human serum albumin (HSA), because HSA is the most abundant plasma protein (35–50 g/L) and because endogenous *S*-nitrosothiols in the human plasma are largely associated with HSA (Marley et al. 2000). *S*-Nitrosated HSA (SNO-HSA) is significantly more stable than low-molecular-weight *S*-nitrosothiols in human circulation (Katsumi et al. 2007). In addition, others have attempted to produce NO delivery systems using a NO-albumin conjugate. Marks et al. (1995) and Ewing et al. (1997) synthesized a macromolecular *S*-nitrosothiol, poly SNO-BSA, in which several *S*-nitrosothiols are formed in bovine serum albumin (BSA) after reduction of the protein disulfide bonds. Independently, Beak et al. (2002) developed a macromolecular NONOate, diazeniumdiolated BSA, in which several NONOate moieties are conjugated to native BSA. In a porcine coronary angioplasty model, the two NO-traffic BSA forms, poly SNO-BSA and diazeniumdiolated BSA, were applied locally to the site of vascular injury and showed high retention at the administration site and reduced platelet attachment and activation. These effects were due to the strong binding of the modified albumins to the injured vessel.

In the development of targeted NO delivery systems for intravenous use, the tissue distribution characteristics of the NO-carrier conjugate should be evaluated in vivo in order to identify the various obstacles to targeted delivery, such as extensive uptake by mononuclear phagocyte systems and rapid loss by glomerular filtration. Katsumi et al. (2005) also examined the pharmacokinetic properties of SNO-BSA. The results showed that serum albumin is a promising carrier for controlling the pharmacokinetic properties of NO after intravenous injection, because *S*-nitrosated albumin shows a relatively high retention in the blood circulation after intravenous injection into mice. However, targeted NO delivery after intravenous injection using a macromolecular carrier has not been successfully achieved so far (Katsumi et al. 2005). To achieve the targeted NO delivery from SNO-HSA after intravenous injection, we need to understand the different functions and structure of HSA and its biological fate in detail. Therefore, we first summarized the effective factors on the *S*-nitrosation and *S*-denitrosation of HSA both in vitro and in biological systems. Secondly, we introduce the biological characteristics of two SNO-HSA analogs, Poly-SNO-HSA, which possesses about seven SNO groups per HSA molecule, and SNO-HSA dimer. Finally, we focused on possible future therapeutic applications of the *S*-nitroso adducts of albumin analogs in this chapter.

8.2 Relationship of *S*-Nitrosation of HSA and Endogenous Ligands of HSA

HSA is a non-glycosylated protein with a molecular weight of 66.5 kDa. The polypeptide is organized in the form of a heart-shaped protein with approximately 67 % α-helix but no β-sheet structure (Peters 1996). It is composed of three homologous domains (I–III) each of which can be subdivided into subdomains (A and B) with distinct helical folding patterns connected by flexible loops (Fig. 8.1). All but 1 (Cys-34) of the 35 cysteine residues are involved in the formation of stabilizing disulfide bonds. In the circulation, the protein has an average half-life of approximately 20 days, and normally about 70 % of the Cys-34 residues are freely accessible; i.e., they are not oxidized or involved in ligand binding and represent the largest fraction of free thiols in the blood (Peters 1996).

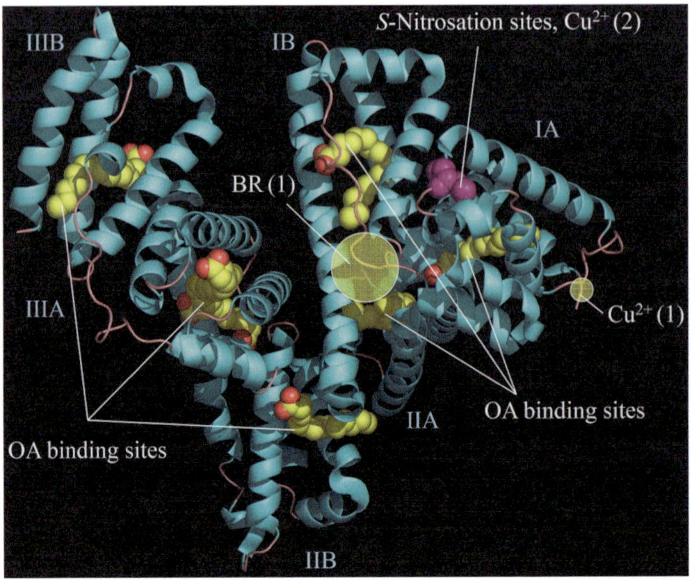

Fig. 8.1 Crystal structure of HSA showing the locations of Cys-34 and the OA, BR, and Cu^{2+} binding sites

In the blood, HSA functions as a transport and depot protein for numerous endogenous and exogenous compounds (Peters 1996; Kragh-Hansen 1981; Kragh-Hansen et al. 2002). HSA purified from serum contains bound endogenous ligands, in particular fatty acids and perhaps also other exogenous ligands. The possible effects of these ligands on the *S*-nitrosation of HSA were examined by incubating non-defatted and charcoal-treated albumin, with GS-NO. The *S*-nitroso moiety of the former preparation was significantly higher ($P<0.01$) than that of the latter (Ishima et al. 2007a). Thus, the presence of ligands greatly enhances the efficiency of *S*-nitrosation. Therefore, we studied the effects of the strongly bound ligands, oleate (OA) (C18:1), bilirubin (BR), and Cu^{2+}, and weakly bound ligands, L-tryptophan, progesterone, ascorbate, Zn^{2+}, and Fe^{2+}, on in vitro *S*-nitrosation of HSA at Cys-34 by *S*-nitrosoglutathione (GS-NO) and 1-hydroxy-2-oxo-3-(N-3-methyl-aminopropyl)-3-methyl-3'-triazene (NOC-7). In these experiments, two types of *S*-nitrosating agents were used, namely, GS-NO which *S*-transnitrosates via NO^+ and NOC-7 which *S*-nitrosates mainly via NO and N_2O_3.

In addition to the seven OA binding sites, the locations of the high-affinity binding site for BR (BR(1)) and the high-affinity binding site (Cu^{2+}(1)) and a secondary binding site for Cu^{2+} (Cu^{2+}(2)) are indicated. Very recently, a crystallographic analysis of HSA complexed with BR has shown that the ligand is bound with a high affinity in subdomain IB rather than in subdomain IIA (Zunszain et al. 2008). The subdivision of HSA into domains (I–III) and subdomains (A and B) is shown. The structure was simulated on the basis of X-ray crystallographic data for HSA-OA (PDB ID code 1gni) and modified with the use of Rasmol (downloaded from http://www.openrasmol.org).

Table 8.1 Effect of ligands on S-nitrosylation of HSA by GS-NO and NOC-7

	Oleic acid	Bilirubin	Cu	Zn	Fe	Tryptophan	Ascorbate	Progesterone
GS-NO	↑	↑↑	↑	→	→	→	→	→
NOC-7	→	→	↑↑	→	→	→	→	→

↑: $P<0.05$, ↑↑: $P<0.01$ as compared with HSA alone

The results obtained with equimolar amounts of protein and ligand are indicated in Table 8.1. It can be seen that OA and BR enhance the efficiency of GS-NO, but not that of NOC-7, whereas Cu^{2+} increases the *S*-nitrosation by NOC-7 but not that caused by GS-NO. In contrast, no significant effect was observed when adding L-tryptophan, progesterone, ascorbate, $(CH_3COO)_2Zn$, or $FeCl_2$. We studied the positive effects of OA, BR, and Cu^{2+}, which bind to different high-affinity sites of HSA, in more detail (Fig. 8.1).

8.2.1 *Effect of OA Binding*

We investigated the effect of increasing OA binding on the *S*-nitrosation of HSA by GS-NO in more detail. The increment in SNO-HSA formation was found to be dose dependent up to a OA:HSA molar ratio of 3; increasing the molar ratio further to 4 or 5 did not result in additional *S*-nitrosation. Because OA does not bind to Cys-34 (Fig. 8.1), the observed effect is most probably due to binding-induced conformational changes of HSA making Cys-34 more accessible to GS-NO (Ishima et al. 2007a). This proposal is supported by the finding that OA binding results in an almost linear increment in binding of the test compound 5,5′-dithiobis-2-nitrobenzoic acid (DTNB) to Cys-34 (Ishima et al. 2007a). The proposal is also supported by crystal structure analysis (Gryzunov et al. 2003; Narazaki et al. 1997; Petitpas et al. 2001). According to that type of analysis, the reactive SH group of Cys-34 is located in a crevice on the surface of HSA. The binding of OA induces conformational changes in the protein, leading to a slight opening of the interface between the two halves of the albumin molecule (subdomains IA-IIA and IIB-IIIB, respectively) and the rotation of domain I. These structural changes result in a greater opening of the crevice that contains Cys-34, thus rendering the SH group of Cys-34 more readily available (Narazaki et al. 1997).

8.2.2 *Effect of BR Binding*

The effect of BR binding on *S*-nitrosation by GS-NO was also studied at different molar ratios of ligand to protein (Ishima et al. 2007a). The binding of 1 mol of BR resulted in a significantly higher level of *S*-nitrosation ($P<0.01$); increasing the molar ratio to 3 or 5 did not cause detectable additional *S*-nitrosation. Thus, only high-affinity BR binding increases *S*-nitrosation. Because this kind of binding takes place in another region of HSA than that containing Cys-34 (Fig. 8.1), the improving effect must be due to conformational changes in the protein related to accommodating the large BR molecule.

In contrast to GS-NO, the high-affinity binding of BR did not influence *S*-nitrosation by NOC-7 (Table 8.1). To test whether this lack of effect could be caused by interaction between NO and HSA-bound BR, we performed spectrophotometric experiments. The experiments showed that the exposure of HSA-BR to NOC-7, but not to GS-NO, resulted in a rapid decrease of absorbancy at 470 nm (representing λ_{max} for HSA-BR) and a concomitant and pronounced increase at 650 nm (representing λ_{max} for HSA-biliverdin). It therefore appears that the following reaction takes place: (^{34}Cys-SH)-HSA-BR + ·NO → (^{34}Cys-SH)-HSA-BV + NO_2^-. Thus, the lack of effect of BR is due to its conversion to biliverdin (BV), and neither the ligand nor the NO_2^- formed improves *S*-nitrosation.

8.2.3 *Effect of Cu^{2+} Binding*

In contrast to the *S*-nitrosating effect of GS-NO, the effect of NOC-7 was significantly increased by Cu^{2+} (Table 8.1). The increasing effect was the same, irrespective of whether the molar ratio of Cu^{2+} to protein was 1:1, 3:1, or 5:1 (Ishima et al. 2007a). Cu^{2+} binds with a very high affinity to a specific site in the N-terminal region of HSA, and His-3 is an essential element at that site (Peters 1996) (Fig. 8.1). To test whether the high-affinity binding of Cu^{2+}, which takes place at a distance from Cys-34, is responsible for the improving effect of NOC-7, or whether the effect is caused by other means, e.g., secondary binding, we mutated His-3 for an alanine. The positive effect of Cu^{2+} disappeared when the His-3 had been replaced. This finding strongly suggests that high-affinity binding explains the improved effect of Cu^{2+} on *S*-nitrosation by NOC-7. The positive effect of high-affinity Cu^{2+} binding is most probably caused by conformational changes induced in the HSA molecule, which renders the SH group of Cys-34 more reactive. Such a mechanism is supported by the results of Zhang and Wilcox (Zhang and Wilcox 2002). These authors, using isothermal titration calorimetry and different spectroscopic techniques, found evidence for an interaction between the first Cu^{2+} binding site and Cys-34 in BSA. However, the conformational changes are different from those caused by OA, because in contrast to the binding of OA, the binding of Cu^{2+} does not affect the accessibility of Cys-34 (Ishima et al. 2007a). In contrast to the above studies, Stubauer et al. (1999) found no effect of a high-affinity-bound Cu^{2+} on RS-NO formation. RS-NO formation was only initiated when that binding site had been saturated, and the authors proposed *S*-nitrosation of Cys-34, when Cu^{2+} binds with a low affinity to the same residue. However, they used BSA and NO gas in their studies.

8.2.4 Methodological Aspects

Commercial HSA preparations for clinical use contain high concentrations of octanoate and *N*-acetyl-L-tryptophanate, because they stabilize the protein during the pasteurization process and also protect it against oxidative stress (Anraku et al. 2004). The presence of these ligands greatly facilitates the formation of SNO-HSA, and a fast and simple method for making such preparations has been developed using GS-NO as the *S*-nitrosylating agent. Interestingly, this method for preparing SNO-HSA is simple, fast, and straightforward; the cytoprotective effect against hepatic ischemia/reperfusion injury was better than that of a preparation carried out in a more traditional way (Ishima et al. 2010). Based on these results, we surmise that this cytoprotective activity of SNO-HSA is enhanced by bound fatty acids.

8.3 Effect of Fatty Acids on *S*-Denitrosation from HSA

The results presented above show that *S*-nitrosation of HSA is facilitated by simultaneous high-affinity ligand binding. The data discussed below show that the opposite is also the case, i.e., high-affinity ligand binding can ease *S*-denitrosation from SNO-HSA to other targets. This effect was observed using endogenous fatty acids as an example, and the effect was observed in both an animal model and in cultured cells. Most experiments were performed with OA, because quantitatively it is the most important fatty acid in human depot fat and because it is a major contributor to the albumin-bound fatty acids. As in the studies on *S*-nitrosation, we used OA:HSA molar ratios up to 5:1. This was done in order to investigate the potential effect of physiological and pathological fatty acid concentrations. Thus, HSA usually carries a total amount of 1–2 M equivalents of fatty acids. However, this value can rise to 4 or more after maximal exercise or other adrenergic stimulation (Peters 1996).

8.3.1 Cytoprotection Against Ischemia/Reperfusion Liver Injury in Rats

To determine the effect of OA binding on *S*-denitrosation from SNO-HSA in vivo, we used an ischemia/reperfusion liver injury model (Ishima et al. 2008). To evaluate liver injury, the extracellular release of the liver enzymes aspartate aminotransferase and alanine aminotransferase was measured via plasma enzyme values. The administration of SNO-HSA diminished, to a significant extent ($P<0.01$), the enzyme concentrations measured at 60 min and 120 min. The protection of the liver cells by SNO-HSA was more pronounced, if the protein also carried OA. The effect of OA on SNO-HSA-mediated cytoprotection appeared to depend on the OA content; e.g., the binding of 5 mol OA had a more pronounced effect than binding of 3 mol. We also found that caprylate (C8:0), a short-chain and saturated fatty acid, potentiated the cytoprotective effect of SNO-HSA. In addition, the pharmacokinetic characteristics of SNO-HSA have been studied in mice. The study showed that binding of as much as 5 mol of OA per mol of SNO-HSA does not affect either the plasma half-life of the protein nor its uptake by the liver, kidney, or spleen (Ishima et al. 2008).

8.3.2 Cytoprotection of HepG2 Cells Exposed to an Anti-Fas Antibody

The advantageous effects of fatty acids on cytoprotection found in the ischemia/reperfusion model may involve multiple mechanisms, including the maintenance of tissue blood flow, the induction of heme oxygenase-1 (a cytoprotective enzyme), the suppression of neutrophil infiltration, and a reduction of apoptosis (Ikebe et al. 2000). Therefore, we investigated the effect of fatty acid binding in a simpler system, a cell line. We examined the influence of OA binding on the antiapoptotic effect of SNO-HSA on HepG2 cells treated with an anti-Fas antibody (Ishima et al. 2008). The results showed that SNO-HSA induced protection of the cells in a concentration-dependent manner. This protection was greatly increased by the binding of 5 mol of OA per mol of SNO-HSA. Thus, fatty acid binding also improves the cytoprotective effect of SNO-HSA in an in vitro system.

The above findings clearly indicate that SNO-HSA *S*-denitrosates in HepG2 cells. We studied this aspect in a direct way, namely, by measuring the concentration of SNO-HSA. We found that the presence of the cells caused a decrease in SNO-HSA and that the decrease was faster and quantitatively more pronounced when OA was present. The effect increased with increasing OA concentration: from 1:1 to 3:1 to 5:1 molar ratios. The OA-mediated promotion of SNO-HSA decay can be explained by an increased accessibility to the *S*-nitroso moiety of HSA and/or by an intensified interaction between SNO-HSA and cell surface thiols. Furthermore, we also examined the issue of whether the improving effect of OA binding on *S*-denitrosation is unique for that fatty acid or whether the effect can also be exerted by a mixture of endogenous fatty acids. We used HSA preparations isolated from hemodialysis patients, because such treatment increases the fatty acid concentrations in the blood. The results showed a good linear correlation between $T_{1/2}$ and the amount of fatty acid bound to SNO-HSA. Thus, in addition to OA, a mixture of endogenous fatty acids facilitates the decay of SNO-HSA by HepG2 cells. This finding has biological and clinical implications, because the plasma concentrations of nonesterified fatty acids can be increased in a number of situations. In addition to hemodialysis, an increase in fatty acids is seen in connection with exercise and other adrenergic stimulation and in pathological conditions such as the metabolic syndrome and diabetes mellitus (Ishima et al. 2008).

8.3.3 Cytoprotective NO Uptake of HepG2 Cells

As mentioned in Sect. 1.3.2, fatty acid binding accelerates SNO-HSA decomposition by HepG2 cells. We investigated whether this decomposition is accompanied by the uptake of NO by the cells using intracellular DAF-FM DA fluorescence (fluorescence of diaminofluorescein-FM diacetate) (Ishima et al. 2008). The intracellular NO concentration increased with incubation time and with increasing OA/SNO-HSA molar ratios. To clarify the *S*-transnitrosation properties of a saturated fatty acid, we tested the effect of stearate (C18:0) on the uptake of NO by the HepG2 cells. Stearate had an effect on the *S*-transnitrosation of SNO-HSA, which was very similar to that of OA.

The fatty acid-induced increment in the transfer of the NO from SNO-HSA into hepatocytes is completely blocked by the addition of filipin III. However, a basal mechanism, not affected by the addition of filipin III addition, shows the transfer of small amounts of NO or modifications thereof (e.g., NO^+). Both systems may involve a membrane protein and operate by transferring NO^+ from one thiol to another.

We examined the issue of whether NO uptake involves contact between albumin and the cell membrane or components thereof using FITC fluorescence (fluorescence of fluorescein isothiocyanate). SNO-HSA was labeled with FITC and the interaction with the HepG2 cells was analyzed by fluorescence microscopy (Ishima et al. 2008). FITC-SNO-HSA was found to bind to the cells, and the degree of binding increased in a dose-dependent manner by the cobinding of OA. Similar results were obtained with HSA that was not *S*-denitrosated. Adding filipin III had no effect on the protein-cell interaction. Thus, OA is proposed to enhance the interaction between SNO-HSA and HepG2 cells. Figure 8.2 proposes a model for the *S*-transnitrosation of HepG2 cells by SNO-HSA. The binding of fatty acids introduces conformational changes in HSA that render the SH group, without or with *S*-nitrosation, more accessible. Fatty acid binding also facilitates the binding of albumin to a receptor on hepatocytes, perhaps the albumin binding adaptor protein gp60 (Minshall et al. 2003). When bound to the receptor, SNO-HSA-OA *S*-transnitrosates to HepG2 cells via two (or more) systems. Because OA-induced transnitrosation is completely blocked by filipin III, an inhibitor of caveolae (Pohl et al. 2002), it is proposed that caveolae are important for this type of *S*-transnitrosation. These findings strongly suggest that OA and NO of SNO-HSA-OA are transported by caveolae-associated proteins. Further studies will be needed to identify and clarify the mechanism for the caveolae-associated proteins.

It is widely assumed that *S*-transnitrosation from SNO-HSA to cells takes place solely or mainly via low-molecular-weight thiols (Simon et al. 1993; Shah et al. 2007; Crane et al. 2002). However, it should be noted that all the experiments with HepG2 cells were performed in the absence of GSH and other low-molecular-weight thiols.

Thus, fatty acid binding improves the cytoprotective effect of SNO-HSA in vivo, and reinforcement of an antiapoptotic effect by fatty acid binding contributes to this.

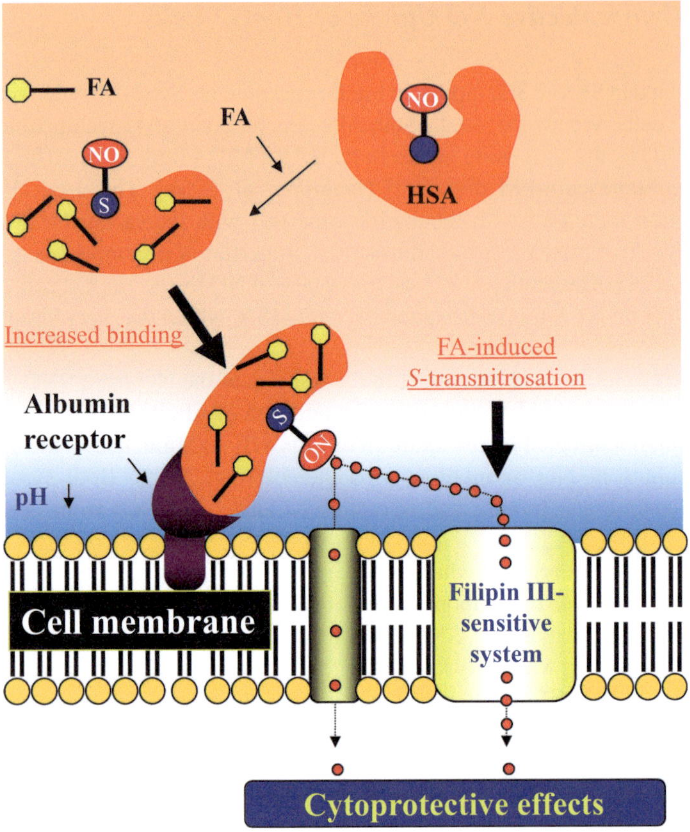

Fig. 8.2 Proposed model for fatty acid-mediated increase in *S*-transnitrosation of HepG2 cells by SNO-HSA. The model operates with two types of *S*-transnitrosation reactions. A basal one and a much more pronounced process, caused by albumin binding of fatty acid, which can be blocked by filipin III. The model also proposes that *S*-transnitrosation takes place without involving low-molecular-weight thiols (We used a modified figure to that proposed in Ref. (Ishima et al. 2008))

A fatty acid bound to SNO-HSA enhances the interaction between SNO-HSA and HepG2 cells and the *S*-transnitrosation of SNO-HSA. This enhances NO transfer from SNO-HSA into hepatocytes and the antiapoptotic effect. We found a novel filipin III-sensitive mechanism for the transfer of NO from SNO-HSA into hepatocytes.

8.4 Poly-SNO-HSA Has Multiple Anticancer Effects

To construct more efficient SNO-HSA preparations, SNO-HSA with *ca.* 7 conjugated SNO groups (Poly-SNO-HSA) was prepared by means of chemical modification using Traut's reagent. The properties of this preparation were compared to those of SNO-HSA, which had, on average, 0.3 SNO group per HSA molecule

(Ishima et al. 2011). For these comparative studies, we used cell culture systems using HepG2 cells and murine colon 26 (C26) tumor cells. The results showed that the cellular uptake of NO from SNO-HSA partly takes place via low-molecular-weight thiols and that it results in cytoprotective effects by the induction of heme oxygenase-1 (HO-1). A cytoprotective effect of SNO-HSA was also observed for HepG2 cells (Sect. 8.3.3). In contrast, the transfer of NO from Poly-SNO-HSA into the cells was faster and more pronounced. The influx mainly takes place by the cell surface protein disulfide isomerase (csPDI). Surprisingly, the considerable NO inflow results in apoptotic cell death by ROS induction and caspase-3 activation and not in cytoprotection. Thus, increasing the number of SNO groups on HSA does not simply intensify the cellular responses to the product but can result in very different effects that are summarized in Fig. 8.3.

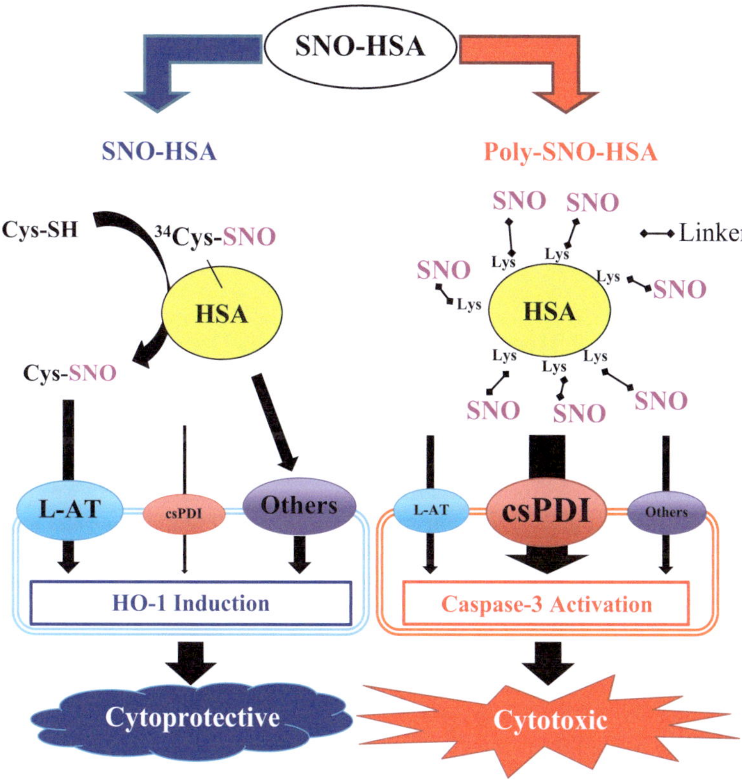

Fig. 8.3 Differences in the mechanisms and consequences of NO traffic from SNO-HSA and Poly-SNO-HSA to cells. NO transfer from the SNO group of Cys-34 on SNO-HSA to the cell is partly mediated by the L-amino acid transporter (L-AT) via *S*-transnitrosation to a free low-molecular-weight thiol. In contrast, the transfer of NO from Poly-SNO-HSA is mainly mediated by a cell surface protein disulfide isomerase (csPDI) without *S*-transnitrosation to a free low-molecular-weight thiol. The relatively slow transfer of NO from SNO-HSA avoids the presence of high intracellular NO concentrations and leads to cytoprotective activity through HO-1 induction. On the other hand, the NO influx from Poly-SNO-HSA is very fast and pronounced and leads to cell death caused by apoptosis (We used a modified figure to that proposed in Ref. (Ishima et al. 2011))

Interestingly, NO donors such as nitroglycerin have been reported to reverse the resistance to anticancer agents (Yasuda 2008). Therefore, we have evaluated the effect of Poly-SNO-HSA on the resistance of human myelogenous leukemic cells (K562 cells) to doxorubicin (Ishima et al. 2012a). The results showed that treatment with Poly-SNO-HSA increased the accumulation of doxorubicin in doxorubicin-resistant K562 cells (K562/dx cells). Furthermore, Poly-SNO-HSA enhanced the anticancer effect of doxorubicin in K562/dx cell-bearing mice. Poly-SNO-HSA reverts doxorubicin resistance by decreasing the expression of P-glycoprotein 1 and HIF-1α (Ishima et al. 2012a).

We also investigated the inhibitory effect of NO on autophagy by using Poly-SNO-HSA. Autophagy is one of the major causes of drug resistance. For example, the angiogenesis inhibitor bevacizumab shows only transient and short-term therapeutic effects, while long-term therapeutic benefits are rarely observed, probably due to hypoxia-induced autophagy. In a C26 cell culture system, SNO-HSA significantly suppressed hypoxia-induced autophagy via inhibiting the phosphorylation of JNK and the expression of its downstream Beclin1. The effect of SNO-HSA was also confirmed in vivo by combining it with bevacizumab (Ishima et al. 2015).

These data may open alternate strategies for cancer chemotherapy by taking advantage of the ability of SNO-HSA to suppress autophagy-mediated drug resistance and enhance the efficacy of chemotherapy (Fig. 8.4) (Ishima and Maruyama 2016).

8.5 *S*-Nitrosated Human Serum Albumin Dimer as a Novel Nano-EPR Enhancer

We synthesized a recombinant HSA dimer and found that its *S*-nitrosated form (SNO-HSA dimer) caused cell death to C26 tumor cells, and the effect was NO dose dependent (Ishima et al. 2012b). Intriguingly, *S*-nitrosation improved the uptake of the HSA dimer in tumor tissue through augmenting the enhanced permeability and retention (EPR) effect. The EPR effect is a unique phenomenon associated with solid tumors, and it can serve as a basis for the development of macromolecular anticancer therapy. These data suggest that the SNO-HSA dimer behaves not only as an anticancer therapeutic drug but also as an enhancer of the EPR effect. Therefore, the SNO-HSA dimer would be a very appealing carrier for utilization of the EPR effect in the future development of cancer therapeutics.

To elaborate this idea further, we investigated the influence of the SNO-HSA dimer on the antitumor effect of two types of macromolecular antitumor drugs, namely, an N-(2-hydroxypropyl) methacrylamide polymer conjugated with zinc protoporphyrin (HPMA-ZnPP), which forms micelles and can be used for fluorescence studies. The other drug was PEGylated liposomal doxorubicin (Doxil), a typical example of a stealth liposome that has been approved for medical usage. In mice bearing C26 tumors with a highly permeable vasculature (high endogenous

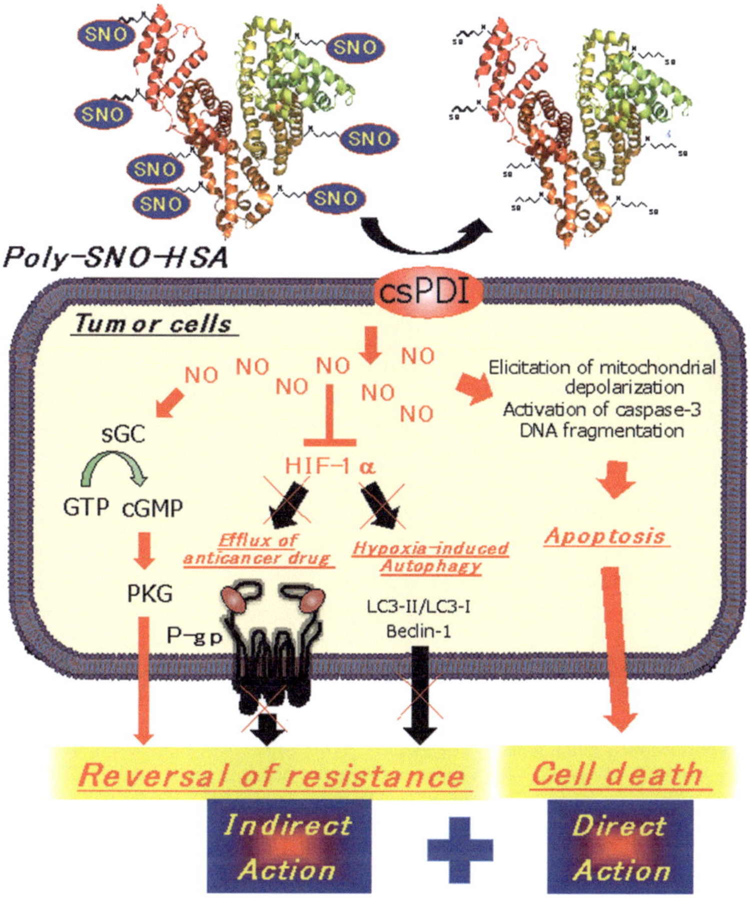

Fig. 8.4 Mechanisms of Poly-SNO-HSA as a safe and strong multiple antitumor agent. The fast and pronounced transfer of NO from Poly-SNO-HSA into the cell mainly takes place via csPDI. Within the cell, a high concentration of NO induces apoptosis and reverts dx resistance or hypoxia-induced autophagy by activating a cGMP-dependent pathway and HIF-1α (We used a modified figure to that proposed in Ref. (Ishima et al. 2012a))

EPR effect), the SNO-HSA dimer increased the tumor accumulation of the drugs by a factor 3-4 and thereby their antitumor effects. Furthermore, the SNO-HSA dimer improved the anti-metastatic effects of Doxil and reduced its minor uptake in non-tumor organs such as the liver and kidney. The tumor accumulation of Doxil in B16 tumors, which are characterized by a low permeable vasculature (low endogenous EPR effect), increased even more (sixfold) in the presence of SNO-HSA-dimer, and the improved accumulation led to a decreased tumor volume and an increased survival of the animals. The administration of the SNO-HSA dimer itself is safe, because it has no effect on blood pressure, heart rate, or on several other biochemical parameters. The present findings indicate that SNO-HSA dimer is promising for

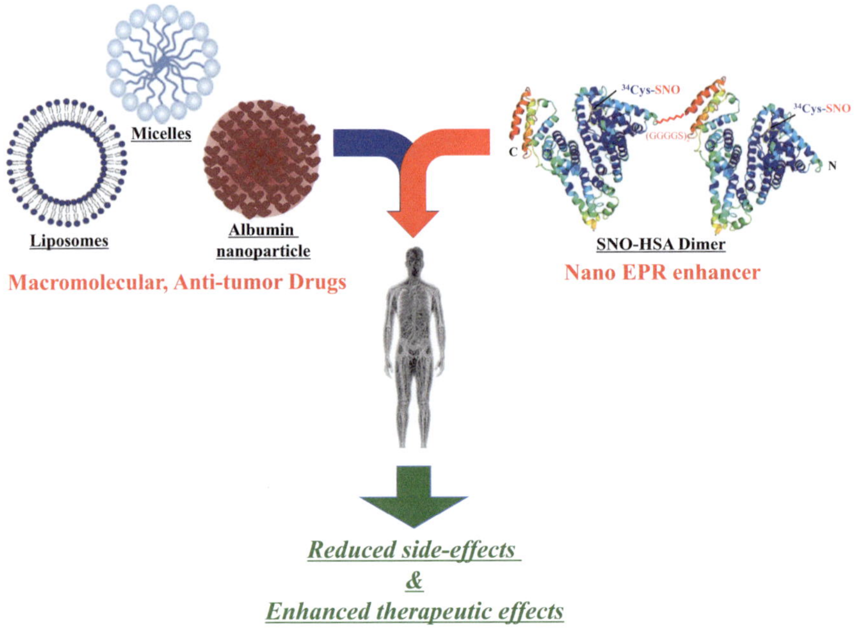

Fig. 8.5 Novel nano-EPR enhancer: SNO-HSA dimer. Administration of the SNO-HSA dimer results in a remarkable increase in vascular permeability that allows nanosized particles (micelles, liposomes, and albumin nanoparticle) to rapidly extravasate into tumor tissue, leading to super-enhanced permeability and retention. The increase in the EPR effect is highest in the first 0.5–1 h after administration of the SNO-HSA dimer and reaches baseline levels by 24 h. In contrast, the distribution of the macromolecular drugs to other organs such as the liver and kidney is decreased in the presence of SNO-HSA dimer

enhancing the EPR effect and consequently the specific, therapeutic effects of macromolecular anticancer drugs (Fig. 8.5) (Kinoshita et al. 2015).

8.6 Concluding Remarks

SNO-HSA possesses cytoprotective effects both in cell cultures and in vivo, and endogenous fatty acids, and perhaps also other high-affinity bound ligands, may serve as novel types of mediators for *S*-denitrosation. Furthermore, SNO-HSA preparations have potent antibacterial effects against bacteria such as *Salmonella typhimurium* (Ishima et al. 2007b). In contrast, Poly-SNO-HSA appears to be a useful candidate as an NO-traffic protein for cancer therapy (Fig. 8.6). In this respect, the SNO-HSA dimer is perhaps an even better candidate, because in addition to its NO trafficking properties, it has tumor targeting potential and superior blood retention properties (Ishima et al. 2012b). However, further studies are warranted to explore the mechanism responsible for the cellular uptake of NO in the pharmacological benefits of *S*-nitrosated HSA and to improve the cell targeting properties of the system.

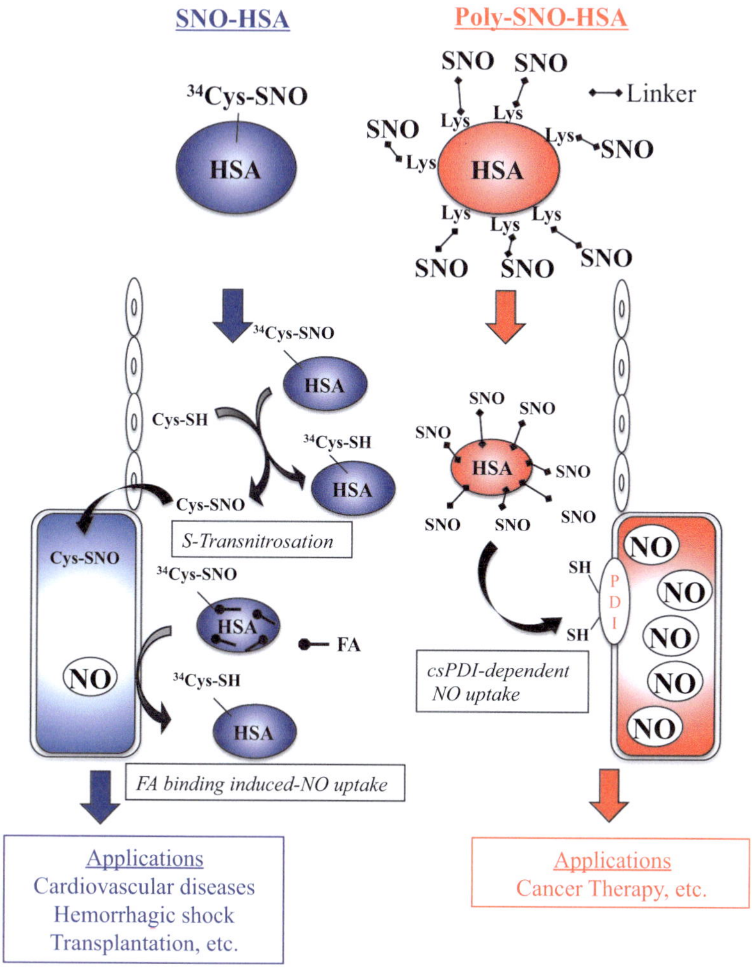

Fig. 8.6 Biological effects and potential therapeutic applications of SNO-HSAs. The figure above indicates some of the beneficial effects of SNO-HSA and Poly-SNO-HSA, which appears to be potentially clinical useful. Our work shows that *S*-denitrosation from SNO-HSA is greatly improved by the binding of fatty acids, and the NO entering the cells from SNO-HSA results in cytoprotection via HO-1 induction. The situation with respect to Poly-SNO-HSA is quite different. The great majority of the NO-moieties enters the cells via csPDI without *S*-transnitrosation to free low-molecular-weight thiol groups, and the pronounced and fast inflow of NO results in cytotoxicity that is exerted via caspase-3 activation

Acknowledgment We wish to thank Dr. H. Maeda, Institute of DDS Research, Sojo University, and Dr. T. Akaike, Graduate School of Medicine Tohoku University for their valuable advice and kind consideration.

References

Anraku M, Tsurusaki Y, Watanabe H, Maruyama T, Kragh-Hansen U, Otagiri M (2004) Stabilizing mechanisms in commercial albumin preparations: octanoate and N-acetyl-L-tryptophanate protect human serum albumin against heat and oxidative stress. Biochim Biophys Acta 1702:9–17

Baek S, Hrabie J, Keefer L, Hou D, Fineberg N, Rhoades R, March K (2002) Augmentation of intrapericardial nitric oxide level by a prolonged-release nitric oxide donor reduces luminal narrowing after porcine coronary angioplasty. Circulation 105:2779–2784

Clancy R, Abramson S (1995) Nitric oxide: a novel mediator of inflammation. Proc Soc Exp Biol Med 210:93–101

Clancy R, Levartovsky D, Leszczynska-Piziak J, Yegudin J, Abramson S (1994) Nitric oxide reacts with intracellular glutathione and activates the hexose monophosphate shunt in human neutrophils: evidence for S-nitrosoglutathione as a bioactive intermediary. Proc Natl Acad Sci U S A 91:3680–3684

Crane M, Ollosson R, Moore K, Rossi A, Megson I (2002) Novel role for low molecular weight plasma thiols in nitric oxide-mediated control of platelet function. J Biol Chem 277:46858–46863

Drapier J, Pellat C, Henry Y (1991) Generation of EPR-detectable nitrosyl-iron complexes in tumor target cells cocultured with activated macrophages. J Biol Chem 266:10162–10167

Ewing J, Young D, Janero D, Garvey D, Grinnell T (1997) Nitrosylated bovine serum albumin derivatives as pharmacologically active nitric oxide congeners. J Pharmacol Exp Ther 283:947–954

Furchgott R, Zawadzki J (1980) The obligatory role of endothelial cells in the relaxation of arterial smooth muscle by acetylcholine. Nature 288:373–376

Gaston B, Reilly J, Drazen J, Fackler J, Ramdev P, Arnelle D, Mullins M, Sugarbaker D, Chee C, Singel D (1993) Endogenous nitrogen oxides and bronchodilator S-nitrosothiols in human airways. Proc Natl Acad Sci U S A 90:10957–10961

Geng Y, Petersson A, Wennmalm A, Hansson G (1994) Cytokine-induced expression of nitric oxide synthase results in nitrosylation of heme and nonheme iron proteins in vascular smooth muscle cells. Exp Cell Res 214:418–428

Gryglewski R, Palmer R, Moncada S (1986) Superoxide anion is involved in the breakdown of endothelium-derived vascular relaxing factor. Nature 320:454–456

Gryzunov Y, Arroyo A, Vigne J, Zhao Q, Tyurin V, Hubel C, Gandley R, Vladimirov Y, Taylor R, Kagan V (2003) Binding of fatty acids facilitates oxidation of cysteine-34 and converts copper-albumin complexes from antioxidants to prooxidants. Arch Biochem Biophys 413:53–66

Hibbs JJ, Taintor R, Vavrin Z, Rachlin E (1988) Nitric oxide: a cytotoxic activated macrophage effector molecule. Biochem Biophys Res Commun 157:87–94

Ignarro L (2000) The unique role of nitric oxide as a signaling molecule in the cardiovascular system. Ital Heart J 1:28–29

Ignarro L, Wei L (2002) Visiting professorial lecture: nitric oxide in the regulation of vascular function: an historical overview. J Card Surg 17:301–306

Ikebe N, Akaike T, Miyamoto Y, Hayashida K, Yoshitake J, Ogawa M, Maeda H (2000) Protective effect of S-nitrosylated alpha(1)-protease inhibitor on hepatic ischemia-reperfusion injury. J Pharmacol Exp Ther 295:904–911

Ishima Y, Akaike T, Kragh-Hansen U, Hiroyama S, Sawa T, Maruyama T, Kai T, Otagiri M (2007a) Effects of endogenous ligands on the biological role of human serum albumin in S-nitrosylation. Biochem Biophys Res Commun 364:790–795
Ishima Y, Sawa T, Kragh-Hansen U, Miyamoto Y, Matsushita S, Akaike T, Otagiri M (2007b) S-Nitrosylation of human variant albumin Liprizzi (R410C) confers potent antibacterial and cytoprotective properties. J Pharmacol Exp Ther 320:969–977
Ishima Y, Akaike T, Kragh-Hansen U, Hiroyama S, Sawa T, Suenaga A, Maruyama T, Kai T, Otagiri M (2008) S-nitrosylated human serum albumin-mediated cytoprotective activity is enhanced by fatty acid binding. J Biol Chem 283:34966–34975
Ishima Y, Hiroyama S, Kragh-Hansen U, Maruyama T, Sawa T, Akaike T, Kai T, Otagiri M (2010) One-step preparation of S-nitrosated human serum albumin with high biological activities. Nitric Oxide 23:121–127
Ishima Y, Yoshida F, Kragh-Hansen U, Watanabe K, Katayama N, Nakajou K, Akaike T, Kai T, Maruyama T, Otagiri M (2011) Cellular uptake mechanisms and responses to NO transferred from mono- and poly-S-nitrosated human serum albumin. Free Radic Res 45:1196–1206
Ishima Y, Hara M, Kragh-Hansen U, Inoue A, Suenaga A, Kai T, Watanabe H, Otagiri M, Maruyama T (2012a) Elucidation of the therapeutic enhancer mechanism of poly-S-nitrosated human serum albumin against multidrug-resistant tumor in animal models. J Control Release 164:1–7
Ishima Y, Chen D, Fang J, Maeda H, Minomo A, Kragh-Hansen U, Ka T, Maruyama T, Otagiri M (2012b) S-nitrosated human serum albumin dimer is not only a novel anti-tumor drug but also a potentiator for anti-tumor drugs with augmented EPR effects. Bioconjug Chem 23:264–271
Ishima Y, Inoue A, Fang J, Kinoshita R, Ikeda M, Watanabe H, Maeda H, Otagiri M, Maruyama T (2015) Poly-S-nitrosated human albumin enhances the antitumor and antimetastasis effect of bevacizumab, partly by inhibiting autophagy through the generation of nitric oxide. Cancer Sci 106:194–200
Ishima Y, Maruyama T (2016) Human serum albumin as carrier in drug delivery systems. Yakugaku Zasshi 136:39–47
Iyengar R, Stuehr D, Marletta M (1987) Macrophage synthesis of nitrite, nitrate, and N-nitrosamines: precursors and role of the respiratory burst. Proc Natl Acad Sci U S A 84:6369–6373
Katsumi H, Nishikawa M, Yamashita F, Hashida M (2005) Development of polyethylene glycol-conjugated poly-S-nitrosated serum albumin, a novel S-Nitrosothiol for prolonged delivery of nitric oxide in the blood circulation in vivo. J Pharmacol Exp Ther 314:1117–1124
Katsumi H, Nishikawa M, Hashida M (2007) Development of nitric oxide donors for the treatment of cardiovascular diseases. Cardiovasc Hematol Agents Med Chem 5:204–208
Kinoshita R, Ishima Y, Ikeda M, Kragh-Hansen U, Fang J, Nakamura H, Chuang VT, Tanaka R, Maeda H, Kodama A, Watanabe H, Maeda H, Otagiri M, Maruyama T (2015) S-Nitrosated human serum albumin dimer as novel nano-EPR enhancer applied to macromolecular anti-tumor drugs such as micelles and liposomes. J Control Release 217:1–9
Kragh-Hansen U (1981) Molecular aspects of ligand binding to serum albumin. Pharmacol Rev 33:17–53
Kragh-Hansen U, Chuang VTG, Otagiri M (2002) Practical aspects of the ligand-binding and enzymatic properties of human serum albumin. Biol Pharm Bull 25:695–704
Lancaster JJ (1994) Simulation of the diffusion and reaction of endogenously produced nitric oxide. Proc Natl Acad Sci U S A 91:8137–8141
Lancaster JJ, Hibbs JJ (1990) EPR demonstration of iron-nitrosyl complex formation by cytotoxic activated macrophages. Proc Natl Acad Sci U S A 87:1223–1227
Marks D, Vita J, Folts J, Keaney JJ, Welch G, Loscalzo J (1995) Inhibition of neointimal proliferation in rabbits after vascular injury by a single treatment with a protein adduct of nitric oxide. J Clin Invest 96:2630–2638
Marley R, Feelisch M, Holt S, Moore KA (2000) chemiluminescence-based assay for S-nitrosoalbumin and other plasma S-nitrosothiols. Free Radic Res 32:1–9
Minshall R, Sessa W, Stan R, Anderson R, Malik A (2003) Caveolin regulation of endothelial function. Am J Physiol Lung Cell Mol Physiol 285:1179–1183

Mizutani T, Layon A (1996) Clinical applications of nitric oxide. Chest 110:506–524
Moncada S, Palmer R, Higgs E (1991) Nitric oxide: physiology, pathophysiology, and pharmacology. Pharmacol Rev 43:109–142
Mülsch A, Mordvintcev P, Vanin A, Busse R (1993) Formation and release of dinitrosyl iron complexes by endothelial cells. Biochem Biophys Res Commun 196:1303–1308
Myers P, Minor RJ, Guerra RJ, Bates J, Harrison D (1990) Vasorelaxant properties of the endothelium-derived relaxing factor more closely resemble S-nitrosocysteine than nitric oxide. Nature 345:161–163
Narazaki R, Maruyama T, Otagiri M (1997) Probing the cysteine 34 residue in human serum albumin using fluorescence techniques. Biochim Biophys Acta 1338:275–281
Nathan C (1992) Nitric oxide as a secretory product of mammalian cells. FASEB J 6:3051–3064
Palmer R, Ferrige A, Moncada S (1987) Nitric oxide release accounts for the biological activity of endothelium-derived relaxing factor. Nature 327:524–526
Peters T Jr (1996) All about albumin: biochemistry, genetics, and medical applications. Academic, San Diego
Petitpas I, Grüne T, Bhattacharya A, Curry S (2001) Crystal structures of human serum albumin complexed with monounsaturated and polyunsaturated fatty acids. J Mol Biol 314:955–960
Pohl J, Ring A, Stremmel W (2002) Uptake of long-chain fatty acids in HepG2 cells involves caveolae: analysis of a novel pathway. J Lipid Res 43:1390–1399
Ródenas J, Mitjavila M, Carbonell T (1998) Nitric oxide inhibits superoxide production by inflammatory polymorphonuclear leukocytes. Am J Physiol 274:827–830
Shah C, Bell S, Locke I, Chowdrey H, Gordge M (2007) Interactions between cell surface protein disulphide isomerase and S-nitrosoglutathione during nitric oxide delivery. Nitric Oxide 16:135–142
Simon D, Stamler J, Jaraki O, Keaney J, Osborne J, Francis S, Singel D, Loscalzo J (1993) Antiplatelet properties of protein S-nitrosothiols derived from nitric oxide and endothelium-derived relaxing factor. Arterioscler Thromb 13:791–799
Stadler J, Bergonia H, Di Silvio M, Sweetland M, Billiar T, Simmons R, Lancaster JJ (1993) Nonheme iron-nitrosyl complex formation in rat hepatocytes: detection by electron paramagnetic resonance spectroscopy. Arch Biochem Biophys 302:4–11
Stamler J, Singel D, Loscalzo J (1992a) Biochemistry of nitric oxide and its redox-activated forms. Science 258:1898–1902
Stamler J, Jaraki O, Osborne J, Simon D, Keaney J, Vita J, Singel D, Valeri C, Loscalzo J (1992b) Nitric oxide circulates in mammalian plasma primarily as an S-nitroso adduct of serum albumin. Proc Natl Acad Sci U S A 89:7674–7677
Stamler J, Osborne J, Jaraki O, Rabbani L, Mullins M, Singel D, Loscalzo J (1993) Adverse vascular effects of homocysteine are modulated by endothelium-derived relaxing factor and related oxides of nitrogen. J Clin Invest 91:308–318
Stubauer G, Giuffrè A, Sarti P (1999) Mechanism of S-nitrosothiol formation and degradation mediated by copper ions. J Biol Chem 274:28128–28133
Yasuda H (2008) Solid tumor physiology and hypoxia-induced chemo/radio-resistance: novel strategy for cancer therapy: nitric oxide donor as a therapeutic enhancer. Nitric Oxide 19:205–216
Zhang Y, Wilcox D (2002) Thermodynamic and spectroscopic study of Cu(II) and Ni(II) binding to bovine serum albumin. J Biol Inorg Chem 7:327–337
Zunszain PA, Ghuman J, McDonagh AF, Curry S (2008) Crystallographic analysis of human serum albumin complexed with 4Z,15E-bilirubin-IXα. J Mol Biol 381:394–406

Chapter 9
Hemoglobin–Albumin Clusters as a Red Blood Cell Substitute

Teruyuki Komatsu

Abstract Core–shell protein cluster comprising bovine hemoglobin (HbBv) in the core and human serum albumin (HSA) at the shell was created as an artificial O_2 carrier designed for use as a red blood cell (RBC) substitute. The protein cluster was prepared by covalent linkage between the Cys-34 residue of HSA and the surface Lys amino groups of HbBv using heterobifunctional cross-linker. The average HSA/HbBv ratio of one cluster was determined as 3.0 ± 0.2; therefore we indicated this hemoglobin–albumin cluster as HbBv–HSA_3. Human Hb A (HbA) can be also used for a core protein to synthesize HbA–HSA_3 cluster. The isoelectric point of HbBv–HSA_3 (pI = 5.1) was markedly lower than that of HbBv and almost identical to the value of HSA. SFM and TEM measurements revealed a triangular shape of HbBv–HSA_3. The complete 3D structure based on TEM data was reconstructed. The clusters showed moderately higher O_2 affinities than the native HbBv and HbA. Viscosity and blood cell counting measurements demonstrated that HbBv–HSA_3 has good compatibility with whole blood. Intravenous administration of HbBv–HSA_3 into anesthetized rats elicited no unfavorable increase in systemic blood pressure by vasoconstriction. The half-life of ^{125}I-labeled cluster in circulating blood is longer than that of HSA. All results indicate that HbBv–HSA_3 has sufficient preclinical safety as an alternative material for RBC transfusion. Interestingly, clusters prepared under N_2 atmosphere showed low O_2 affinity resembling human RBC. Furthermore, the exterior HSA units possess a remarkable ability to bind antioxidant agent, such as Pt nanoparticle (PtNP). The peripheral HSA–PtNP shell prevents oxidation of the core HbBv, which enables the formation of an extremely stable O_2 complex even in H_2O_2 solution. This chapter reviews the synthesis, structure, O_2-binding property, and preclinical safety of hemoglobin–albumin cluster as a promising RBC substitute for practical use.

Keywords Artificial oxygen carrier • Protein cluster • Oxygen-binding property • Blood pressure • Blood retention

T. Komatsu (✉)
Department of Applied Chemistry, Faculty of Science and Engineering, Chuo University, 1-13-27 Kasuga, Bunkyo-ku, Tokyo 112-8551, Japan
e-mail: komatsu@kc.chuo-u.ac.jp

M. Otagiri, V.T.G. Chuang (eds.), *Albumin in Medicine*,
DOI 10.1007/978-981-10-2116-9_9

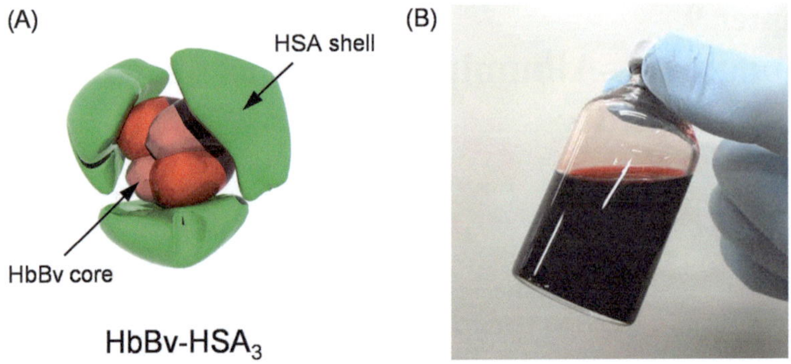

Fig. 9.1 (**A**) Illustration of molecular structure of HbBv–HSA_3 in which an HbBv core is wrapped covalently by three HSAs (Tomita et al. 2013). (**B**) HbBv–HSA_3 solution (20 g/dL) in PBS (pH 7.4)

9.1 Introduction

Over the last few decades, hemoglobin (Hb)-based O_2 carriers (HBOCs) of many kinds have been designed and developed as red blood cell (RBC) substitutes (Squires 2002; Jahr et al. 2011; Kluger and Lui 2013; Mondery-Pawlowski et al. 2013), such as intramolecularly cross-linked Hb (Snyder et al. 1987; Nagababu et al. 2002), polymerized Hb (Buehler et al. 2005; Pearce et al. 2006; Kluger and Zhang 2003; Hu and Kluger 2008), poly(ethylene glycol)-decorated Hb (Vandegriff et al. 2003; Manjula et al. 2003; Li et al. 2008, 2009), enzyme-conjugated Hb (D'Agnilloo and Chang 1998; Alagic et al. 2005), saccharide-linked Hb (Zhang et al. 2008), and nano-/microparticle-encapsulated Hb (Sakai 2012; Xiong et al. 2013). In any period, the social requests have promoted the development of the RBC substitute, for example, a need in battlefield, a concern to virus diffusion, and a primary measure for crisis management. Currently, the motive of the artificial O_2 carrier is moving to a medical measure to supplement blood transfusion treatment. The declining birthrate and aging population make it difficult to retain a stable blood transfusion system. The number of old people will continue to increase, although the population of blood donors is expected to decrease. In fact, the Japanese Red Cross Society predicts a blood shortage equivalent to 890,000 people per year in 2027 (Ministry of Health, Labor and Welfare, Japan 2014). However, no HBOC product has been assigned yet for medical use (Jahr et al. 2011; Pearce et al. 2006; Natanson et al. 2008; Kluger 2010). The major concern of the Hb derivatives is vasoconstriction, which causes a mild increase in systemic blood pressure. This pressor response is inferred to be due to quick scavenging of nitric oxide (NO), the endothelial-derived relaxing factor, by Hb leaked into the extravascular space (Shultz et al. 1993; Rohlfs et al. 1998; Doherty et al. 1998).

Recently, we prepared a covalent core–shell structured protein cluster composed of Hb in the core and human serum albumin (HSA) at the shell as a unique HBOC (Fig. 9.1) (Tomita et al. 2013; Hosaka et al. 2014; Haruki et al. 2015). The average

HSA/Hb ratio of one cluster was 3.0±0.2. We indicate this hemoglobin–albumin cluster as Hb–HSA$_{3}$. It is noteworthy that intravenous transfusion of Hb–HSA$_{3}$ does not elicit the acute increase in blood pressure (Haruki et al. 2015). This is attributed to the fact that Hb–HSA$_{3}$ is not eliminated from the vasculature walls because of the electrostatic repulsion between the negative surface net charges of the cluster and the glomerular basement membrane around the endothelial cells. This chapter reviews the synthesis, structure, O_2-binding property, and preclinical safety of Hb–HSA$_{3}$ as a promising RBC substitute for practical use.

9.2 Synthesis and Structure of Hemoglobin–Albumin Cluster

HSA is a heart-shaped monomeric protein bearing one free sulfhydryl group of Cys at position 34 (Curry et al. 1998). Therefore, we used a heterobifunctional cross-linking agent, *N*-succinimidyl-4-(*N*-maleimidomethyl)cyclohexane-1-carboxylate (SMCC), as a connector between the Cys-34 residue of HSA and the surface Lys amino groups of Hb to create a core–shell cluster of Hb and HSA (Fig. 9.2) (Tomita et al. 2013; Hosaka et al. 2014). First, SMCC was reacted with bovine Hb (HbBv) in phosphate-buffered saline (PBS, pH 7.4) solution. Then the resulting maleimide-activated HbBv was added dropwise into the HSA solution, followed by stirring at 4 °C. Size-exclusion chromatography (SEC) of the resultant mixture showed distinct three peaks at the high molecular weight region. Native PAGE also exhibited three new bands above HSA (Tomita et al. 2013; Hosaka et al. 2014). Using gel filtration chromatography (GFC), unreacted HSA was removed and the major products were collected. Based on the Hb assay and protein assay, the HSA/HbBv ratio (mol/mol) of the cluster was determined to be 3.0±0.2. We indicate this product as Hb–HSA$_{3}$ with italicized subscript 3. Reaction with human adult Hb (HbA) also generated similar protein cluster, HbA–HSA$_{3}$ (Kimura et al. 2015).

The CD spectrum of HbBv–HSA$_{3}$ fit perfectly with the sum of the HbBv spectrum and a threefold-enlarged HSA spectrum (Tomita et al. 2013; Hosaka et al. 2014). This result implies that (i) the HbBv:HSA ratio in one cluster is 1:3 (mol:mol) on average and (ii) the secondary structure of the individual protein unit remains constant after the cluster formation. The isoelectric points of HbBv–HSA$_{3}$ (*p*I: 5.1) were markedly lower than the value of native HbBv (*p*I, 7.0) and resembled to that of HSA (*p*I: 4.9). These results supported that the HbBv core is covalently wrapped by HSAs (Tomita et al. 2013; Hosaka et al. 2014).

Scanning probe microscopy (SPM) images of HbBv–HSA$_3$ on a mica surface in PBS solution depicted clearly triangular shape of several entities (Fig. 9.3A) (Tomita et al. 2013). We were convinced of a triangular core–shell structure with HbBv in the center and three exterior HSAs are formed.

Furthermore, the 3D reconstruction of HbBv–HSA$_3$ based on transmission electron microscopy (TEM) images revealed a complete triangular structure (Tomita

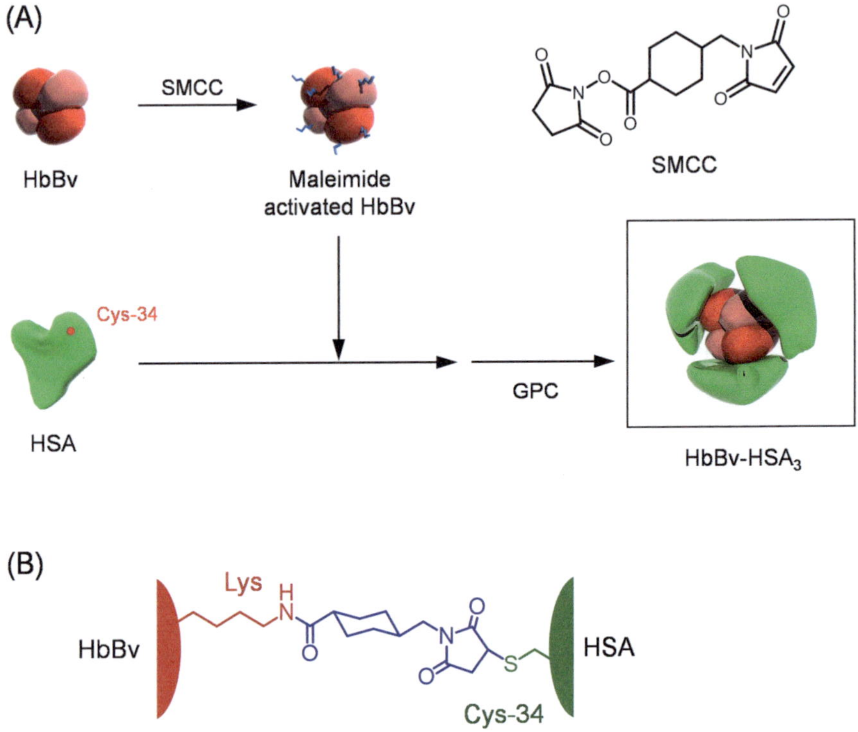

Fig. 9.2 (**A**) Schematic illustration of the synthetic route of HbBv–HSA_3 using heterobifunctional cross-linker (SMCC). (**B**) The covalent linkage structure between the Lys residue of HbBv and the Cys-34 residue of HSA

et al. 2013). The original TEM pictures of HbBv–HSA_3 showed individual particles (diameter: approximately 10 nm), but detailed structure information was unavailable owing to the low contrast. Then we used additional image processing procedure (single-particle analysis). From the obtained class sum images with an enhanced signal-to-noise ratio (Fig. 9.3B) (Tomita et al. 2013; Kimura et al. 2015), the 3D volume of HbBv–HSA_3 was reconstructed. We calculated a presentation of the protein moieties by fitting their PDB data into the reconstructed volume. The proposed geometries conferred a possible spatial arrangement of the HbBv interior and three HSA exterior. The fitting of three HSAs defined an arrangement of the Cys-34 of HSA, which suggested the potential binding Lys partners on HbBv (Fig. 9.3C). The HSA binding sites on HbA in HbA–HSA_3 are almost the same as those of HbBv–HSA_3 (Kimura et al. 2015).

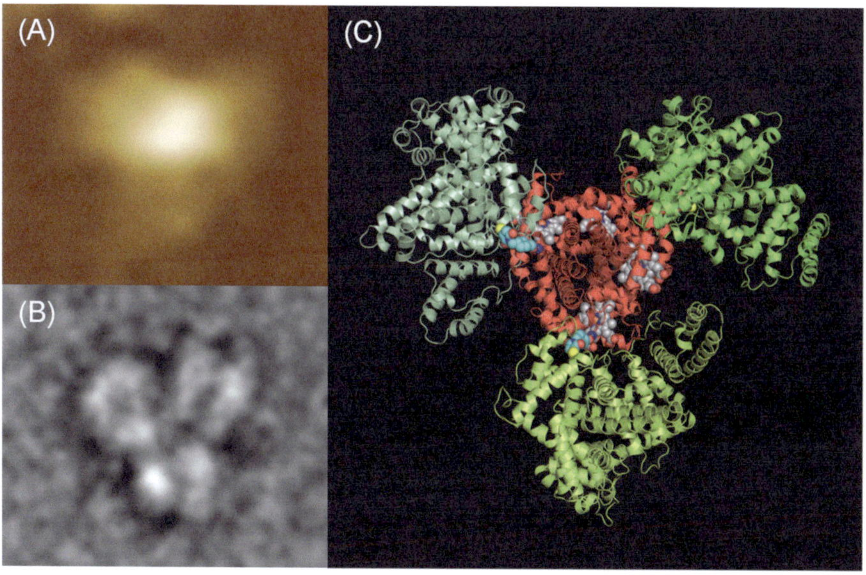

Fig. 9.3 (**A**) SPM image of HbBv–HSA_3 bound on mica surface in PBS (pH 7.4) solution at 25 °C. (**B**) A selected class sum image of HbBv–HSA_3. (**C**) Spatial molecular view of HbBv–HSA_3 derived from a 3D volume reconstruction. Color code: HbBv, *red*; HSAs, *chartreuse*, *lemon*, and *pale green*. Hemes (**C**, *gray*) in HbBv, Cys-34 (S, *yellow*) in HSA

9.3 O_2-Binding Property of Hemoglobin–Albumin Cluster

The deoxy, oxy, and carbonyl forms of HbBv–HSA_3 in PBS solution under N_2, O_2, and CO atmospheres, respectively, showed identical absorption spectra to the corresponding forms of naked HbBv (Fig. 9.4) (Hosaka et al. 2014; Antonini and Brunori 1971). We reasoned that the electronic states of the prosthetic heme groups in HbBv were unaltered by the covalent linkages of HSAs.

The O_2 affinity (P_{50}: O_2 partial pressure where Hb is half-saturated with O_2) and the cooperativity coefficient (Hill coefficient: n) of the HbBv–HSA_3 were measured using an automatic recording system for blood O_2 equilibrium curve (Hemox Analyzer). The P_{50} of native HbBv was 23 Torr at 37 °C, whereas the value of HbBv–HSA_3 was 9 Torr (Table 9.1) (Hosaka et al. 2014). The n value decreased from 2.6 to 1.5. The results imply that the HbBv–HSA_3 shows higher O_2 affinity than HbBv does. The P_{50} and n value reductions were also seen in HbA–HSA_3 (P_{50} = 8 Torr, n = 1.4) (Table 9.1) (Kimura et al. 2015). There are two possible explanations for the increase of O_2 affinity and decrease in cooperativity. The first is the binding of the maleimide terminal of SMCC to Cys-93(β) in HbBv and HbA. Modification of the sulfhydryl group of Cys-93(β), which is located nearby the proximal His-92(β) coordinated to the heme (Mueser et al. 2000), is known to enhance the O_2 affinity (Manjula et al. 2003; Li et al. 2009; Zhang et al. 2008). Furthermore, it reduces the available motion of the $\alpha_1\beta_1/\alpha_2\beta_2$ interface and induces

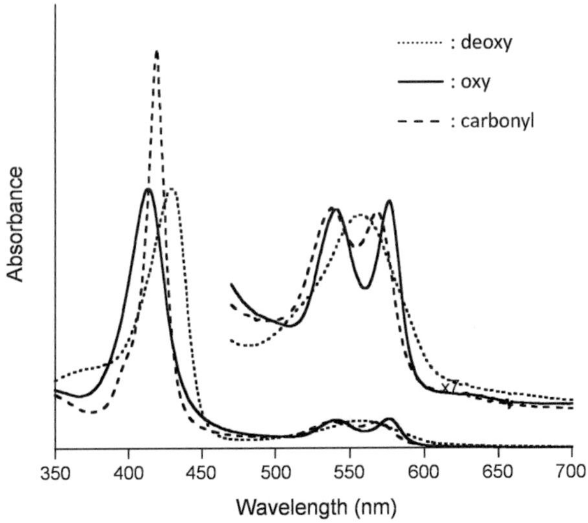

Fig. 9.4 Visible absorption spectral changes of the HbBv–HSA_3 in PBS solution (pH 7.4) at 25 °C

Table 9.1 O_2-binding parameters of HbBv–HSA_3 and HbA–HSA_3 in PBS solution (pH 7.4) at 37 °C

	P_{50} (Torr)	n	Ref.
HbBv–HSA_3	9	1.5	Hosaka et al. (2014)
HbBv–HSA_4	9	1.5	Yamada et al. (2016)
HbA–HSA_3	8	1.4	Kimura et al. (2015)
HbA(T)–HSA_3	26	1.2	Kimura et al. (2015)
$\alpha\alpha$HbA(T)–HSA_3	35	1.4	Kimura et al. (2015)
HbBv	23	2.6	Hosaka et al. (2014)
HbA	12	2.4	Kimura et al. (2015)
	12[a]	2.4[a]	Elmer et al. (2012)
RBC	25	2.5	Kimura et al. (2015)

[a]In Hemox buffer (pH 7.4), 37 °C

disturbance of the quaternary structure of Hb from the Tense (T)-state to the Relaxed (R)-state (Zhang et al. 2008). In fact, the number of cysteinyl thiols per HbBv decreased from 2.0 to 0.2 after the SMCC reaction, indicating that two Cys-93(β) of HbBv are blocked by SMCC maleimide. The second reason is the modifications of surface Lys groups of HbBv or HbA by succinimide terminal of SMCC. They are needed to create the cluster, but the chemical modifications of Lys groups on Hb influence the O_2 affinity (Kluger and Zhang 2003; Hu and Kluger 2008; Vandegriff et al. 2003). In particular, Lys-82(β) plays a key role to modulate the quaternary structural change from the T-state to R-state of Hb. Our 3D reconstruction suggested that Lys-82(β) is a binding partner of Cys-34 of HSA (Tomita et al. 2013;

Kimura et al. 2015). It can be concluded that (i) the masking of Cys-93(*β*) increases the O_2 affinity and (ii) modification of surface Lys groups locks the R-state configuration of the central HbBv or HbA, resulting in the decrease of O_2-binding cooperativity. Interestingly, HbBv–HSA_4, which is hemoglobin–albumin cluster bearing four HSA units (large-size variant), showed the same O_2-binding parameters as HbBv–HSA_3 (Yamada et al. 2016). It implies that the central Hb conformation is independent of the binding number of HSA.

9.4 Preclinical Safety of Hemoglobin–Albumin Cluster

The viscosity of the HbBv–HSA_3 solution (20 g/dL, [Hb] = 5.0 g/dL) is dependent on the shear rate, namely, a Newtonian fluid (Haruki et al. 2015). The viscosity at 230 s^{-1}, the shear rate in the human arterial wall, was 2.8 cP, which is lower than that of blood (3.8 cP). A mixture solution of freshly drawn whole blood and HbBv–HSA_3 (1/1, v/v) showed non-Newtonian viscosity, which obeyed a nonlinear correlation to the shear rate. The viscosity was reasonably high: 3.3 cP at 230 s^{-1}. Furthermore, we counted the number of blood cell components [RBC, white blood cell (WBC), and platelet (PLT)] of the blood/HbBv–HSA_3 mixture solution in vitro (Haruki et al. 2015). The numbers of RBC, WBC, and PLT decreased in proportion to their respective dilution ratios. These results indicate that HbBv–HSA_3 has good compatibility with whole blood.

The HbBv–HSA_3 solution (20 g/dL) was injected into anesthetized rats (6 mL/kg) and observed their mean arterial pressure (MAP) (Haruki et al. 2015). Notably, a small transient alternation in MAP was observed after administration of HbBv–HSA_3 (Fig. 9.5). The slight elevation of ΔMAP (25.3 ± 2.9 mmHg) from the basal value was followed by a decrease to 10 mmHg and retained constant during the monitoring time. The response is almost identical to that observed after infusion of HSA (20 g/dL). On the contrary, the administration of ββ-cross-linked HbBv (XLHbBv, 5 g/dL) is associated with an acute increase in ΔMAP (55.5 ± 5.9 mmHg) and urinary excretion of Hb from 10 min after the injection.

This non-vasopressor response of HbBv–HSA_3 is attributed to the negative surface net charge and high molecular weight of the cluster. HSA shows low vascular permeability of less than 1/100 Hb because of the electrostatic repulsion between the albumin surface and the glomerular basement membrane around the endothelial cells (Haraldsson et al. 2008). The isoelectric point of HbBv–HSA_3 (pI = 5.1) is close to that of HSA. Furthermore, the molecular weight of HbBv–HSA_3 (26.4 kDa) is much greater than that of HSA (66.5 kDa). Thus, the release of HbBv–HSA_3 into the extravascular space was attenuated. In contrast, the small XLHbBv having neutral surface charge passes through the vascular endothelium and contributes to the consumption of NO. Moreover, XLHbBv passes through the renal glomerulus, thereby inducing excretion of Hb in urine.

The ^{125}I-labeled HbBv–HSA_3 was injected into rats to evaluate blood retention (Haruki et al. 2015). The ^{125}I-labeled native HbBv was cleared rapidly from

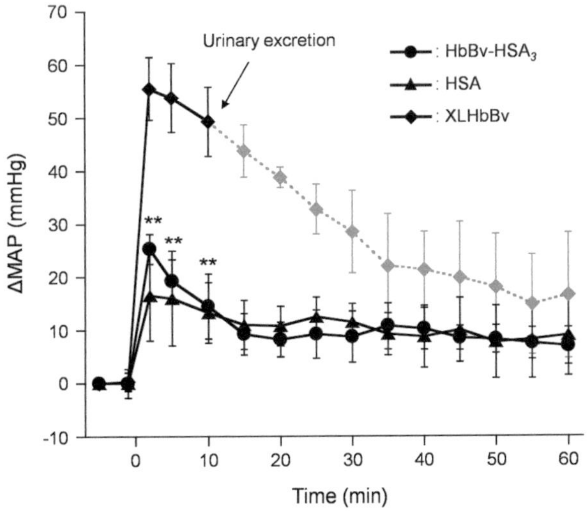

Fig. 9.5 Difference of mean arterial pressure (ΔMAP) from the basal value after intravenous administration of HbBv–HSA$_3$, HSA, and XLHbBv solutions to rats. Each data point represents the mean ± SD (n=4). **p<0.01 vs. XLHbBv. Basal values are 84.5±4.4 mmHg in the HbBv–HSA$_3$ group, 85.0±7.3 mmHg in the HSA group, and 87.5±5.1 mmHg in the XLHbBv group

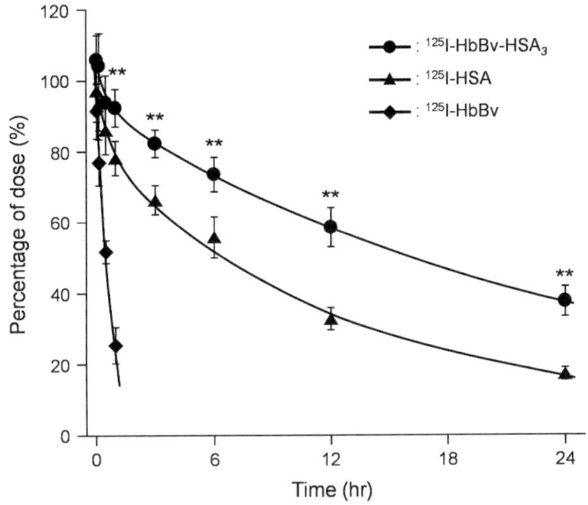

Fig. 9.6 Relative plasma concentration of ^{125}I–HbBv–HSA$_3$, ^{125}I–HSA, and ^{125}I–HbBv after intravenous administration to rats. Each data point represents the mean ± SD (n=6). **p<0.01 vs. ^{125}I–HSA

circulation with the half-life ($T_{1/2}$) of 0.53 h (Fig. 9.6). On the one hand, the time course of HbBv–HSA$_3$ demonstrated very slow kinetics. The $T_{1/2}$ of HbBv–HSA$_3$ was significantly long (18.5 h) and 1.7-fold greater than that of HSA ($T_{1/2}$ = 11.0 h). The negative surface net charge and large molecular size of HbBv–HSA$_3$ prevent filtration by the renal glomerulus. We reasoned that the superior blood retention property of HbBv–HSA$_3$ is attributable to suppression of movement to the extravascular space and renal filtration. All parameters of HbBv–HSA$_4$ were comparable to those of HbBv–HSA$_3$ (Yamada et al. 2016). The HSA-binding number on Hb is ineffective to extend the circulation persistence.

All animals injected with HbBv–HSA$_3$ solution (20 g/dL, 6 mL/kg) were alive for 7 days (Haruki et al. 2015). No remarkable change was found in their appearance or behavior during the measurement time. The body weight increased gradually thereafter. The 26 analytes of the serum biochemical tests after 7 days from the administration showed almost identical data to those of the control groups (HSA injection group, sham-operated group). Microscopic observations of the stained specimens of major organs (liver, kidney, spleen, lungs, and heart) showed no histopathologic disorder in their tissues.

9.5 Various Hemoglobin–Albumin Cluster Derivatives

9.5.1 Low O_2 Affinity Model

In general, HBOCs possess high O_2 affinity compared to RBC (Nagababu et al. 2002; Kluger and Zhang 2003; Hu and Kluger 2008; Li et al. 2009). High O_2 affinity is inferred (i) to prevent the transport of a sufficient amount of O_2 to tissues under physiological conditions, but (ii) to avoid early O_2 offloading on the arterial side of circulation, which may be beneficial for targeted O_2 delivery to the hypoxic regions (Rohlfs et al. 1998; Intaglietta 2004; Winslow 2003; Zhang and Palmer 2010). One of the interesting challenges of artificial O_2 carrier is to prepare a novel HBOC with controllable O_2-binding affinity. It could become a promising RBC substitute and O_2-providing therapeutic reagent for clinical situations.

The heterobifunctional cross-linker SMCC binds not only to the amino groups of Lys on HbA but also to the sulfhydryl group of Cys-93(β). As a result, the HbA–HSA$_3$ showed high O_2 affinity as described before. In our synthesis, the HbA is kept in the carbonyl form to prevent the autoxidation of the hemes. It is known that the Cys-93(β) in deoxygenated T-state HbA is less accessible to cross-linking agents (Buehler et al. 2006). We found that SMCC cannot bind to the Cys-93(β) of deoxy HbA under N_2 atmosphere. As expected, the cluster prepared in N_2, HbA(T)–HSA$_3$, showed lower O_2 affinity (P_{50} = 26 Torr) than the native HbA (12 Torr) does (Table 9.1) (Kimura et al. 2015). The n value of HbA(T)–HSA$_3$ was 1.2, indicating a loss of O_2-binding cooperativity. We reasoned that the HbA center was locked in the T-state conformation by the binding of SMCC under N_2 atmosphere. Interestingly,

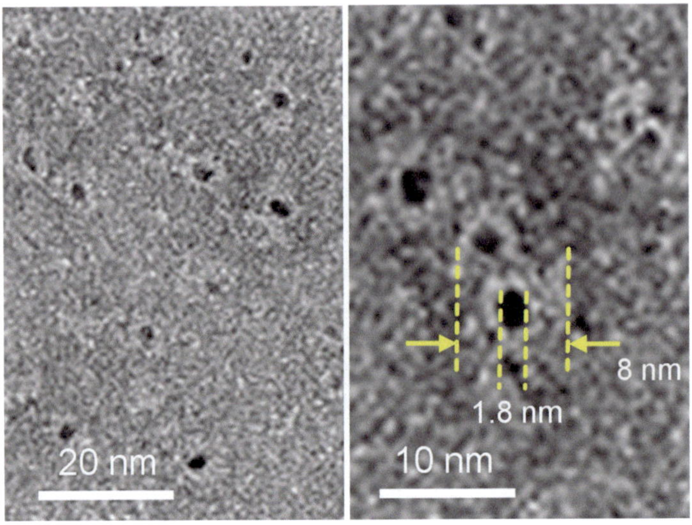

Fig. 9.7 TEM images of HSA–PtNP complexes

the cluster including an $\alpha\alpha$-cross-linked Hb with bis(3,5-dibromosalicyl)fumarate (DBBF), $\alpha\alpha$HbA(T), demonstrated markedly low O_2 affinity (P_{50} = 35 Torr): lower than that of human RBC (Table 9.1) (Kimura et al. 2015). We inferred that the T-state conformation of the $\alpha\alpha$HbA(T) core was preserved strongly by chemical modification of the surface Lys groups. These HbA–HSA_3 clusters with different O_2 affinities can support a new generation of RBC substitute that is better tuned to a role in O_2 delivery.

9.5.2 *Antioxidation Model*

If one can confer an additional functionality to the external HSA unit of Hb–HSA_3, it would become a promising O_2 carrier with high performance. In this context, we designed to add antioxidant property to HbBv–HSA_3. Pt nanoparticle (PtNP) is known to act as an effective catalysis for both $O_2^{\bullet-}$ and H_2O_2 dismutations (Kajita et al. 2007; Hamasaki et al. 2008; San et al. 2012) and shows almost no cytotoxicity against cells (Hamasaki et al. 2008). We found that small PtNP (diameter: approximately 1.8 nm) is incorporated into HSA. TEM images demonstrated the formation of equivalent complex of HSA and PtNP (Fig. 9.7) (Hosaka et al. 2014). Close inspections of TEM micrographs revealed that each PtNP is incorporated in the center of the protein. We reasoned that one PtNP binds to the positively charged cleft of HSA, forming a 1:1 HSA–PtNP complex. The obtained HSA–PtNP complex showed superoxide dismutase (SOD) ($O_2^{\bullet-}$ dismutation) activity and catalase (H_2O_2 dismutation) activity with high efficiency (Hosaka et al. 2014). The IC_{50}

value (the concentration of enzyme necessary to attain 50 % inhibition of the Cyt. *c* reduction) of the HSA–PtNP complex was 0.16 μM, which resembled the value of native Cu, Zn-SOD (Weser and Schubotz 1981). The HSA–PtNP complex possesses a strong capability to catalyze the dismutation of $O_2^{\bullet -}$. The catalase activity of the HSA–PtNP complex was determined by measuring the H_2O_2 decomposition. The T_{50} value (time required for quenching half of H_2O_2) of HSA–PtNP was 19 min, which is two orders of magnitude larger than that of native catalase.

The HbBv–HSA$_3$ also possesses the capability of binding PtNP into the HSA shells (Hosaka et al. 2014). The *K* value and binding number of PtNP with the exterior HSA unit were 1.1×10^7 M^{-1} and 1.1. The resultant HbBv–HSA$_3$(PtNP) cluster forms a very stable O_2 adduct, even in aqueous H_2O_2 (20 μM) solution. We can conclude that the HSA–PtNP shell acts as an efficient scavenger for external H_2O_2 and achieves protection of the core HbBv.

The similar HbBv–HSA$_3$ derivative with high resistance toward oxidation reactions was prepared by incorporation of Mn(II)-protoporphyrin IX into the exterior HSA units (Daijima and Komatsu 2014). These artificial O_2 carriers having triple functionalities (O_2 transport, $O_2^{\bullet -}$ dismutation, H_2O_2 dismutation) might be useful in clinical conditions with ischemia–reperfusion.

9.6 Conclusion

Covalently wrapping of HbBv or HbA with the most abundant plasma protein, HSA, generated a core–shell structured protein cluster as a promising O_2 carrier for RBC substitute. The cluster was prepared by covalent linkage between Hb's Lys and HSA's Cys-34 using heterobifunctional cross-linker. Major products were isolated using gel filtration chromatography, and the average HSA/Hb ratio of the product was 3.0 ± 0.2. We designated the clusters as HbBv–HSA$_3$ and HbA–HSA$_3$. The low isoelectric points (*p*I = 5.1) of the clusters were almost equal to that of HSA, proving the covering of Hb core by negatively charged HSA. The 3D reconstruction of HbBv–HSA$_3$ based on TEM images revealed a complete triangular structure. The possible spatial arrangement of the HbBv center and HSA exteriors was determined. The HbBv–HSA$_3$ and HbA–HSA$_3$ showed higher O_2 affinity (P_{50} = 9 Torr) than the native Hbs. The viscosity measurements and blood cell counting measurements of the mixture solution of whole blood and HbBv–HSA$_3$ revealed the high blood compatibility of this O_2-carrier protein. The administration of HbBv–HSA$_3$ to anesthetized rats caused a slight change in MAP, which is identical to that observed in the control group with HSA. This hemodynamic response contrasts against the acute hypertension occurred after infusion of XLHbBv. The $T_{1/2}$ of HbBv–HSA$_3$ was 1.7-fold longer than that of HSA. The non-vasopressor response and superior blood retention property of HbBv–HSA$_3$ are attributable to the negative surface net charge and larger molecular weight of the cluster. The serum biochemical parameters resembled those of the control groups. Histopathologic inspections proved that HbBv–HSA$_3$ gave no negative side effects in any major organ. These results support the preclinical safety of the HbBv–HSA$_3$ solution. Clusters prepared under N_2 atmosphere showed low O_2 affinity (P_{50} = 26 Torr). Moreover, the cluster containing an $\alpha\alpha$-cross-linked HbA possessed markedly low O_2 affinity (P_{50} = 35 Torr). A PtNP binds within a cleft of HSA, yielding a stable HSA–PtNP complex. This platinated protein showed high $O_2^{-\bullet}$ and H_2O_2 dismutation activities. The HbBv–HSA$_3$ also captured PtNP into the HSA units. The obtained HbBv–HSA$_3$(PtNP) cluster formed very stable O_2 complex even in aqueous H_2O_2 solution. All the results indicate that a series of hemoglobin–albumin clusters can be of tremendous medical importance as an alternative material to RBCs for transfusion in many clinical situations.

Acknowledgments This work was supported by a Grant-in-Aid for Scientific Research on Innovative Area ("Coordination Programming" Area 2107, No. 21108013) from MEXT Japan, a Chuo University Grant for Special Research, and a Joint Research Grant from the Institute of Science and Engineering, Chuo University. The author acknowledges Prof. Mitsutomo Kohno (Tokai University), Dr. Kazuaki Taguchi, Prof. Masaki Otagiri (Sojo University), and Prof. Toru Maruyama (Kumamoto University) for their great supports and valuable comments on animal experiments, and Dr. Christoph Böttcher (Freie Universität Berlin) for his skillful experiments related to TEM measurements and 3D reconstruction.

References

Alagic A, Koprianiuk A, Kluger R (2005) J Am Chem Soc 127:8036–8043

Antonini E, Brunori M (1971) Hemoglobin and myoglobin in their reactions with ligands. In: Neuberger A, Tatum EL (eds) North-Holland research monographs, vol 21, Frontiers of Biology. North-Holland, Amsterdam, pp 13–39

Buehler PW, Boyskins RA, Jia Y, Norris S, Freedberg DI, Alayash AI (2005) Structural and functional characterization of glutaraldehyde-polymerized bovine hemoglobin and its isolated fractions. Anal Chem 77:3466–3478

Buehler PW, Boykins RA, Norris S, Alayash AL (2006) Anal Chem 78:4634–4641

Curry S, Mandelkow H, Brick P, Franks N (1998) Crystal structure of human serum albumin complexed with fatty acid reveals an asymmetric distribution of binding sites. Nat Struct Biol 5:827–835

D'Agnilloo F, Chang TMS (1998) Nat Biotechnol 16:667–671

Daijima Y, Komatsu T (2014) Haemoglobin wrapped covalently by human serum albumin mutants containing Mn(III) protoporphyrin IX: an O_2 complex stable in H_2O_2 solution. Chem Commun 50:14716–14719

Doherty DH, Doyle MP, Curry SR, Vali RJ, Fattor TJ, Olson JS, Lemon DD (1998) Rate of reaction with nitric oxide determines the hypertensive effect of cell-free hemoglobin. Nat Biotechnol 16:672–676

Elmer J, Zorc K, Rameez S, Zhou Y, Cabrales P, Palmer AF (2012) Hypervolemic infusion of *Lumbricus terrestris* erythrocruorin purified by tangential-flow filtration. Transfusion 52:1729–1740

Hamasaki T, Kashiwagi T, Imada T, Nakamichi N, Aramaki S, Toh K, Morisawa S, Shimakoshi H, Hisaeda Y, Shirahata S (2008) Kinetic analysis of superoxide radical-scavenging and hydroxyl radical-scavenging activities of platinum nanoparticles. Langmuir 24:7354–7364

Haraldsson B, Nyström J, Deen WM (2008) Properties of the glomerular barrier and mechanisms of proteinuria. Physiol Rev 88:451–487

Haruki R, Kimura T, Iwsaki H, Yamada K, Kamiyama I, Kohno M, Taguchi K, Nagao S, Maruyama T, Otagiri M, Komatsu T (2015) Safety evaluation of hemoglobin-albumin cluster "HemoAct" as a red blood cell substitute. Sci Rep 5:12778, 1–9

Hosaka H, Haruki R, Yamada K, Böttcher C, Komatsu T (2014) Hemoglobin–albumin cluster incorporating a Pt nanoparticle: artificial O_2 carrier with antioxidant activities. PLoS ONE 9:e110541, 1–9

Hu D, Kluger R (2008) Functional cross-linked hemoglobin bis-tetramers: geometry and cooperativity. Biochemistry 47:12551–12561

Intaglietta M (2004) Microvascular transport factors in the design of effective blood substitutes. In: Messmer K, Burhop KE, Hutter J (eds) Microcirculatory effects of hemoglobin solutions. Karger AG, Basel, pp 8–15

Jahr JS, Sadighi A, Doherty L, Li A, Kim HW (2011) Hemoglobin-based oxygen carriers: history, limits, brief summary of the state of the art, including clinical trials. In: Bettati S, Mozzarelli A (eds) Chemistry and biochemistry of oxygen therapeutics: from transfusion to artificial blood. Wiley, West Sussex, pp 301–316

Kajita M, Hikosaka K, Iitsuka M, Kanayama A, Toshima N, Miyamoto Y (2007) Platinum nanoparticle is a useful scavenger of superoxide anion and hydrogen peroxide. Free Radic Res 41:615–626

Kimura T, Shinohara R, Böttcher C, Komatsu T (2015) Core-shell clusters of human haemoglobin A and human serum albumin: artificial O_2-carriers having various O_2-affinities. J Mater Chem B 3:6157–6164

Kluger R (2010) Red cell substitutes from hemoglobin – do we start all over again? Curr Opin Chem Biol 14:538–543

Kluger R, Lui FE (2013) HBOCs from chemical modification of Hb. In: Kim HW, Greenburg AG (eds) Hemoglobin-based oxygen carriers as red cell substitutes and oxygen therapeutics. Springer, Berlin, pp 159–183

Kluger R, Zhang J (2003) Hemoglobin dendrimers: functional protein clusters. J Am Chem Soc 125:6070–6071

Li D, Hu T, Manjula BN, Acharya SA (2008) Non-conservative surface decoration of hemoglobin: influence of neutralization of positive charges at PEGylation sites on molecular and functional properties of PEGylated hemoglobin. Biochim Biophys Acta 1784:1395–1401

Li D, Hu T, Manjula BN, Acharya SA (2009) Extension arms facilitated pegylation of αα-hemoglobin with modifications targeted exclusively to amino groups: functional and structural advantages of free Cys-93(β) in the PEG-Hb adduct. Bioconjug Chem 20:2062–2070

Manjula BN, Tsai A, Upadhya R, Perumalsamy K, Smith K, Malavalli A, Vandegriff K, Winslow RM, Intaglietta M, Prabhakaran M, Friedman JM, Acharya AS (2003) Site-specific PEGylation of hemoglobin at Cys-93(β): correlation between the colligative properties of the PEGylated protein and the length of the conjugated PEG chain. Bioconjug Chem 14:464–472

Ministry of Health, Labor and Welfare, Japan (2014) Proceedings of blood donation promotion committee, pharmaceutical affairs and food sanitation council on 2 December, 2014. http://www.mhlw.go.jp/file/05-Shingikai-11121000-Iyakushokuhinkyoku-Soumuka/0000067177.pdf. Accessed 30 Dec 2015

Mondery-Pawlowski CL, Tian LL, Pan V, Gupta AS (2013) Synthetic approaches to RBC mimicry and oxygen carrier systems. Biomacromolecules 14:939–948

Mueser TC, Rogers PH, Arnone A (2000) Interface sliding as illustrated by the multiple quaternary structures of liganded hemoglobin. Biochemistry 39:15353–15364

Nagababu E, Ramasamy S, Rifkind JM, Jia Y, Alayash AI (2002) Site-specific cross-linking of human and bovine hemoglobins differentially alters oxygen binding and redox side reactions producing rhombic heme and heme degradation. Biochemistry 41:7407–7415

Natanson C, Kern SJ, Lurie P, Banks SM, Wolfe SM (2008) Cell-free hemoglobin-based blood substitutes and risk of myocardial infarction and death. J Am Med Assoc 299:2304–2312

Pearce LB, Gawryl MS, Rentko VT, Moon-Massat PF, Rausch CW (2006) HBOC-201 (hemoglobin glutamer-250) (bovine), hemopure®): clinical studies. In: Winslow RM (ed) Blood substitutes. Elsevier, San Diego, pp 437–450

Rohlfs RJ, Bruner E, Chiu A, Gonzales A, Gonzales ML, Magde D, Magde MD Jr, Vandegriff KD, Winslow RM (1998) Arterial blood pressure responses to cell-free hemoglobin solutions and the reaction with nitric oxide. J Biol Chem 273:12128–12134

Sakai H (2012) Present situation of the development of cellular-type hemoglobin-based oxygen carrier (hemoglobin-vesicles). Curr Drug Discov Technol 9:188–193

San BH, Moh SH, Kim KK (2012) The effect of protein shells on the antioxidation activity of protein-encapsulated platinum nanoparticles. J Mater Chem 22:1774–1780

Shultz SC, Grady B, Cole F, Hamilton I, Burhop K, Malcolm DS (1993) A role for endothelin and nitric oxide in the pressor response to diaspirin cross-linked hemoglobin. J Lab Clin Med 122:301–308

Snyder SR, Welty EV, Walder RY, Williams LA, Walder JA (1987) HbXL99α: a hemoglobin derivative that is cross-linked between the α subunits is useful as a blood substitute. Proc Natl Acad Sci U S A 84:7280–7284

Squires JE (2002) Artificial blood. Science 295:1002–1005

Tomita D, Kimura T, Hosaka H, Daijima Y, Haruki R, Böttcher C, Komatsu T (2013) Covalent core–shell architecture of hemoglobin and human serum albumin as an artificial O_2 carrier. Biomacromolecules 14:1816–1825

Vandegriff KD, Malavalli A, Wooldbridge J, Lohman W, Winslow RM (2003) MP4, a new nonvasoactive PEG-Hb conjugate. Transfusion 43:509–516

Weser U, Schubotz LM (1981) Imidazole-bridged copper complexes as Cu_2Zn_2-superoxide dismutase models. J Mol Catal 13:249–261

Winslow RM (2003) Current status of blood substitute research: towards a new paradigm. J Intern Med 253:508–517

Xiong Y, Liu ZZ, Georgieva R, Smuda K, Steffen A, Sendeski M, Voigt A, Patzak A, Bäumler H (2013) Nonvasoconstrictive hemoglobin particles as oxygen carriers. ACS Nano 7:7454–7461

Yamada K, Yokomaku K, Haruki R, Taguchi K, Nagao S, Maruyama T, Otagiri M, Komatsu T (2016) Influence of molecular structure on O_2-binding properties and blood circulation of hemoglobin-albumin clusters. PLoS ONE 11:e0149526, in press

Zhang N, Palmer AF (2010) Polymerization of human hemoglobin using the crosslinker 1,11-bis(maleimido)triethylene glycol for use as an oxygen carrier. Biotechnol Prog 26:1481–1485

Zhang Y, Bhatt VS, Sun G, Wang PG, Palmer AF (2008) Site-selective glycosylation of hemoglobin on Cys 93. Bioconjug Chem 19:2221–2230

Chapter 10
The Influence of FcRn on Albumin-Fused and Targeted Drugs

Jeannette Nilsen, Kine Marita Knudsen Sand, Malin Bern, Peng Lei, Inger Sandlie, and Jan Terje Andersen

Abstract Albumin escapes intracellular degradation by binding to the neonatal Fc receptor (FcRn), which results in a very long serum half-life of nearly 3 weeks in humans. The broadly expressed FcRn is unique in that it binds both its ligands, immunoglobulin G (IgG) and albumin, in a strictly pH-dependent fashion, and this has proven to be fundamental for rescue from degradation. Further, elucidation of the biology of FcRn as well as its relationship with albumin is necessary to obtain a better understanding of how albumin homeostasis is regulated. This will be of great importance for optimal applications of albumin as a therapeutic molecule. Indeed, albumin is attracting increasing interest as it is utilized to extend the serum half-life of drugs and improve pharmacokinetics. We review the current status of albumin-based therapeutics in light of FcRn biology and the prospect of a new generation of albumin molecules with improved binding to FcRn.

Keywords FcRn • Albumin • The FcRn-albumin interaction • FcRn recycling • Albumin-based therapeutics • Albumin engineering • Half-life

10.1 The Discovery That FcRn Protects Albumin

Long before a mechanistic explanation was provided, it was recognized that albumin features a half-life that is exceptionally long relative to that of other serum proteins and that also correlates with its serum concentration. This was demonstrated in studies conducted in the 1950–1970s, when the half-life of radiolabeled albumin injected

J. Nilsen
Centre for Immune Regulation (CIR) and Department of Immunology, Oslo University Hospital Rikshospitalet, Oslo, Norway

Institute of Clinical Medicine, University of Oslo, Oslo, Norway

K.M.K. Sand • M. Bern • P. Lei • I. Sandlie • J.T. Andersen (✉)
Centre for Immune Regulation (CIR) and Department of Immunology, Oslo University Hospital Rikshospitalet, Oslo, Norway

Department of Biosciences, University of Oslo, Oslo, Norway
e-mail: j.t.andersen@medisin.uio.no

M. Otagiri, V.T.G. Chuang (eds.), *Albumin in Medicine*,
DOI 10.1007/978-981-10-2116-9_10

into humans with abnormally low albumin levels was estimated to be 50–100 days (Bennhold and Kallee 1959; Cormode et al. 1975; Gordon et al. 1959). The concentration-catabolism relationship was first revealed for IgG (Fahey and Robinson 1963), the only serum protein to share this property. In 1964, Brambell and colleagues proposed the now well-accepted hypothesis that IgG is protected from degradation by a saturable receptor-mediated mechanism (Brambell et al. 1964). Then, 30 years later, the receptor, named FcRn, identified by Simister and Rees (Simister and Rees 1985) was shown to protect IgG from degradation, when reduced levels of circulating IgG were observed in mice deficient for the β2-microglobulin (β_2m) subunit of FcRn (Israel et al. 1996; Junghans and Anderson 1996).

In 1966, Schultze and Heremans postulated that the same mechanism is operating for albumin, based on studies of the relative catabolic rates of IgG and albumin in patients suffering from agammaglobulinemia and analbuminemia (Schultze and Heremans 1966). Yet, further progress followed first in 2003 when Anderson and co-workers reported that FcRn actually binds albumin (Chaudhury et al. 2003). The interaction was discovered by chance when bovine serum albumin was co-eluted with recombinant soluble human FcRn (hFcRn) from an IgG affinity column. In the same report, mice genetically engineered to lack expression of the receptor, either having a defective FcRn heavy chain (HC) or β2m gene, were used to provide evidence for the involvement of FcRn in regulating albumin homeostasis. In both mouse strains, the serum concentration of albumin were measured to be about 40 % of normal levels, and in line with this, the half-life of injected albumin in these mice was shorter than in wild-type mice (Chaudhury et al. 2003). The role of FcRn in protection from degradation, and maintaining high serum levels, was further supported by an observation made in two siblings diagnosed with familial hypercatabolic hypoproteinemia, which showed low expression of functional FcRn due to a single-point mutation in the β2m gene, resulting in very low serum concentrations of both albumin and IgG (Waldmann and Terry 1990; Wani et al. 2006).

10.2 FcRn Gene Expression and Tissue Distribution

FcRn is a heterodimer that consists of a major histocompatibility complex class I-like HC, which has three extracellular domains (α1, α2, and α3) followed by a segment that traverses the membrane and a cytoplasmic tail. The HC is non-covalently associated with the soluble β2m subunit, which is constitutively expressed (Burmeister et al. 1994a, b; Simister and Mostov 1989). The HC-encoding *FCGRT* gene is located on chromosome 19q13 in humans and 7 in mice (Ahouse et al. 1993; Kandil et al. 1996).

There is limited knowledge about the mechanisms that regulate the expression of *FCGRT*. Computational analysis of the *FCGRT* gene cloned from rat, mouse, and human has revealed several putative transcription factor binding sites in the promoter regions (Jiang et al. 2004; Mikulska et al. 2000; Mikulska and Simister 2000; Tiwari and Junghans 2005), and the relevance of these sites in the human promoter region was then demonstrated by their ability to bind their corresponding transcription factors, including stimulating protein 1 (Sp1), Sp2, Sp3, c-Fos, c-Jun, and YY1

(Mikulska 2015). Furthermore, site-directed mutagenesis of the binding motifs altered promoter activity. For example, mutating the Sp1 binding site at nucleotide-313 reduced promoter activity, as did mutating the AP-1 motif in position-276, in both epithelial and endothelial cells, supporting their crucial role in human *FCGRT* gene regulation (Mikulska 2015).

Studies have also shown that FcRn expression may be modulated by the presence of pro-inflammatory substances and cytokines. More specifically, tumor necrosis factor-α (TNF-α) or interleukin-1 (IL-1)-β stimulation of human epithelial cells, macrophage-like THP-1 cells and monocytes was shown to upregulate *FCGRT* gene transcription. Upregulation was also observed when the two latter cell types, both expressing toll-like receptor (TLR) 4 and TLR9, were exposed to lipopolysaccharide or CpG-containing DNA from bacteria. Importantly, the increase in FcRn expression was dependent on nuclear factor-kappaB (NF-kB) activation (Liu et al. 2007). Three identified binding sequences in intron 2 and 4 of the human *FCGRT* gene were shown to interact with the NF-kB p65 and p50 subunits and to modulate the promoter activity in a luciferase gene reporter system, further supporting NF-kB as a regulator of FcRn expression (Liu et al. 2007). Another study revealed that stimulating the same cells (human epithelial cells, macrophage-like THP-1 cells, and monocytes) with interferon (INF)-γ results in downregulation of *FCGRT* expression via the JAK-STAT signaling pathway (Liu et al. 2008).

In rodents, FcRn expression is developmentally regulated in the intestine. During the first weeks of life, the receptor is highly expressed and mediates IgG transfer from mother's milk. It is then rapidly downregulated at the time of weaning (Martin et al. 1997; Israel et al. 1995). Interestingly, administration of thyroxine or corticosteroids to suckling rat pups decreases intestinal FcRn expression and IgG uptake in a dose-dependent manner (Martin et al. 1993). FcRn continues to be expressed in other tissues in adult life (Akilesh et al. 2007, 2008; Borvak et al. 1998; Kim et al. 2008; Montoyo et al. 2009; Schlachetzki et al. 2002; Spiekermann et al. 2002). Indeed, FcRn is expressed by a wide range of tissues and in many different species. More specifically, hFcRn expression has been detected in endothelial cells of the microvasculature of the skin (Ober et al. 2004b), retina (Powner et al. 2014), and placenta (Antohe et al. 2001) and in epithelial cells in the liver (Andersen et al. 2012b), kidneys (Haymann et al. 2000), lung (Spiekermann et al. 2002), eye (Powner et al. 2014), intestine (Dickinson et al. 1999; Israel et al. 1997), epidermis (Cauza et al. 2005; Cianga et al. 2007), mammary gland (Cianga et al. 2003), placenta (Simister et al. 1996), and the female genital tract (Li et al. 2011). Hematopoietic cells, such as monocytes, macrophages, dendritic cells, and neutrophils, have also been reported to express the receptor (Vidarsson et al. 2006; Zhu et al. 2001).

Endothelial cells lining the interior of blood vessels were initially presumed to be the main site for FcRn-mediated rescue, until a study using bone marrow chimeras revealed that both non-hematopoietic and hematopoietic cells are of great importance (Akilesh et al. 2007; Kobayashi et al. 2009). The half-life of injected IgG in wild-type mice transplanted with bone marrow from FcRn-deficient mice was found to be 2.3 days only, which was shorter compared to 5 days measured in wild-type mice with a bone morrow transplant from other wild-type mice. Moreover, in FcRn-deficient mice with transplanted bone marrow from wild-type mice or other FcRn-deficient mice, the half-life of injected IgG was found to be 2.8 days and 1.3 days, respectively (Akilesh et al. 2007). These results demonstrate the involvement of hematopoietic cells in regulating IgG homeostasis. Furthermore, mice conditionally deleted for FcRn in endothelial cells and hematopoietic cells were shown to have four- and twofold reduced serum levels of IgG and albumin, respectively, compared to wild-type mice (Montoyo et al. 2009). This is similar to the observation made in mice globally deleted for FcRn expression (Chaudhury et al. 2003). Thus, in line with the functional expression of FcRn in a wide variety of hematopoietic cell types, it is clear that a significant fraction of IgG and albumin protection is due to the activities of these cells. However, further studies are needed to fully understand the contribution of FcRn in various tissues and organs on homeostatic regulation of IgG and albumin such as the kidneys and the liver. In addition, it would be important to address the regulation of FcRn expression under different conditions.

10.3 FcRn-Mediated Transport in the Cell

Insight into the mechanisms of FcRn-mediated intracellular transport of IgG has been obtained through studies using advanced live cell fluorescence microscopy of epithelial and endothelial cells (Gan et al. 2013; Ober et al. 2004a, b; Prabhat et al. 2007; Ward et al. 2003, 2005; Claypool et al. 2004; Tesar et al. 2006). These studies support a trafficking model for FcRn that includes two main pathways, recycling and transcytosis. Following internalization of IgG and albumin, by fluid-phase pinocytosis or receptor-mediated endocytosis, and endosome acidification, the ligands bind FcRn in endosomal compartments. The complexes are then sorted to the cell surface, where exposure to the neutral pH of the extracellular milieu causes dissociation and release of the ligands. Depending on the cell type, the complexes may be recycled to the cell surface of the side of entry or trancytosed to the opposite side of the cell. In contrast, molecules that do not have affinity for FcRn, or that find FcRn already occupied, are delivered to the lysosomes for degradation. Thus, once FcRn becomes saturated, unbound ligands progress through the lysosomal pathway. Indeed, the fate of endocytosed ligands depends on their concentration in the endosomes, which is directly proportional to the concentration in the serum, and in this way, FcRn acts as a homeostatic regulator. Illustrations of the FcRn-mediated transport pathways are shown in Fig. 10.1a, b.

Fig. 10.1 FcRn mediated recycling and transcytosis. (**a**) The model of FcRn-mediated recycling in an endothelial cell. (1) IgG and albumin enter the cell from the bloodstream through pinocytosis or receptor-mediated endocytosis and are sorted into early endosomes. (2) The ligands bind to FcRn in recycling endosomes at acidic pH. (3) FcRn recycles its ligands back to the plasma membrane. (4) Release of the ligands back into the circulation is triggered by the neutral pH of the blood. (5) Proteins that are not recycled are degraded in lysosomes. (**b**) The model of FcRn mediated transcytosis in an epithelial cell. (1) Acidic pH at epithelial surfaces may trigger binding of IgG and albumin to FcRn at the cell surface. (2) The ligands are sorted into endosomes and to the basolateral side of the cell. (3) Release of the ligands at the basolateral side of the cell is triggered by the neutral pH of the extracellular milieu (The figure is modified from (Sand et al. 2015))

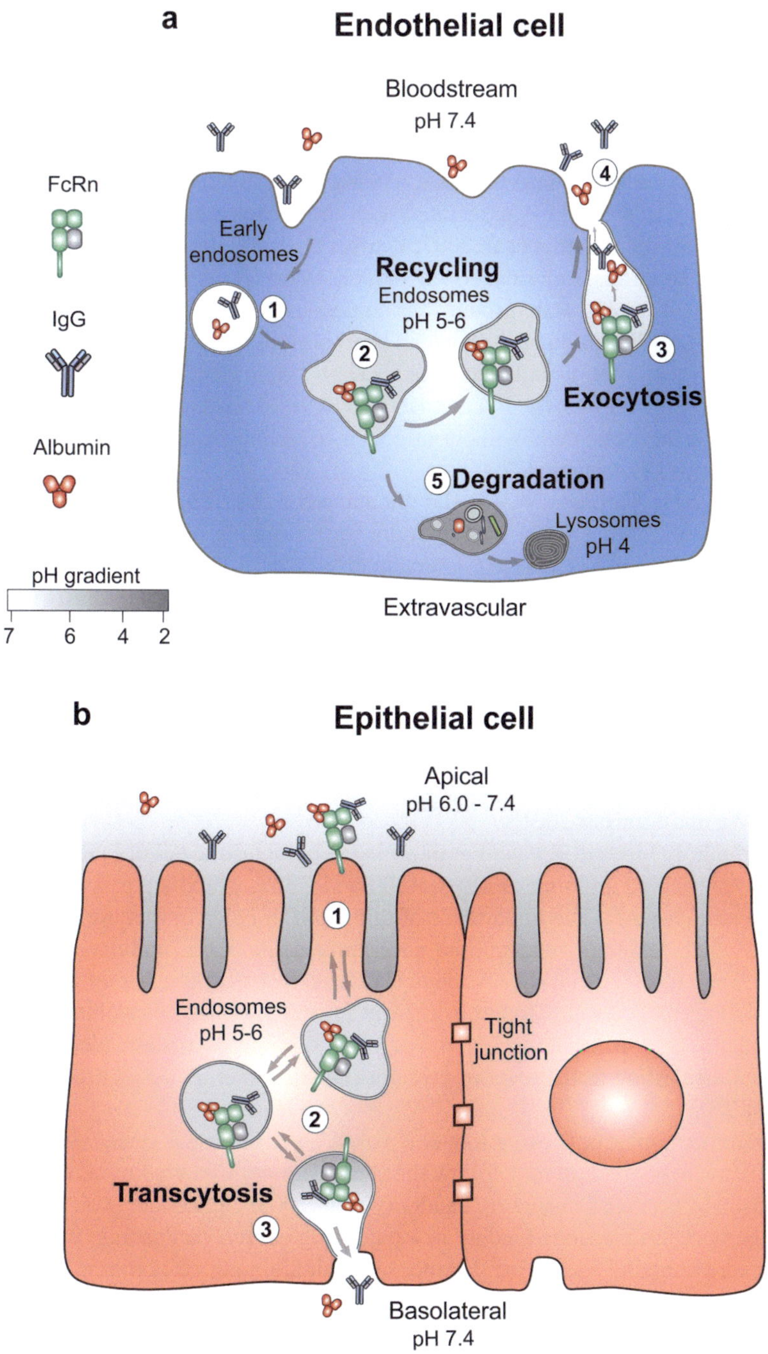
a
Endothelial cell
Bloodstream
pH 7.4
FcRn
IgG
Albumin
Early endosomes
1
Recycling
Endosomes
pH 5-6
2
4
3
Exocytosis
5 Degradation
Lysosomes
pH 4
pH gradient
7
6
4
2
Extravascular
b
Epithelial cell
Apical
pH 6.0 - 7.4
1
Endosomes
pH 5-6
Tight junction
2
Transcytosis
3
Basolateral
pH 7.4

Intracellular transport is regulated by small Ras-like GTPases, the Rab proteins, and in human endothelial cells, Rab4, Rab5, and Rab11 are present on FcRn-containing endosomes (Ward et al. 2005). Rab5 is found in early endosomes, whereas Rab4 and Rab11 are involved in recycling from sorting endosomes to the plasma membrane (Zhen and Stenmark 2015). Ward and colleagues have shown that FcRn can be sorted into tubulovesicular compartments positive for Rab4 and Rab11, while only Rab11 is associated with the receptor during exocytosis at the plasma membrane (Prabhat et al. 2007; Ward et al. 2005). Importantly, these studies support the so-called Rab conversion model, where Rabs are gradually lost and replaced by different Rab proteins as endosomes mature (Rink et al. 2005). Moreover, Rab11 has also been shown to be important for recycling of hFcRn to the basolateral membrane of polarized Madin-Darby canine kidney cells, while being dispensable for transcytosis. In contrast, Rab25 was found to regulate transcytosis of hFcRn in the same cells without affecting recycling from the same endosomal compartments (Tzaban et al. 2009).

Conserved tryptophan (W311) and dileucine (L322/L323) sorting motifs within the cytoplasmic tail of FcRn are important for trafficking of the receptor (Wu and Simister 2001). Rat FcRn W311 interacts directly with the μ2-subunit of adaptor protein-2 (AP-2) (Wernick et al. 2005), while the dileucine motif interacts with the σ- and γ-subunits of the same adaptor protein, which links the membrane protein to the clathrin-coated pits forming in the initial phase of endocytosis (Edeling et al. 2006). Furthermore, substitution of residues in the two motifs with other amino acids affected the efficiency of endocytosis in polarized rat cells, as well as subcellular distribution, revealing their role in rapid endocytosis of the receptor from the plasma membrane into endosome and also in basolateral targeting (Newton et al. 2005; Wu and Simister 2001).

Calmodulin binds to yet another motif in the cytoplasmic tail of FcRn, including the two conserved arginine residues, R300 and R302, which are part of a putative amphipathic α-helix. The importance of this motif in FcRn trafficking was demonstrated when targeted mutagenesis, abolishing calmodulin binding, reduced the half-life and transcytosis of the receptor (Dickinson et al. 2008). Moreover, calmodulin binding to FcRn was found to depend on the presence of calcium, and thus, calmodulin may control intracellular sorting through reversible binding to the receptor, according to the level of calcium (Dickinson et al. 2008; McMahon and Gallop 2005).

The intracellular trafficking model is based on experiments done with IgG as a ligand; however albumin may follow the same pathways, as binding of each ligand occurs at distinct binding sites (Chaudhury et al. 2003; Oganesyan et al. 2014). Yet, evidence for simultaneous binding in a physiological relevant setting, where FcRn is membrane bound, is lacking. Furthermore, albumin has one binding site for the receptor, whereas IgG has two that could potentially crosslink FcRn molecules in the membrane and in turn have an impact on trafficking (Chaudhury et al. 2003; West and Bjorkman 2000). In addition, the presence of other albumin-binding receptors may affect the route taken by albumin in different cell types (Bern et al. 2015). Further studies are needed to fully understand how cellular trafficking of the ligands is orchestrated in different types of cells.

The first set of data that supports FcRn-mediated transcytosis of albumin comes from studies in the kidneys. Proximal tubule cells (PTC) lining the proximal convoluted tubule reabsorbs albumin that enters the glomerular filtrate (Birn and Christensen 2006; Maunsbach 1966). These cells express FcRn (Haymann et al. 2000) as well as the megalin-cubilin complex (Kozyraki et al. 2001), another albumin-binding receptor, and several lines of evidence support a role of both receptors in retrieval of albumin (Sarav et al. 2009; Tenten et al. 2013; Amsellem et al. 2010; Aseem et al. 2014; Birn et al. 2000; Cui et al. 1996). They may work in concert, such that albumin is taken up by receptor-mediated endocytosis after binding to the megalin-cubilin complex (Amsellem et al. 2010; Aseem et al. 2014; Birn et al. 2000; Cui et al. 1996), and delivered to acidified endosomes where binding to FcRn takes place. From here, albumin in complex with FcRn is transcytosed to the basolateral surface of the PTC and directed back to the circulation (Tenten et al. 2013).

In mice genetically modified to express mouse serum albumin (MSA) in the kidney podocytes upon induction with doxycycline, MSA secreted into the filtrate was taken up by PTC and delivered intact into the circulation in an FcRn-dependent manner (Tenten et al. 2013). In yet another study, wild-type mice with a transplanted kidney from FcRn-deficient mice developed hypoalbuminemia, whereas FcRn-deficient mice that received an FcRn-expressing kidney had increased serum albumin levels (Sarav et al. 2009). Furthermore, dogs that have a defective cubilin gene and mice with 90 % reduced expression of cubilin in the kidneys due to conditional Cre-loxP knockdown show reduced renal proximal tubular uptake, increased urinary loss, and decreased albumin serum levels (Amsellem et al. 2010; Birn et al. 2000). As cubilin associates with megalin, both mice and humans deficient in megalin expression also show markedly reduced tubular reabsorption of albumin (Birn et al. 2000; Moestrup et al. 1998; Storm et al. 2013).

Notably, this pathway may function as a selective process where only albumin with intact receptor binding properties are returned to the circulation, while modified albumin ends up in the urine or in intracellular compartments destined for degradation.

10.4 The Nature of the FcRn-Albumin Interaction

The FcRn-albumin interaction has been studied extensively since its first discovery. With the first report came the clue that albumin and IgG bind to recombinant soluble FcRn simultaneously, as bovine serum albumin was co-eluted with hFcRn from an human IgG-coupled affinity column, and like the FcRn-IgG interaction, the FcRn-albumin interaction was remarkably pH dependent (Chaudhury et al. 2003). Since then, site-specific mutagenesis and interactions assays have been central in mapping of the interaction (Andersen et al. 2006, 2010b, 2012a; Chaudhury et al. 2006). Data obtained through such studies, together with close scrutiny of available FcRn and albumin crystal structures, were used to build a docking model of hFcRn in complex with human serum albumin (HSA) (Andersen et al. 2012a), which guided further characterization.

Albumin consists of three homologous domains, named DI, DII, and DIII, where each is composed of A and B subdomains (DIA, DIB, DIIA, DIIB, DIIIA, and DIIIB) (Dockal et al. 1999). Using site-directed mutagenesis, three conserved histidines (H464, H510, and H535) and a lysine in position 500 (K500) within DIII were identified as key residues involved in FcRn-binding (Andersen et al. 2012a). That the C-terminal DIII contains the main binding site for FcRn was first revealed when an HSA variant lacking this domain failed to bind the receptor (Andersen et al. 2010b). In addition, a recombinant DIII was shown to bind FcRn, although with tenfold weaker affinity than full-length HSA, which also suggested that other parts of the molecule contribute to the interaction (Andersen et al. 2012a). Indeed, two surface-exposed loops in the N-terminal DI were found in close proximity to the receptor in the docking model of the hFcRn-HSA complex, and their involvement was subsequently demonstrated when alanine substitution of selected loop residues gave rise to altered FcRn-binding affinity (Andersen et al. 2012a; Sand et al. 2014a). An illustration of a crystal structure of HSA is shown in Fig. 10.2a.

The HC of FcRn has three extracellular domains, α1 and α2, which sits on top of α3 and the non-covalently associated β2m subunit (Burmeister et al. 1994a, b). In the α2-domain of FcRn, a fully conserved histidine at position 166 (H166) was found to be crucial for binding to albumin (Andersen et al. 2006). This was first revealed by the complete loss of binding upon alanine substitution, while a structural explanation that H166 stabilizes a loop in the α1-domain when protonated at low pH was proposed after inspection of two crystal structures of hFcRn (Mezo et al. 2010; West and Bjorkman 2000). Moreover, the fundamental role of four conserved tryptophan residues (W51, W53, W59 and W61) within this loop were demonstrated when targeted mutagenesis resulted in abolished or reduced binding (Sand et al. 2014b; Schmidt et al. 2013). This explained the need for a stabilized loop and emphasized the regulatory role of H166. Interestingly, this confirmed that the interaction has a hydrophobic character, which had been previously predicted using isothermal titration calorimetry (Chaudhury et al. 2006). An illustration of a crystal structure of hFcRn is shown in Fig. 10.2b.

Further insights into the molecular details of the interaction were provided with two reported co-crystal structures of hFcRn in complex with HSA, solved with either wild-type HSA or HSA13, a variant with four amino acid substitutions (V418M, T420A, E505G, and V547A) (Oganesyan et al. 2014; Schmidt et al. 2013).

Fig. 10.2 The crystal structures of HSA, hFcRn and the hFcRn-HSA complex. (**a**) The crystal structure of HSA with its three domains indicated. The major FcRn-interacting amino acids in DI, DIIIA, and DIIIB are highlighted as *ball* and *sticks* (Sugio et al. 1999). (**b**) The crystal structure of the hFcRn HC and the β2m subunit (Mezo et al. 2010). The IgG and the albumin-binding sites are highlighted. Some of the central residues involved in albumin binding are W53, W59, and H166. (**c**) The hFcRn-HSA co-crystal structure (Oganesyan et al. 2014). The main sites of the interaction are shown as close-ups. Residues in HSA DIIIA, T422, V426, L460, L463, H464, and T467 create a hydrophobic pocket for FcRn W59. FcRn W53 makes hydrophobic stacking with F507, F509 and F551 in DIIIB. HSA DI N111 and N109 in loop 2 interact with FcRn residues S58 and K63, respectively, and HSA DI R81 and D89 in loop 1 form an intramolecular hydrogen bond and R81 interacts with FcRn T153 (Oganesyan et al. 2014). The figures were made using PyMol and the crystal structure data of HSA (PDB 1AO6), hFcRn (PDB 3M1B) and hFcRn in complex with HSA (PDB 4N0F)

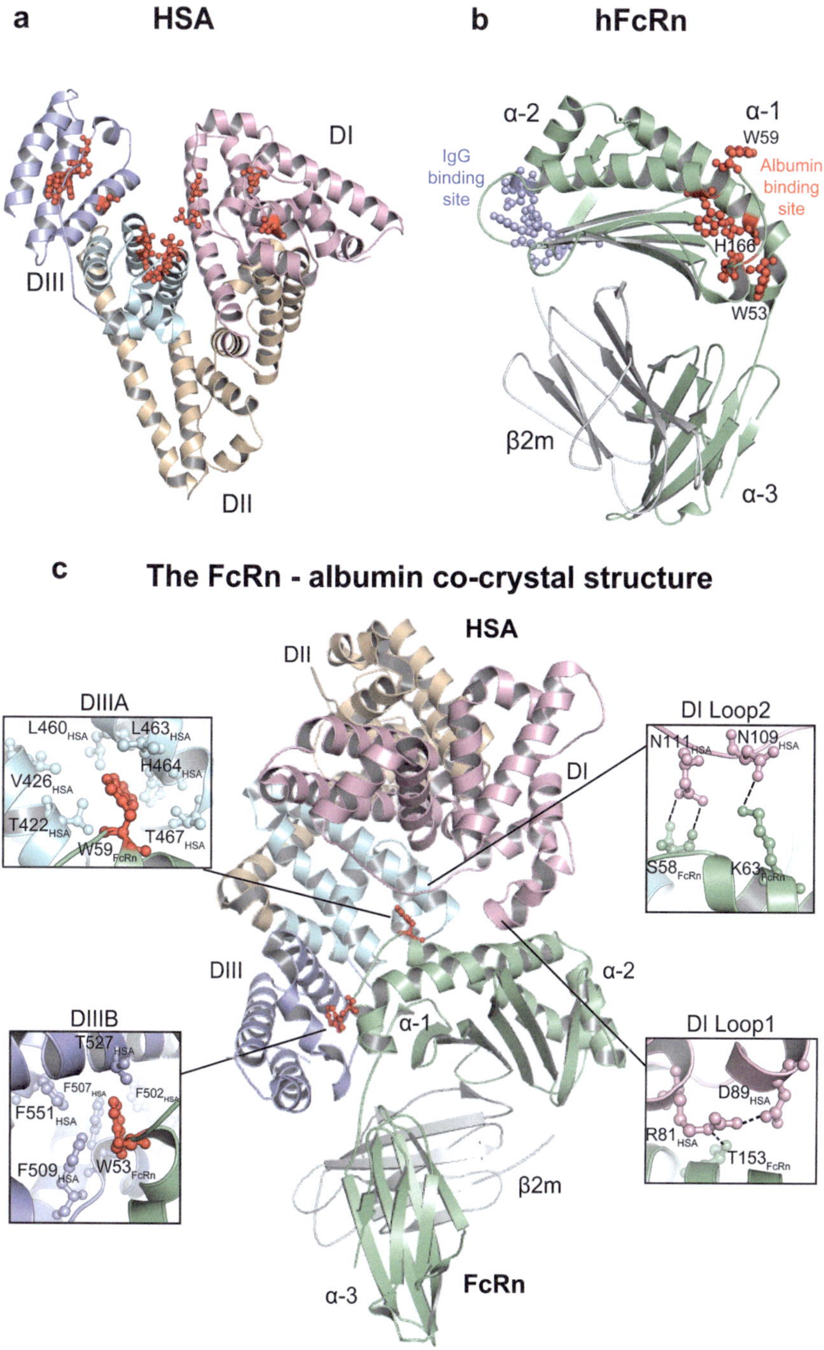
a
HSA
DI
DIII
DII
b
hFcRn
α-2
α-1
W59
IgG binding site
Albumin binding site
H166
W53
β2m
α-3
c
The FcRn - albumin co-crystal structure
HSA
DII
DIIIA
L460HSA
L463HSA
H464HSA
V426HSA
T422HSA
T467HSA
W59FcRn
DI Loop2
N111HSA
N109HSA
S58FcRn
K63FcRn
DI
DIII
α-2
α-1
DIIIB
T527HSA
F507HSA
F502HSA
F551HSA
F509HSA
W53FcRn
DI Loop1
D89HSA
R81HSA
T153FcRn
β2m
α-3
FcRn

The latter was developed using yeast display and features improved binding affinity for FcRn both at acidic and neutral pH (Oganesyan et al. 2014). Both co-crystals confirmed the involvement of both DI and DIII in FcRn binding, and despite some differences between the two co-crystal structures, probably due to the introduced mutations in HSA13, the overall binding modes and the interaction cores are very similar. Notably, an additional crystal structure of hFcRn in complex with wild-type HSA and an engineered fragment crystallizable (Fc) fragment with improved binding to the receptor was also reported, which revealed that the mode of albumin binding is not altered by the presence of IgG (Schmidt et al. 2013). Both co-crystal structures show the two exposed DI-loops in contact with residues of the α1–α2 helices of FcRn (Oganesyan et al. 2014; Schmidt et al. 2013). Furthermore, FcRn-W53 and FcRn-W59 make hydrophobic interactions with residues located within pockets of the subdomains DIIIB (F507, F509, F551, and T527) and DIIIA (T422, V426, L460, L463, and T467), respectively. Engagement of FcRn-W53 requires a conformational change in a loop comprising residue 500–510 that connects DIIIA and DIIIB, which is only stabilized upon protonation of H510 and H535 of albumin at acidic pH, while FcRn-W59 inserts into a hydrophobic area that is sustained by the nearby H464 of HSA (Oganesyan et al. 2014; Schmidt et al. 2013). Collectively, these findings demonstrate that DIII is crucial for pH-dependent binding to FcRn, while DI modulates and stabilizes the interaction. An illustration of the hFcRn-HSA co-crystal structure with close-ups of the interacting residues is shown in Fig. 10.2c

10.5 Albumin in Therapy

Small therapeutic peptides, proteins and chemical drugs have poor therapeutic efficacy due to rapid clearance via the kidney and liver. This may be overcome by taking advantage of the long serum half-life of IgG or albumin (Kratz 2014; Czajkowsky et al. 2012; Elsadek and Kratz 2012; Sockolosky and Szoka 2015). One example of the use of IgG is etanercept (Enbrel®), which is a genetic fusion between the TNF receptor and the Fc region of human IgG1, which blocks binding of TNF-α to cellular TNF receptors and inhibits pro-inflammatory activity in autoimmune disease (Ducharme and Weinberg 2008). Strategies developed to exploit albumin include covalent association via genetic fusion to the N- or C-terminal end, chemical conjugation and encapsulation of drugs into albumin nanoparticles. Moreover, non-covalent association strategies target endogenous albumin via albumin-binding molecules (Elsadek and Kratz 2012). These strategies were all established before the FcRn-albumin relationship was appreciated, and whether the strategies interfere with receptor binding and transport of albumin has not been fully addressed. This aspect is important to consider prior to design of new albumin-based therapeutics. Illustrations of drugs targeting albumin via different strategies are given in Fig. 10.3.

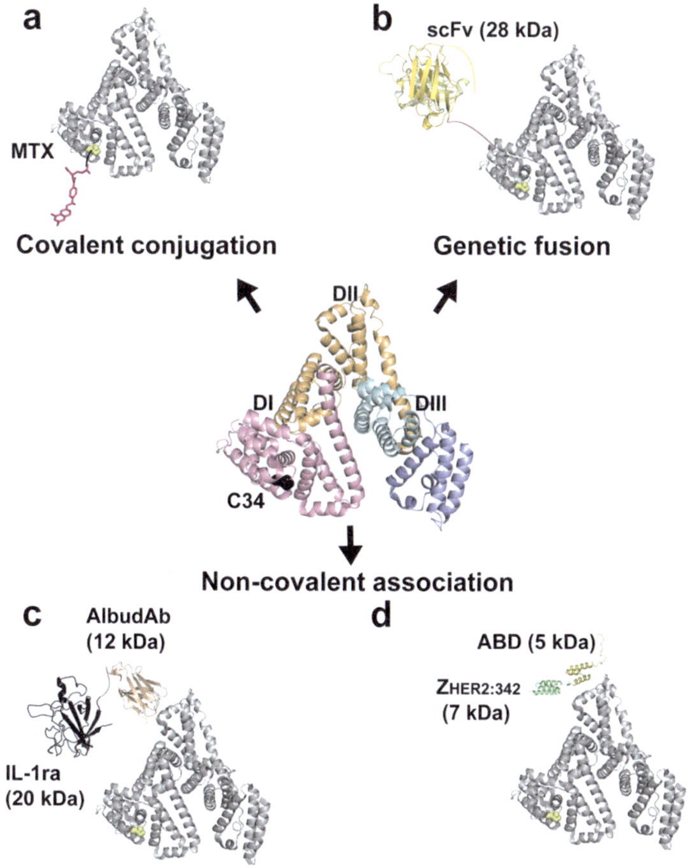

Fig. 10.3 Albumin targeting of drugs by different strategies. (*Middle*) The crystal structure of HSA with marked domains (DI, DII, and DIII) and its free C34 indicated in *black*. (**a**) Covalent conjugation of the chemical drug MTX to C34 of HSA. (**b**) Genetic fusion of a scFv fragment to the N-terminal end of HSA. Association of HSA with (**c**) an AlbudAb-IL-1ra fusion and (**d**) an ABD-$Z_{HER2:342}$ fusion. The figures were made using PyMOL with the following PDB files: HSA (1E7H), MTX (4OCX), scFv (3JUY), IL-1ra (1ILR), AlbudAb (1OP9), ABD (1GJT), and Affibody (2KZJ) (The figure is modified from (Bern et al. 2015))

10.5.1 Covalent Association with Albumin

10.5.1.1 Genetic Fusion

The genetic fusion strategy is attractive in that the therapeutic is synthesized as one transcript without the need for further in vitro processing. There are a number of examples of therapeutic proteins that when genetically fused to wild-type albumin show improved pharmacokinetics. Examples are hirudin (Syed et al. 1997),

CD4 (Yeh et al. 1992), insulin (Duttaroy et al. 2005), growth hormone (Osborn et al. 2002), granulocyte colony-stimulating factor (Halpern et al. 2002), α- and β-interferons (Bain et al. 2006; Subramanian et al. 2007; Sung et al. 2003), and antibody fragments (Evans et al. 2010; McDonagh et al. 2012; Muller et al. 2007; Smith et al. 2001; Yazaki et al. 2008).

Recombinant INF-α2b, used in therapy for chronic hepatitis C virus infections, has a serum half-life of only 4 h in humans; however, upon genetic fusion to HSA, the half-life increases by 35-fold to 141 h (Bain et al. 2006). Another example is Tanzeum®/Eperzan®, a fusion of glucagon-like peptide-1 (GLP-1) to wild-type HSA, which was approved by the FDA in 2014 and is the first albumin fusion that has entered the market. It is used for treatment of type II diabetes and has a half-life of 5 days compared to 2 min for the unfused drug (Poole and Nowlan 2014; Baggio et al. 2004).

Furthermore, a range of other wild-type HSA fusions is now in preclinical or clinical trials. For instance, genetic fusion of recombinant coagulation factors to HSA has shown promising results. Commercially available recombinant coagulation factor IX (rIX/BeneFIX®) has a half-life of 22 h in humans, and patients require multiple infusions per week to minimize the number of bleeding episodes (BeneFIX 2015; Santagostino et al. 2012). Development of a fusion protein linking rIX to the N-terminal end of albumin (rIX-FP) via a cleavable linker, derived from the endogenous activation peptide in native FIX, allows for in vivo cleavage of activated FIX from HSA when required during blood coagulation (Metzner et al. 2009). This fusion has been extensively evaluated in clinical trials for treatment of hemophilia B (Metzner et al. 2009; Nolte et al. 2012; Santagostino et al. 2012, 2016). It is well tolerated, and a phase III trial was recently completed, which showed that the half-life was extended by sixfold and the pharmacodynamic activity was considerably improved compared to rIX (Santagostino et al. 2016). Idelvion® (rIX-FP) was approved by the FDA in March of 2016.

Whether fusion to albumin interferes with pH-dependent binding to FcRn has not been given much attention. To our knowledge this has only been addressed in one study, where a peptide or an antibody single-chain variable fragment (scFv) was genetically fused via a glycine-serine linker to the N- or C-terminal end of HSA (Andersen et al. 2013). The results revealed that C-terminal fusion had a negative effect on binding to FcRn, which at most reduced the affinity by twofold, compared to only minor effects observed upon N-terminal fusion (Andersen et al. 2013). No cellular studies have been conducted where the effects of N- and C-terminal fusions have been compared directly with regard to FcRn-mediated cellular transport. Although the fusions still bind FcRn well, an effect may be seen in vivo in the presence of high levels endogenous albumin competing for binding to the receptor. Binding to FcRn should be addressed for each individual HSA fusion as the nature of the fusion partners may affect receptor binding differently. Importantly, this potential limitation may be overcome by the use of HSA variants engineered for improved FcRn binding that has shown extended serum half-life (Andersen et al. 2014; Schmidt et al. 2013). One example is the single-point mutation variant,

K573P, which has been shown to maintain its favorable effect on receptor binding post fusion of a scFv molecule either to the N or C-terminal end and even to both (Andersen et al. 2014).

10.5.1.2 Covalent Conjugation

Albumin accumulates at tumor sites and inflamed tissues, and this was early utilized for the delivery of antitumor agents by chemical conjugation to albumin (Kratz et al. 2007; Stehle et al. 1997a; Wunder et al. 2003). Methotrexate (MTX)-albumin conjugates for treatment of renal carcinomas and autoimmune diseases is one example (Bolling et al. 2006; Stehle et al. 1997b; Wunder et al. 2003). Random conjugation of drugs to surface-exposed amino acid residues on albumin may negatively affect FcRn binding and clearance, and a more controlled approach would be to target a free cysteine (C34) in DI of albumin. The Drug Affinity Complex (DAC®) technology performs such site-specific conjugation of drugs to either exogenous or endogenous albumin (Kratz et al. 2000). Exendin-4, a GLP-1 homolog (CJC–1131) for treatment of type 2 diabetes, bound to C34 (Baggio et al. 2008; Kim et al. 2003; Leger et al. 2004) has shown increased half-life of 9–15 days in human compared to a few hours for GLP-1 analogs (Giannoukakis 2003). Another example is aldoxorubicin, a prodrug of doxorubicin, which rapidly binds to C34 after intravenous administration via an acid-sensitive linker (Kratz et al. 2000). Aldoxorubicin is in clinical trials for treatment of sarcoma and glioblastoma, and the strategy relies on the acidic environment at tumor sites to allow cleavage of the linker and release of the drug from albumin (Chawla et al. 2015; Kratz 2014). Whether the linker is protected or cleaved during FcRn-mediating transport remains to be investigated. In addition, little is known about the expression of FcRn in different types of cancer tissues. Such knowledge will be important to understand how FcRn handles albumin fusions in cancerous tissues. In a recent study, a cytotoxin-conjugated and tumor targeting designed ankyrin repeat protein (DARPin) was conjugated to C34 of MSA, which extended the serum half-life of the DARPin from 11 min to 17.4 h in mice (Simon et al. 2013). In addition, the conjugate was shown to bind mouse FcRn (mFcRn) with expected pH dependency (Simon et al. 2013).

Another albumin-based approach that has been used to target drugs to tumors involves assembly of albumin and drugs into nanoparticles. Nanoparticle albumin-bound (nab)-paclitaxel (Abraxane®) consists of the lipophilic drug paclitaxel, which is encapsulated with HSA molecules under high pressure. Following administration, the nanoparticles dissolve and release HSA-bound paclitaxel into the bloodstream. The drug was first approved for treatment of metastatic breast cancer and is currently in clinical trials for treatment of non-small lung cancer, pancreatic cancer, and melanoma (Gradishar et al. 2005; Gupta et al. 2014; Hersh et al. 2010; Kottschade et al. 2011, 2013; Miele et al. 2009; Petrelli et al. 2010; Von Hoff et al. 2011). Whether albumin-based nanoparticles are capable of interacting with FcRn has not been investigated.

The C34 residue is not located near the FcRn-binding site in DI (Oganesyan et al. 2014); however, a recent study demonstrated that conjugation of polyethylene glycol (PEG) polymers to C34 of HSA lowered binding to hFcRn by two- to three-fold at acidic pH (Petersen et al. 2015). Notably, a larger drop in binding affinity was observed when a 5 kDa PEG was attached compared to larger polymers of 10 kDa and 30 kDa (Petersen et al. 2015). This suggests that covalent attachment to C34 may introduce conformational changes or steric hindrance that influences the interaction to FcRn, but to different degrees depending on the nature of the payload. This example highlights the need for case-specific assessment of potential impact on FcRn binding. In addition, the authors demonstrated that an engineered albumin variant (K573P) with improved affinity for FcRn may be used to compensate for the decrease in binding caused by the C34-conjugated payload (Petersen et al. 2015).

10.5.2 Non-covalent Association with Albumin

Non-covalent association with albumin has been explored as an alternative strategy to improve the pharmacokinetic properties of therapeutics and may be achieved by genetic fusion or conjugation of the drug to an albumin-binding molecule. Such albumin-binding molecules exist in various formats and albumin's own ligands have also been used for this purpose (described in Chap. 1). Anti-albumin-binding antibody fragments are an alternative. One example is a bispecific antigen-binding $(Fab)_2$ fragment with one arm targeting TNF and the other albumin, which show fivefold longer half-life compared to a monospecific anti-TNF $F(ab)_2$ in rats (Smith et al. 2001). The estimated half-life of 42.5 h was close to that measured for endogenous rat serum albumin (RSA), which was 49.1 h, supporting that the anti-albumin arm of the molecule does not interfere with binding to FcRn (Smith et al. 2001). Further, camelid-derived anti-albumin-binding nanobodies (Tijink et al. 2008) and variable domains of shark new antigen receptor have also been exploited (Muller et al. 2012).

To benefit from FcRn-mediated half-life extension, it is important that binding of targeting molecules to albumin does not negatively affect receptor binding. Moreover, binding to albumin needs to be retained at neutral pH, as well as through the lower pH encountered in the endosomes. In addition to controlling these aspects, it is also necessary to consider cross-species reactivity of albumin-binding molecules to albumin when choosing a preclinical animal model. When optimal binding to albumin is achieved, the only limitation is that the half-life cannot be extended beyond that of endogenous albumin.

10.5.2.1 Albumin-Binding Peptides

Dennis and co-workers have revealed the potential of albumin-binding peptides in improving pharmacokinetic properties of short-lived molecules (Dennis et al. 2002). First, albumin-specific peptides sharing a core sequence (DICLPRWGCLW) were identified using phage display technology. Many of these could recognize albumin from mouse, rat, rabbit, and human. One of the selected peptides (SA21) was measured to have a half-life of 2.3 h in rabbit compared to only 7.3 min of an unrelated peptide of the same size (Dennis et al. 2002). Furthermore, a Fab molecule was shown to acquire albumin-binding properties upon attachment of a related peptide (SA06) while retaining its ability to bind antigen. This gave rise to 37-fold and 26-fold increased half-life in rabbit and mouse, respectively, achieving 25–45 % of the half-life of albumin in these animals (Dennis et al. 2002).

In a follow-up study, the length of the SA06-peptide was modified to generate a panel of peptide variants with different affinities for albumin. These were fused to a Fab molecule, derived from the clinically approved antibody trastuzumab (Herceptin®), with specificity for human epidermal growth factor receptor 2 (HER2), denoted AB.Fab4D5 (Nguyen et al. 2006). When injected into rodents, it was demonstrated that albumin-binding affinity correlated with serum half-life, as fusions with low-affinity peptides were more rapidly eliminated than high-affinity peptide fusions (Nguyen et al. 2006).

Furthermore, AB.Fab4D5 with the original SA06 peptide was compared to Fab4D5 lacking an albumin-binding peptide and to trastuzumab, the full-length parental antibody, in targeting of HER2-positive tumors in allograft mouse models (Dennis et al. 2007). AB.Fab4D5 showed more rapid targeting of the tumor compared to trastuzumab and thus improved tumor to normal tissue ratio. In addition, both greater penetration and more even distribution within the tumor were observed for Fab4D5 and AB.Fab4D5 compared to the full-length antibody, with the latter being superior. Although both molecules localized rapidly to the tumor, Fab4D5 was also quickly removed and found to accumulate in the kidneys. In contrast, AB.Fab4D5 showed prolonged tumor retention and did not accumulate in the kidneys, showing the importance of the albumin-binding peptide (Dennis et al. 2007).

The favorable tumor targeting properties of AB.Fab4D5 may be related to having a smaller size (120 kDa when complexed with albumin) compared to trastuzumab (150 kDa). However, it is also interesting that albumin accumulates at tumor sites and inflamed tissues (Kratz et al. 2007; Stehle et al. 1997a; Wunder et al. 2003). Albumin-binding receptors, such as glycoprotein60 (gp60) and the secreted protein acidic and rich in cysteine (SPARC), have been suggested to mediate transport and accumulation of albumin from blood to the tumor site (Bern et al. 2015). Thus, receptor-mediated mechanisms may in part explain the more efficient and even uptake of AB.Fab4D5 when in complex albumin. However, further characterization of the different albumin receptors in regard to tumor expression is needed to understand how they contribute to the distribution of albumin-binding fusions.

10.5.2.2 The Albumin-Binding Domain

The concept of improving the pharmacokinetic properties of therapeutics via albumin-targeting was first demonstrated using a naturally existing albumin-binding domain (ABD), derived from the bacterial *Streptococcal* protein G (SpG), which when fused to a soluble CD4, extended its serum half-life in mice (Nygren et al. 1991). Since then, a minimal three-helical domain within SpG, referred to as the ABD, which binds albumin from multiple species with low nM affinity, has been widely used as a fusion partner (Johansson et al. 2002). For example, ABD has been genetically fused to the anti-HER2 Fab4D5, and in mice, the half-life was increased by tenfold to 20.9 h for the ABD fusion compared to 2.1 h for the Fab molecule alone (Schlapschy et al. 2007).

Similarly, ABD has also been fused to a dimeric anti-HER2 Affibody ($Z_{HER2:342}$) molecule. The Affibody scaffold is a small domain, derived from the IgG binding domain of *Staphylococcal* protein A (7 kDa), which has been used to make phage display libraries, enabling selection of binders to various targets (Nord et al. 1997). When ABD-$Z_{HER2:342}$ fusion molecules were radiolabeled and given to HER2-positive microxenograft mice, high tumor uptake and reduced accumulation in the kidneys were observed compared to $Z_{HER2:342}$ not fused to ABD (Tolmachev et al. 2007). Thus, non-covalent association with albumin led to redistribution of the drug, similar to that observed for AB.Fab4D5. Notably, the ABD-$Z_{HER2:342}$ fusion was shown to exhibit similar biodistribution profile as RSA in wild-type rats (Andersen et al. 2011). Moreover, binding of ABD or the ABD-$Z_{HER2:342}$ fusion to albumin did not affect pH-dependent binding to FcRn in vitro (Andersen et al. 2011). Thus, ABD binds to a site of albumin that does not interfere with pH-dependent FcRn binding.

Furthermore, the recently reported ABDCon is an engineered ABD molecule with high thermal stability that was developed from consensus analysis of ABD derived from SpG and close homologs from other bacterial proteins. The domain was shown to bind albumin from mouse, monkey, and human with affinities in the pM range (Jacobs et al. 2015). Genetic fusion of ABDCon to a 10 kDa scaffold protein, which does not bind to any known target proteins, resulted in serum half-lives of 60 h in mice and 182 h in cynomolgus monkeys. Furthermore, it was demonstrated that by introduction of single-point mutations in ABDCon, altering binding affinity for albumin, the serum half-life of the fusion protein could be adjusted and fine-tuned in mice (Jacobs et al. 2015). In another study, the naturally occurring ABD was put through affinity maturation using phage display, which gave rise to a variant with extremely high affinity for HSA in the fM range (Jonsson et al. 2008). The Albumod™ platform uses such high-affinity binding ABDs to extend the half-life of biopharmaceuticals (Elsadek and Kratz 2012).

10.5.2.3 The AlbudAb™ Platform

The AlbudAb™ technology exploits albumin-binding domain antibodies (dAb), to extend the half-life of short-lived drugs. A dAb consists of a single variable antibody domain, corresponding to variable region of the heavy (V_H) or light chain (V_L) of human antibodies and has a molecular weight of 11–13 kDa (Ward et al. 1989). The dAb was initially identified when Ward and co-workers discovered that the V_H chain domain could be expressed in the absence of a V_L chain domain and still bind its antigen (Ward et al. 1989). Furthermore, highly stable and aggregation-resistant human dAbs have been selected using phage display technology (Jespers et al. 2004). Holt and colleagues were the first to examine the potential of human dAbs as fusion partners to increase serum half-life of therapeutic drugs (Holt et al. 2008). To obtain dAbs with specificity for albumin, phage display selections were performed, which gave rise to a large collection of unique human V_H and V_K dAbs with affinities in the low nM range. Notably, promiscuous dAb variants with the ability to bind albumin from mouse, rat, and human were obtained by altering the albumin target during selection, which facilitates the evaluation of fusion candidates in preclinical animal models (Holt et al. 2008).

One of the variants (dAbr3), with an affinity of 13 nM for RSA, was shown to exhibit a half-life of 49 h, being close to that measured for RSA, while another variant (dAbr16), with an affinity of 1 μM, was estimated to have a shorter half-life of 43 h, thus also demonstrating that binding affinity for endogenous albumin correlates with serum half-life (Holt et al. 2008). Moreover, the half-life of a dAbm16 with an affinity of 70 nM for MSA was measured to be 24 h when injected into mice, which was 51-fold longer than that of a dAb without albumin-binding affinity (Holt et al. 2008). Next, dAbm16 was fused to an IL-1 receptor antagonist (IL-1ra), which is a potent inhibitor of IL-1 signaling and approved for treatment of rheumatoid arthritis. The resulting fusion protein exhibited a half-life of 4.3 h in mice compared to 2 min for IL-1ra. Furthermore, in a mouse model of collagen-induced arthritis, the increase in half-life was demonstrated to improve the in vivo efficacy of IL-1ra in a dose-dependent manner (Holt et al. 2008). Moreover, IL-1ra fused to a dAb (dAbh8) adopted a biodistribution profile very close to that of MSA, while IL1-ra alone was primarily found in the renal cortex or in the bladder 24 h after injection (Holt et al. 2008).

The AlbudAb approach has also been tested with human INF-α2b fused to an albumin-binding dAb, denoted DOM7 h14 (Walker et al. 2010). In rat, the pharmacokinetics of INF-α2b-DOM7 h14 was compared to recombinant INF-α2b, as well as to INF-α2b fused to HSA (HSA-INF-α2b), and both fusion-proteins showed serum half-life extension by more than 12-fold compared to the 1.5 h of INF-α2b. In an antiviral assay using A549 human lung carcinoma cells challenged with encephalomyocarditis virus, the antiviral activity of INF-α2b-DOM7 was significantly increased relative to HSA-INF-α2b. Moreover, INF-α2b-DOM7 was superior in suppressing tumor growth in a human melanoma xenograft mouse model (Walker et al. 2010). However, the result of this comparison may well be influenced by cross-species differences in FcRn binding as discussed in Sect. 10.7.

Moreover, the first evaluation of an AlbudAb-based drug was recently performed in human. Extendin-4 genetically fused to an AlbudAb (GSK2374697) was tested in healthy individuals in a phase I clinical trial (O'Connor-Semmes et al. 2014). Results from this study show that the AlbudAb improved the pharmacokinetic profile of extendin-4, increasing the half-life from 2.5 h to 6–10 days. Furthermore, the drug showed the anticipated effects on glucose and insulin levels and gastric emptying after a meal, as well as a tolerability profile expected for a GLP-1 agonist (O'Connor-Semmes et al. 2014; Ko et al. 2014).

The in vivo data showing that the serum half-life of AlbudAbs is close to that of endogenous albumin suggests that the FcRn-albumin interaction is not affected. In line with this, Herring and Schon have provided data indicating that the AlbudAbs selected by phage display are primarily domain 2 binders (Herring and Schon 2012). However, the fusion partner may still influence the interaction, which should be controlled in each case.

10.6 Engineering of Albumin for Improved Binding to FcRn

Increased insight into FcRn biology and its relationships with IgG and albumin provides new opportunities in therapy. Importantly, engineering can increase serum half-life and function beyond that of endogenous albumin or IgG. So far, engineering of the FcRn-IgG interaction to optimize IgG effector functions and in vivo efficacy has received the most attention. The major challenge has been to improve binding without disrupting the pH dependency of the interaction. However, there are several examples of IgG variants, with amino acid substitutions at the core or near the FcRn-binding site, that show improved pH-dependent binding and increased serum half-life (Dall'Acqua et al. 2006; Ghetie et al. 1997; Hinton et al. 2004, 2006; Mi et al. 2008; Zalevsky et al. 2010), which also have been demonstrated to benefit treatment of infections and diseases in preclinical models (Zalevsky et al. 2010; Ko et al. 2014).

Engineering of albumin is also being increasingly explored, and the first examples of albumin variants with altered FcRn-binding properties have been reported. Development of one such variant was inspired from a cross-species study demonstrating that MSA binds more strongly to hFcRn than HSA (Andersen et al. 2010a). Swapping of DIII from MSA onto DI–DII of HSA gave rise to a chimeric albumin variant with considerably improved binding toward hFcRn. Moreover, when the last C-terminal α-helix in HSA was replaced by the corresponding sequence in MSA, binding was improved by fourfold (Andersen et al. 2013).

Another study based on cross-species analysis identified a single amino acid substitution (K573P) that gave rise to 12-fold improved binding affinity to hFcRn when introduced in DIII of HSA (Andersen et al. 2014). All species have a proline at position 573, except for humans and orangutans. Notably, when K573 was replaced by any of the other 19 amino acids, improved binding to hFcRn was observed at pH 6 with no or minor effect on binding at pH 7.4 (Andersen et al. 2014). This positive effect on FcRn binding is not easily explained by the available co-crystal structures (Oganesyan et al. 2014; Schmidt et al. 2013). Introduction of K573P in HSA also improved binding to mFcRn by more than 20-fold, which partly explains why wild-type HSA binds poorly to the mouse receptor (Andersen et al. 2010a, 2014). Nevertheless, improved FcRn-binding affinity measured for HSA-K573P translated into 1.4-fold prolonged half-life in both hFcRn transgenic and wild-type mice and extended half-life by 1.6-fold in cynomolgus monkeys (Andersen et al. 2014).

HSA variants with improved affinity for hFcRn have also been identified using yeast display technology (Schmidt et al. 2013). One variant, with two amino acid substitutions in DIII (E505G/V547A), showed more than tenfold improved affinity at pH 6.0 with a minor increase at pH 7.4. This increase in affinity extended the half-life by 1.5-fold and 1.3-fold, in hFcRn transgenic mice and cynomolgus monkeys, respectively (Schmidt et al. 2013). Another two variants, with three (V418M, T420A, and E505G) or four (V418M, T420A, E505G, and V547A) amino acid substitutions, showed further improved binding to FcRn at acidic pH but also at neutral pH, and these were eliminated faster from the circulation of hFcRn transgenic mice than wild-type HSA (Schmidt et al. 2013). The fast clearance could be due to increased binding affinity at pH 7.4 that may disrupt efficient release during FcRn-mediated recycling. In addition, the mice developed antibody responses toward the HSA variants, which may be part of the explanation (Schmidt et al. 2013).

Targeting amino acid residues in DI of HSA by mutagenesis have also generated variants with increased binding to the receptor, and thus combining DIII and DI mutations may potentially give rise to variants with superior FcRn binding (Sand et al. 2014a). Importantly, these examples demonstrate that the serum half-life of albumin can be altered by modifying FcRn affinity, which may be used to benefit the next generation of albumin-based therapeutics.

10.7 Preclinical Development of Albumin-Based Therapeutics

Rodents are commonly used in preclinical evaluation of the pharmacokinetics and pharmacodynamics of IgG- and albumin-based drugs. In this regard, it has become evident that the utility of conventional mice is limited due large cross-species differences in binding to FcRn (Andersen et al. 2010a, 2014). Interaction studies using surface plasmon resonance have revealed that mFcRn has weak affinity for HSA and binds MSA tenfold more strongly. In contrast, hFcRn has stronger affinity for MSA than for HSA (Andersen et al. 2010a). The consequence was demonstrated in wild-type mice when the serum half-life of wild-type HSA was measured to be close to that of a HSA variant (K500A) with more than 30-fold reduced affinity to hFcRn (Andersen et al. 2014). Thus, the injected HSA was not rescued from degradation by FcRn recycling, a result of weak binding affinity in combination with the presence of high levels of competing endogenous MSA. Furthermore, the half-life of HSA in wild-type rats was found to be only 15 h compared to 49 h for RSA (Smith et al. 2001).

Differences in FcRn binding across species were first revealed for IgG, when hFcRn was found to weakly interact with mouse IgG and mFcRn was found to bind more strongly to human IgG than to mouse IgGs (Ober et al. 2001). This provided an explanation for the rapid clearance of the first mouse monoclonal IgG antibodies tested in humans (Frodin et al. 1990; Saleh et al. 1992) and also a reason for why human IgG has longer serum half-life in wild-type mice than mouse IgG (Petkova et al. 2006; Roopenian et al. 2003).

Transgenic mice that express hFcRn and lack the mouse receptor have become valuable models for in vivo pharmacokinetic evaluations of human IgG-based therapeutics (Petkova et al. 2006; Roopenian et al. 2003; Proetzel and Roopenian 2014) and engineered IgG molecules with altered FcRn-binding kinetics, which have been shown to provide good predictions of behavior in monkeys (Zalevsky et al. 2010; Proetzel and Roopenian 2014; Tam et al. 2013). Notably, this is observed despite the lack of competition from endogenous mouse IgG, which is rapidly eliminated from the circulation due to low affinity for the hFcRn. High levels of MSA in these mice make them less appropriate for preclinical use in regard to HSA-based therapeutics, as this give rise to a situation similar to that in wild-type mice, injected HSA being outcompeted for binding to hFcRn.

The first generation of mice that may serve as more suitable models is emerging. It was recently reported that one of the original hFcRn transgenic strains has been further genetically engineered to lack expression of endogenous MSA (Roopenian et al. 2015). In these mice the serum half-life of intravenous administrated HSA was measured to be 24.1 days, as compared to 2.6 days in wild-type mice and 5.8 days in the original hFcRn transgenic strain (Roopenian et al. 2015). Thus, this novel mouse stain allows for studying the behavior of albumin-based therapeutics under the influence of FcRn, both in the absence of competition and potentially also in the presence of preloaded competing HSA.

Furthermore, double transgenic humanized mice, having the genes encoding mFcRn and MSA replaced by the human counterparts, have also been described (Viuff et al. 2016). Unlike the hFcRn transgenic mice developed previously, expression of the receptor, as well as albumin, is regulated by the endogenous mouse promoters, which may give rise to differences in tissue distribution. Nevertheless, high levels of HSA are maintained in these mice, and it was demonstrated that HSA variants with reduced or increased binding affinity for hFcRn showed correlating reduced and increased serum half-life (Viuff et al. 2016). Thus, this novel mouse strain may prove to be a valuable model system for pharmacokinetic evaluation of recombinant HSA-based therapeutics, as well as for drugs whose action is based on reversible binding to endogenous albumin.

10.8 Concluding Remarks

The long serum half-life of albumin is due to protection from intracellular degradation by the broadly expressed cellular receptor, FcRn, which transports internalized albumin back to the circulation through a mechanism that is strictly pH dependent. This feature of albumin has been greatly exploited therapeutically to extend the serum half-life of drugs that would otherwise have poor pharmacokinetics. A number of strategies involving albumin fusion, conjugation, or targeting have been employed, and a range of albumin-based drugs have shown improved therapeutic outcome in clinical trials, and several have been approved for human use. Increasing knowledge about the FcRn-albumin relationship is likely to pave the way for the next generation of albumin-based therapeutics, as this offers new opportunities for design and tailoring. Indeed, the first examples of engineered albumin variants with altered FcRn-binding kinetics have already been reported, and recent results show that improved pH-dependent binding to FcRn translates into prolonged serum half-life. Importantly, novel transgenic mice specifically designed for preclinical evaluation of engineered HSA and HSA-based therapeutics may prove to be valuable assets in this rapidly growing area.

References

Ahouse JJ, Hagerman CL, Mittal P, Gilbert DJ, Copeland NG, Jenkins NA, Simister NE (1993) Mouse MHC class I-like Fc receptor encoded outside the MHC. J Immunol 151:6076–6088

Akilesh S, Christianson GJ, Roopenian DC, Shaw AS (2007) Neonatal FcR expression in bone marrow-derived cells functions to protect serum IgG from catabolism. J Immunol 179:4580–4588

Akilesh S, Huber TB, Wu H, Wang G, Hartleben B, Kopp JB, Miner JH, Roopenian DC, Unanue ER, Shaw AS (2008) Podocytes use FcRn to clear IgG from the glomerular basement membrane. Proc Natl Acad Sci U S A 105:967–972

Amsellem S, Gburek J, Hamard G, Nielsen R, Willnow TE, Devuyst O, Nexo E, Verroust PJ, Christensen EI, Kozyraki R (2010) Cubilin is essential for albumin reabsorption in the renal proximal tubule. J Am Soc Nephrol 21:1859–1867

Andersen JT, Dee Qian J, Sandlie I (2006) The conserved histidine 166 residue of the human neonatal Fc receptor heavy chain is critical for the pH-dependent binding to albumin. Eur J Immunol 36:3044–3051

Andersen JT, Daba MB, Berntzen G, Michaelsen TE, Sandlie I (2010a) Cross-species binding analyses of mouse and human neonatal Fc receptor show dramatic differences in immunoglobulin G and albumin binding. J Biol Chem 285:4826–4836

Andersen JT, Daba MB, Sandlie I (2010b) FcRn binding properties of an abnormal truncated analbuminemic albumin variant. Clin Biochem 43:367–372

Andersen JT, Pehrson R, Tolmachev V, Daba MB, Abrahmsen L, Ekblad C (2011) Extending half-life by indirect targeting of the neonatal Fc receptor (FcRn) using a minimal albumin binding domain. J Biol Chem 286:5234–5241

Andersen JT, Dalhus B, Cameron J, Daba MB, Plumridge A, Evans L, Brennan SO, Gunnarsen KS, Bjoras M, Sleep D, Sandlie I (2012a) Structure-based mutagenesis reveals the albumin-binding site of the neonatal Fc receptor. Nat Commun 3:610

Andersen JT, Foss S, Kenanova VE, Olafsen T, Leikfoss IS, Roopenian DC, Wu AM, Sandlie I (2012b) Anti-carcinoembryonic antigen single-chain variable fragment antibody variants bind mouse and human neonatal Fc receptor with different affinities that reveal distinct cross-species differences in serum half-life. J Biol Chem 287:22927–22937

Andersen JT, Cameron J, Plumridge A, Evans L, Sleep D, Sandlie I (2013) Single-chain variable fragment albumin fusions bind the neonatal Fc receptor (FcRn) in a species-dependent manner: implications for in vivo half-life evaluation of albumin fusion therapeutics. J Biol Chem 288:24277–24285

Andersen JT, Dalhus B, Viuff D, Ravn BT, Gunnarsen KS, Plumridge A, Bunting K, Antunes F, Williamson R, Athwal S, Allan E, Evans L, Bjoras M, Kjaerulff S, Sleep D, Sandlie I, Cameron J (2014) Extending serum half-life of albumin by engineering neonatal Fc receptor (FcRn) binding. J Biol Chem 289:13492–13502

Antohe F, Radulescu L, Gafencu A, Ghetie V, Simionescu M (2001) Expression of functionally active FcRn and the differentiated bidirectional transport of IgG in human placental endothelial cells. Hum Immunol 62:93–105

Aseem O, Smith BT, Cooley MA, Wilkerson BA, Argraves KM, Remaley AT, Argraves WS (2014) Cubilin maintains blood levels of HDL and albumin. J Am Soc Nephrol 25:1028–1036

Baggio LL, Huang QL, Brown TJ, Drucker DJ (2004) A recombinant human glucagon-like peptide (GLP)-1-albumin protein (Albugon) mimics peptidergic activation of GLP-1 receptor-dependent pathways coupled with satiety, gastrointestinal motility, and glucose homeostasis. Diabetes 53:2492–2500

Baggio LL, Huang Q, Cao X, Drucker DJ (2008) An albumin-exendin-4 conjugate engages central and peripheral circuits regulating murine energy and glucose homeostasis. Gastroenterology 134:1137–1147

Bain VG, Kaita KD, Yoshida EM, Swain MG, Heathcote EJ, Neumann AU, Fiscella M, Yu R, Osborn BL, Cronin PW, Freimuth WW, McHutchison JG, Subramanian GM (2006) A phase 2 study to evaluate the antiviral activity, safety, and pharmacokinetics of recombinant human albumin-interferon alfa fusion protein in genotype 1 chronic hepatitis C patients. J Hepatol 44:671–678

BeneFIX. BeneFIX (coagulation Factor IX Recombinant) (2015) Electronic medicines compendium. https://www.medicines.org.uk/emc/medicine/20376/SPC/BeneFIX/#PHARMACODINAMIC_PROPS. Accessed 16 Feb 2016

Bennhold H, Kallee E (1959) Comparative studies on the half-life of I 131-labeled albumins and nonradioactive human serum albumin in a case of analbuminemia. J Clin Invest 38:863–872

Bern M, Sand KM, Nilsen J, Sandlie I, Andersen JT (2015) The role of albumin receptors in regulation of albumin homeostasis: implications for drug delivery. J Control Release: Off J Control Release Soc 211:144–162

Birn H, Christensen EI (2006) Renal albumin absorption in physiology and pathology. Kidney Int 69:440–449

Birn H, Fyfe JC, Jacobsen C, Mounier F, Verroust PJ, Orskov H, Willnow TE, Moestrup SK, Christensen EI (2000) Cubilin is an albumin binding protein important for renal tubular albumin reabsorption. J Clin Invest 105:1353–1361

Bolling C, Graefe T, Lubbing C, Jankevicius F, Uktveris S, Cesas A, Meyer-Moldenhauer WH, Starkmann H, Weigel M, Burk K, Hanauske AR (2006) Phase II study of MTX-HSA in combination with cisplatin as first line treatment in patients with advanced or metastatic transitional cell carcinoma. Invest New Drugs 24:521–527

Borvak J, Richardson J, Medesan C, Antohe F, Radu C, Simionescu M, Ghetie V, Ward ES (1998) Functional expression of the MHC class I-related receptor, FcRn, in endothelial cells of mice. Int Immunol 10:1289–1298

Brambell FW, Hemmings WA, Morris IG (1964) A theoretical model of gamma-globulin catabolism. Nature 203:1352–1354

Burmeister WP, Gastinel LN, Simister NE, Blum ML, Bjorkman PJ (1994a) Crystal structure at 2.2 A resolution of the MHC-related neonatal Fc receptor. Nature 372:336–343

Burmeister WP, Huber AH, Bjorkman PJ (1994b) Crystal structure of the complex of rat neonatal Fc receptor with Fc. Nature 372:379–383

Cauza K, Hinterhuber G, Dingelmaier-Hovorka R, Brugger K, Klosner G, Horvat R, Wolff K, Foedinger D (2005) Expression of FcRn, the MHC class I-related receptor for IgG, in human keratinocytes. J Invest Dermatol 124:132–139

Chaudhury C, Mehnaz S, Robinson JM, Hayton WL, Pearl DK, Roopenian DC, Anderson CL (2003) The major histocompatibility complex-related Fc receptor for IgG (FcRn) binds albumin and prolongs its lifespan. J Exp Med 197:315–322

Chaudhury C, Brooks CL, Carter DC, Robinson JM, Anderson CL (2006) Albumin binding to FcRn: distinct from the FcRn-IgG interaction. Biochemistry 45:4983–4990

Chawla SP, Chua VS, Hendifar AF, Quon DV, Soman N, Sankhala KK, Wieland DS, Levitt DJ (2015) A phase 1B/2 study of aldoxorubicin in patients with soft tissue sarcoma. Cancer 121(4):570–579

Cianga P, Cianga C, Cozma L, Ward ES, Carasevici E (2003) The MHC class I related Fc receptor, FcRn, is expressed in the epithelial cells of the human mammary gland. Hum Immunol 64:1152–1159

Cianga P, Cianga C, Plamadeala P, Branisteanu D, Carasevici E (2007) The neonatal Fc receptor (FcRn) expression in the human skin. Virchows Arch 451:859–860

Claypool SM, Dickinson BL, Wagner JS, Johansen FE, Venu N, Borawski JA, Lencer WI, Blumberg RS (2004) Bidirectional transepithelial IgG transport by a strongly polarized basolateral membrane Fc gamma-receptor. Mol Biol Cell 15:1746–1759

Cormode EJ, Lyster DM, Israels S (1975) Analbuminemia in a neonate. J Pediatr 86:862–867

Cui SY, Verroust PJ, Moestrup SK, Christensen EI (1996) Megalin/gp330 mediates uptake of albumin in renal proximal tubule. Am J Physiol-Ren 271:F900–F907

Czajkowsky DM, Hu J, Shao Z, Pleass RJ (2012) Fc-fusion proteins: new developments and future perspectives. EMBO Mol Med 4:1015–1028

Dall'Acqua WF, Kiener PA, Wu H (2006) Properties of human IgG1s engineered for enhanced binding to the neonatal Fc receptor (FcRn). J Biol Chem 281:23514–23524

Dennis MS, Zhang M, Meng YG, Kadkhodayan M, Kirchhofer D, Combs D, Damico LA (2002) Albumin binding as a general strategy for improving the pharmacokinetics of proteins. J Biol Chem 277:35035–35043

Dennis MS, Jin HK, Dugger D, Yang RH, McFarland L, Ogasawara A, Williams S, Cole MJ, Ross S, Schwall R (2007) Imaging tumors with an albumin-binding Fab, a novel tumor-targeting agent. Cancer Res 67:254–261

Dickinson BL, Badizadegan K, Wu Z, Ahouse JC, Zhu XP, Simister NE, Blumberg RS, Lencer WI (1999) Bidirectional FcRn-dependent IgG transport in a polarized human intestinal epithelial cell Line. J Clin Investig 104:903–911

Dickinson BL, Claypool SM, D'Angelo JA, Aiken ML, Venu N, Yen EH, Wagner JS, Borawski JA, Pierce AT, Hershberg R, Blumberg RS, Lencer WI (2008) Ca2+-dependent calmodulin binding to FcRn affects immunoglobulin G transport in the transcytotic pathway. Mol Biol Cell 19:414–423

Dockal M, Carter DC, Ruker F (1999) The three recombinant domains of human serum albumin – structural characterization and ligand binding properties. J Biol Chem 274:29303–29310

Ducharme E, Weinberg JM (2008) Etanercept. Expert Opin Biol Ther 8:491–502

Duttaroy A, Kanakaraj P, Osborn BL, Schneider H, Pickeral OK, Chen C, Zhang GY, Kaithamana S, Singh M, Schulingkamp R, Crossan D, Bock J, Kaufman TE, Reavey P, Carey-Barber M, Krishnan SR, Garcia A, Murphy K, Siskind JK, McLean MA, Cheng S, Ruben S, Birse CE, Blondel O (2005) Development of a long-acting insulin analog using albumin fusion technology. Diabetes 54:251–258

Edeling MA, Smith C, Owen D (2006) Life of a clathrin coat: insights from clathrin and AP structures. Nat Rev Mol Cell Biol 7:32–44

Elsadek B, Kratz F (2012) Impact of albumin on drug delivery – new applications on the horizon. J Control Release: Off J Control Release Soc 157:4–28

Evans L, Hughes M, Waters J, Cameron J, Dodsworth N, Tooth D, Greenfield A, Sleep D (2010) The production, characterisation and enhanced pharmacokinetics of scFv-albumin fusions expressed in Saccharomyces cerevisiae. Protein Expr Purif 73:113–124

Fahey JL, Robinson AG (1963) Factors controlling serum gamma-globulin concentration. J Exp Med 118:845–868

Frodin JE, Lefvert AK, Mellstedt H (1990) Pharmacokinetics of the mouse monoclonal antibody 17-1A in cancer patients receiving various treatment schedules. Cancer Res 50:4866–4871

Gan Z, Ram S, Ober RJ, Ward ES (2013) Using multifocal plane microscopy to reveal novel trafficking processes in the recycling pathway. J Cell Sci 126:1176–1188

Ghetie V, Popov S, Borvak J, Radu C, Matesoi D, Medesan C, Ober RJ, Ward ES (1997) Increasing the serum persistence of an IgG fragment by random mutagenesis. Nat Biotechnol 15:637–640

Giannoukakis N (2003) CJC-1131. Conju Chem Curr Opin Invest Drugs 4:1245–1249

Gordon RS Jr, Bartter FC, Waldmann T (1959) Idiopathic hypoalbuminemias: clinical staff conference at the National Institutes of Health. Ann Intern Med 51:553–576

Gradishar WJ, Tjulandin S, Davidson N, Shaw H, Desai N, Bhar P, Hawkins M, O'Shaughnessy J (2005) Phase III trial of nanoparticle albumin-bound paclitaxel compared with polyethylated castor oil-based paclitaxel in women with breast cancer. J Clin Oncol 23:7794–7803

Gupta N, Hatoum H, Dy GK (2014) First line treatment of advanced non-small-cell lung cancer – specific focus on albumin bound paclitaxel. Int J Nanomed 9:209–221

Halpern W, Riccobene TA, Agostini H, Baker K, Stolow D, Gu ML, Hirsch J, Mahoney A, Carrell J, Boyd E, Grzegorzewski KJ (2002) Albugranin (TM), a recombinant human granulocyte colony stimulating factor (G-CSF) genetically fused to recombinant human albumin induces prolonged myelopoietic effects in mice and monkeys. Pharm Res 19:1720–1729

Haymann JP, Levraud JP, Bouet S, Kappes V, Hagege J, Nguyen G, Xu Y, Rondeau E, Sraer JD (2000) Characterization and localization of the neonatal Fc receptor in adult human kidney. J Am Soc Nephrol 11:632–639

Herring C, Schon O (2012) AlbudAb™ technology platform–versatile albumin binding domains for the development of therapeutics with tunable half-lives. In: Kontermann RE (ed) Therapeutic proteins: strategies to modulate their plasma half-lives. Wiley-VCH Verlag GmbH & Co., Weinheim

Hersh EM, O'Day SJ, Ribas A, Samlowski WE, Gordon MS, Shechter DE, Clawson AA, Gonzalez R (2010) A phase 2 clinical trial of nab-paclitaxel in previously treated and chemotherapy-naive patients with metastatic melanoma. Cancer-Am Cancer Soc 116:155–163

Hinton PR, Johlfs MG, Xiong JM, Hanestad K, Ong KC, Bullock C, Keller S, Tang MT, Tso JY, Vasquez M, Tsurushita N (2004) Engineered human IgG antibodies with longer serum half-lives in primates. J Biol Chem 279:6213–6216

Hinton PR, Xiong JM, Johlfs MG, Tang MT, Keller S, Tsurushita N (2006) An engineered human IgG1 antibody with longer serum half-life. J Immunol 176:346–356

Holt LJ, Basran A, Jones K, Chorlton J, Jespers LS, Brewis ND, Tomlinson IM (2008) Anti-serum albumin domain antibodies for extending the half-lives of short lived drugs. Protein Eng Des Sel 21:283–288

Israel EJ, Patel VK, Taylor SF, Marshakrothstein A, Simister N (1995) Requirement for a beta(2)-microglobulin-associated Fc receptor for acquisition of maternal Igg by fetal and neonatal mice. J Immunol 154:6246–6251

Israel EJ, Wilsker DF, Hayes KC, Schoenfeld D, Simister NE (1996) Increased clearance of IgG in mice that lack beta(2)-microglobulin: possible protective role of FcRn. Immunology 89:573–578

Israel EJ, Taylor S, Wu Z, Mizoguchi E, Blumberg RS, Bhan A, Simister NE (1997) Expression of the neonatal Fc receptor, FcRn, on human intestinal epithelial cells. Immunology 92:69–74

Jacobs SA, Gibbs AC, Conk M, Yi F, Maguire D, Kane C, O'Neil KT (2015) Fusion to a highly stable consensus albumin binding domain allows for tunable pharmacokinetics. Protein Eng Des Sel 28:385–393

Jespers L, Schon O, Famm K, Winter G (2004) Aggregation-resistant domain antibodies selected on phage by heat denaturation. Nat Biotechnol 22:1161–1165

Jiang L, Wang J, Solorzano-Vargas RS, Tsai HV, Gutierrez EM, Ontiveros LO, Kiela PR, Wu SV, Martin MG (2004) Characterization of the rat intestinal Fc receptor (FcRn) promoter: transcriptional regulation of FcRn gene by the Sp family of transcription factors. Am J Physiol Gastrointest Liver Physiol 286:G922–G931

Johansson MU, Frick IM, Nilsson H, Kraulis PJ, Hober S, Jonasson P, Linhult M, Nygren PA, Uhlen M, Bjorck L, Drakenberg T, Forsen S, Wikstrom M (2002) Structure, specificity, and mode of interaction for bacterial albumin-binding modules. J Biol Chem 277:8114–8120

Jonsson A, Dogan J, Herne N, Abrahmsen L, Nygren PA (2008) Engineering of a femtomolar affinity binding protein to human serum albumin. Protein Eng Des Sel 21:515–527
Junghans RP, Anderson CL (1996) The protection receptor for IgG catabolism is the beta(2)-microglobulin-containing neonatal intestinal transport receptor. Proc Natl Acad Sci U S A 93:5512–5516
Kandil E, Egashira M, Miyoshi O, Niikawa N, Ishibashi T, Kasahara M (1996) The human gene encoding the heavy chain of the major histocompatibility complex class I-like Fc receptor (FCGRT) maps to 19q13.3. Cytogenet Cell Genet 73:97–98
Kim JG, Baggio LL, Bridon DP, Castaigne JP, Robitaille MF, Jette L, Benquet C, Drucker DJ (2003) Development and characterization of a glucagon-like peptide 1-albumin conjugate: the ability to activate the glucagon-like peptide 1 receptor in vivo. Diabetes 52:751–759
Kim H, Fariss RN, Zhang C, Robinson SB, Thill M, Csaky KG (2008) Mapping of the neonatal Fc receptor in the rodent eye. Invest Ophthalmol Vis Sci 49:2025–2029
Ko SY, Pegu A, Rudicell RS, Yang ZY, Joyce MG, Chen XJ, Wang KY, Bao S, Kraemer TD, Rath T, Zeng M, Schmidt SD, Todd JP, Penzak SR, Saunders KO, Nason MC, Haase AT, Rao SS, Blumberg RS, Mascola JR, Nabel GJ (2014) Enhanced neonatal Fc receptor function improves protection against primate SHIV infection. Nature 514:642–645
Kobayashi K, Qiao SW, Yoshida M, Baker K, Lencer WI, Blumberg RS (2009) An FcRn-dependent role for anti-flagellin immunoglobulin G in pathogenesis of colitis in mice. Gastroenterology 137:1746–1756, e1741
Kottschade LA, Suman VJ, Amatruda T, McWilliams RR, Mattar BI, Nikcevich DA, Behrens R, Fitch TR, Jaslowski AJ, Markovic SN (2011) A phase II trial of nab-paclitaxel (ABI-007) and carboplatin in patients with unresectable stage IV melanoma a North Central Cancer Treatment Group Study, N057E. Cancer-Am Cancer Soc 117:1704–1710
Kottschade LA, Suman VJ, Perez DG, McWilliams RR, Kaur JS, Amatruda TT, Geoffroy FJ, Gross HM, Cohen PA, Jaslowski AJ, Kosel ML, Markovic SN (2013) A randomized phase 2 study of temozolomide and bevacizumab or nab-paclitaxel, carboplatin, and bevacizumab in patients with unresectable stage IV melanoma a North Central Cancer Treatment Group Study, N0775. Cancer-Am Cancer Soc 119:586–592
Kozyraki R, Fyfe J, Verroust PJ, Jacobsen C, Dautry-Varsat A, Gburek J, Willnow TE, Christensen EI, Moestrup SK (2001) Megalin-dependent cubilin-mediated endocytosis is a major pathway for the apical uptake of transferrin in polarized epithelia. Proc Natl Acad Sci U S A 98:12491–12496
Kratz F (2014) A clinical update of using albumin as a drug vehicle – a commentary. J Control Release 190:331–336
Kratz F, Muller-Driver R, Hofmann I, Drevs J, Unger C (2000) A novel macromolecular prodrug concept exploiting endogenous serum albumin as a drug carrier for cancer chemotherapy. J Med Chem 43:1253–1256
Kratz F, Abu Ajaj K, Warnecke A (2007) Anticancer carrier-linked prodrugs in clinical trials. Expert Opin Invest Drug 16:1037–1058
Leger R, Thibaudeau K, Robitaille M, Quraishi O, van Wyk P, Bousquet-Gagnon N, Carette J, Castaigne JP, Bridon DP (2004) Identification of CJC-1131-albumin bioconjugate as a stable and bioactive GLP-1(7-36) analog. Bioorg Med Chem Lett 14:4395–4398
Li ZL, Palaniyandi S, Zeng RY, Tuo WB, Roopenian DC, Zhu XP (2011) Transfer of IgG in the female genital tract by MHC class I-related neonatal Fc receptor (FcRn) confers protective immunity to vaginal infection. Proc Natl Acad Sci U S A 108:4388–4393
Liu X, Ye L, Christianson GJ, Yang JQ, Roopenian DC, Zhu X (2007) NF-kappaB signaling regulates functional expression of the MHC class I-related neonatal Fc receptor for IgG via intronic binding sequences. J Immunol 179:2999–3011
Liu XD, Ye LL, Bai Y, Mojidi H, Simister NE, Zhu XP (2008) Activation of the JAK/STAT-1 signaling pathway by IFN-gamma can down-regulate functional expression of the MHC class I-related neonatal Fc receptor for IgG. J Immunol 181:449–463

Martin MG, Wu SV, Walsh JH (1993) Hormonal control of intestinal Fc receptor gene expression and immunoglobulin transport in suckling rats. J Clin Invest 91:2844–2849

Martin MG, Wu SV, Walsh JH (1997) Ontogenetic development and distribution of antibody transport and Fc receptor mRNA expression in rat intestine. Dig Dis Sci 42:1062–1069

Maunsbach AB (1966) Albumin absorption by renal proximal tubule cells. Nature 212:546–547

McDonagh CF, Huhalov A, Harms BD, Adams S, Paragas V, Oyama S, Zhang B, Luus L, Overland R, Nguyen S, Gu JM, Kohli N, Wallace M, Feldhaus MJ, Kudla AJ, Schoeberl B, Nielsen UB (2012) Antitumor activity of a novel bispecific antibody that targets the ErbB2/ErbB3 oncogenic unit and inhibits heregulin-induced activation of ErbB3. Mol Cancer Ther 11:582–593

McMahon HT, Gallop JL (2005) Membrane curvature and mechanisms of dynamic cell membrane remodelling. Nature 438:590–596

Metzner HJ, Weimer T, Kronthaler U, Lang W, Schulte S (2009) Genetic fusion to albumin improves the pharmacokinetic properties of factor IX. Thromb Haemost 102:634–644

Mezo AR, Sridhar V, Badger J, Sakorafas P, Nienaber V (2010) X-ray crystal structures of monomeric and dimeric peptide inhibitors in complex with the human neonatal Fc receptor, FcRn. J Biol Chem 285:27694–27701

Mi WT, Wanjie S, Lo ST, Gan Z, Pick-Herk B, Ober RJ, Ward ES (2008) Targeting the neonatal Fc receptor for antigen delivery using engineered Fc fragments. J Immunol 181:7550–7561

Miele E, Spinelli GP, Miele E, Tomao F, Tomao S (2009) Albumin-bound formulation of paclitaxel (Abraxane (R) ABI-007) in the treatment of breast cancer. Int J Nanomed 4:99–105

Mikulska JE (2015) Correction: analysis of response elements involved in the regulation of the human neonatal Fc receptor gene (FCGRT). PLoS One 10:e0139744

Mikulska JE, Simister NE (2000) Analysis of the promoter region of the human FcRn gene. Biochim Biophys Acta 1492:180–184

Mikulska JE, Pablo L, Canel J, Simister NE (2000) Cloning and analysis of the gene encoding the human neonatal Fc receptor. Eur J Immunogenet 27:231–240

Moestrup SK, Kozyraki R, Kristiansen M, Kaysen JH, Rasmussen HH, Brault D, Pontillon F, Goda FO, Christensen EI, Hammond TG, Verroust PJ (1998) The intrinsic factor-vitamin B12 receptor and target of teratogenic antibodies is a megalin-binding peripheral membrane protein with homology to developmental proteins. J Biol Chem 273:5235–5242

Montoyo HP, Vaccaro C, Hafner M, Ober RJ, Mueller W, Ward ES (2009) Conditional deletion of the MHC class I-related receptor FcRn reveals the sites of IgG homeostasis in mice. Proc Natl Acad Sci U S A 106:2788–2793

Muller D, Karle A, Meibburger B, Hofig I, Stork R, Kontermann RE (2007) Improved pharmacokinetics of recombinant bispecific antibody molecules by fusion to human serum albumin. J Biol Chem 282:12650–12660

Muller MR, Saunders K, Grace C, Jin M, Piche-Nicholas N, Steven J, O'Dwyer R, Wu L, Khetemenee L, Vugmeyster Y, Hickling TP, Tchistiakova L, Olland S, Gill D, Jensen A, Barelle CJ (2012) Improving the pharmacokinetic properties of biologics by fusion to an anti-HSA shark VNAR domain. MAbs 4:673–685

Newton EE, Wu Z, Simister NE (2005) Characterization of basolateral-targeting signals in the neonatal Fc receptor. J Cell Sci 118:2461–2469

Nguyen A, Reyes AE, Zhang M, McDonald P, Wong WLT, Damico LA, Dennis MS (2006) The pharmacokinetics of an albumin-binding Fab (AB.Fab) can be modulated as a function of affinity for albumin. Protein Eng Des Sel 19:291–297

Nolte MW, Nichols TC, Mueller-Cohrs J, Merricks EP, Pragst I, Zollner S, Dickneite G (2012) Improved kinetics of rIX-FP, a recombinant fusion protein linking factor IX with albumin, in cynomolgus monkeys and hemophilia B dogs. J Thromb Haemost 10:1591–1599

Nord K, Gunneriusson E, Ringdahl J, Stahl S, Uhlen M, Nygren PA (1997) Binding proteins selected from combinatorial libraries of an alpha-helical bacterial receptor domain. Nat Biotechnol 15:772–777

Nygren PA, Uhlen M, Flodby P, Andersson R, Wigzell H (1991) In vivo stabilization of a human recombinant CD4 derivative by fusion to a serum-albumin-binding receptor. Vaccines 91:363–368

Ober RJ, Radu CG, Ghetie V, Ward ES (2001) Differences in promiscuity for antibody-FcRn interactions across species: implications for therapeutic antibodies. Int Immunol 13:1551–1559

Ober RJ, Martinez C, Lai X, Zhou J, Ward ES (2004a) Exocytosis of IgG as mediated by the receptor, FcRn: an analysis at the single-molecule level. Proc Natl Acad Sci U S A 101:11076–11081

Ober RJ, Martinez C, Vaccaro C, Zhou J, Ward ES (2004b) Visualizing the site and dynamics of IgG salvage by the MHC class I-related receptor, FcRn. J Immunol 172:2021–2029

O'Connor-Semmes RL, Lin J, Hodge RJ, Andrews S, Chism J, Choudhury A, Nunez DJ (2014) GSK2374697, a novel albumin-binding domain antibody (AlbudAb), extends systemic exposure of exendin-4: first study in humans – PK/PD and safety. Clin Pharmacol Ther 96:704–712

Oganesyan V, Damschroder MM, Cook KE, Li Q, Gao CS, Wu HR, Dall'Acqua WF (2014) Structural insights into neonatal Fc receptor-based recycling mechanisms. J Biol Chem 289:7812–7824

Osborn BL, Sekut L, Corcoran M, Poortman C, Sturm B, Chen GX, Mather D, Lin HL, Parry TJ (2002) Albutropin: a growth hormone-albumin fusion with improved pharmacokinetics and pharmacodynamics in rats and monkeys. Eur J Pharmacol 456:149–158

Petersen SS, Klaning E, Ebbesen MF, Andersen B, Cameron J, Sorensen ES, Howard KA (2015) Neonatal Fc receptor binding tolerance toward the covalent conjugation of payloads to cysteine 34 of human albumin variants. Mol Pharm 13:677–82

Petkova SB, Akilesh S, Sproule TJ, Christianson GJ, Al Khabbaz H, Brown AC, Presta LG, Meng YG, Roopenian DC (2006) Enhanced half-life of genetically engineered human IgG1 antibodies in a humanized FcRn mouse model: potential application in humorally mediated autoimmune disease. Int Immunol 18:1759–1769

Petrelli F, Borgonovo K, Barni S (2010) Targeted delivery for breast cancer therapy: the history of nanoparticle-albumin-bound paclitaxel. Expert Opin Pharm 11:1413–1432

Poole RM, Nowlan ML (2014) Albiglutide: first global approval. Drugs 74:929–938

Powner MB, McKenzie JA, Christianson GJ, Roopenian DC, Fruttiger M (2014) Expression of neonatal Fc receptor in the eye. Invest Ophthalmol Vis Sci 55:1607–1615

Prabhat P, Gan Z, Chao J, Ram S, Vaccaro C, Gibbons S, Ober RJ, Ward ES (2007) Elucidation of intracellular recycling pathways leading to exocytosis of the Fc receptor, FcRn, by using multifocal plane microscopy. Proc Natl Acad Sci U S A 104:5889–5894

Proetzel G, Roopenian DC (2014) Humanized FcRn mouse models for evaluating pharmacokinetics of human IgG antibodies. Methods 65:148–153

Rink J, Ghigo E, Kalaidzidis Y, Zerial M (2005) Rab conversion as a mechanism of progression from early to late endosomes. Cell 122:735–749

Roopenian DC, Christianson GJ, Sproule TJ, Brown AC, Akilesh S, Jung N, Petkova S, Avanessian L, Choi EY, Shaffer DJ, Eden PA, Anderson CL (2003) The MHC class I-like IgG receptor controls perinatal IgG transport, IgG homeostasis, and fate of IgG-Fc-coupled drugs. J Immunol 170:3528–3533

Roopenian DC, Low BE, Christianson GJ, Proetzel G, Sproule TJ, Wiles MV (2015) Albumin-deficient mouse models for studying metabolism of human albumin and pharmacokinetics of albumin-based drugs. MAbs 7:344–351

Saleh MN, Khazaeli MB, Wheeler RH, Dropcho E, Liu T, Urist M, Miller DM, Lawson S, Dixon P, Russell CH et al (1992) Phase I trial of the murine monoclonal anti-GD2 antibody 14G2a in metastatic melanoma. Cancer Res 52:4342–4347

Sand KM, Bern M, Nilsen J, Dalhus B, Gunnarsen KS, Cameron J, Grevys A, Bunting K, Sandlie I, Andersen JT (2014a) Interaction with both domain I and III of albumin is required for optimal pH-dependent binding to the neonatal Fc receptor (FcRn). J Biol Chem 289:34583–34594

Sand KM, Dalhus B, Christianson GJ, Bern M, Foss S, Cameron J, Sleep D, Bjoras M, Roopenian DC, Sandlie I, Andersen JT (2014b) Dissection of the neonatal Fc receptor (FcRn)-albumin interface using mutagenesis and anti-FcRn albumin-blocking antibodies. J Biol Chem 289:17228–17239

Sand KMK, Bern M, Nilsen J, Noordzij HT, Sandlie I, Andersen JT (2015) Unraveling the interaction between FcRn and albumin: opportunities for design of albumin-based therapeutics. Front Immunol 5:1–21
Santagostino E, Negrier C, Klamroth R, Tiede A, Pabinger-Fasching I, Voigt C, Jacobs I, Morfini M (2012) Safety and pharmacokinetics of a novel recombinant fusion protein linking coagulation factor IX with albumin (rIX-FP) in hemophilia B patients. Blood 120:2405–2411
Santagostino E, Martinowitz U, Lissitchkov T, Pan-Petesch B, Hanabusa H, Oldenburg J, Boggio L, Negrier C, Pabinger I, von Depka Prondzinski M, Altisent C, Castaman G, Yamamoto K, Alvarez-Roman MT, Voigt C, Blackman N, Jacobs I (2016) Long acting recombinant coagulation factor IX albumin fusion protein (rIX-FP) in hemophilia B: results of a phase 3 trial. Blood 127:1761
Sarav M, Wang Y, Hack BK, Chang A, Jensen M, Bao LH, Quigg RJ (2009) Renal FcRn reclaims albumin but facilitates elimination of IgG. J Am Soc Nephrol 20:1941–1952
Schlachetzki F, Zhu CN, Pardridge WM (2002) Expression of the neonatal Fc receptor (FcRn) at the blood-brain barrier. J Neurochem 81:203–206
Schlapschy M, Theobald I, Mack H, Schottelius M, Wester HJ, Skerra A (2007) Fusion of a recombinant antibody fragment with a homo-amino-acid polymer: effects on biophysical properties and prolonged plasma half-life. Protein Eng Des Sel 20:273–284
Schmidt MM, Townson SA, Andreucci AJ, King BM, Schirmer EB, Murillo AJ, Dombrowski C, Tisdale AW, Lowden PA, Masci AL, Kovalchin JT, Erbe DV, Wittrup KD, Furfine ES, Barnes TM (2013) Crystal structure of an HSA/FcRn complex reveals recycling by competitive mimicry of HSA ligands at a pH-dependent hydrophobic interface. Structure 21:1966–1978
Schultze HE, Heremans JF (1966) Molecular biology of human proteins: with special reference to plasma proteins. Elsevier, New York, 1
Simister NE, Mostov KE (1989) An Fc receptor structurally related to MHC class I antigens. Nature 337:184–187
Simister NE, Rees AR (1985) Isolation and characterization of an Fc receptor from neonatal rat small intestine. Eur J Immunol 15:733–738
Simister NE, Story CM, Chen HL, Hunt JS (1996) An IgG-transporting Fc receptor expressed in the syncytiotrophoblast of human placenta. Eur J Immunol 26:1527–1531
Simon M, Frey R, Zangemeister-Wittke U, Pluckthun A (2013) Orthogonal assembly of a designed ankyrin repeat protein-cytotoxin conjugate with a clickable serum albumin module for half-life extension. Bioconjug Chem 24:1955–1966
Smith BJ, Popplewell A, Athwal D, Chapman AP, Heywood S, West SM, Carrington B, Nesbitt A, Lawson AD, Antoniw P, Eddelston A, Suitters A (2001) Prolonged in vivo residence times of antibody fragments associated with albumin. Bioconjug Chem 12:750–756
Sockolosky JT, Szoka FC (2015) The neonatal Fc receptor, FcRn, as a target for drug delivery and therapy. Adv Drug Deliv Rev 91:109–124
Spiekermann GM, Finn PW, Ward ES, Dumont J, Dickinson BL, Blumberg RS, Lencer WI (2002) Receptor-mediated immunoglobulin G transport across mucosal barriers in adult life: functional expression of FcRn in the mammalian lung. J Exp Med 196:303–310
Stehle G, Sinn H, Wunder A, Schrenk HH, Stewart JCM, Hartung G, MaierBorst W, Heene DL (1997a) Plasma protein (albumin) catabolism by the tumor itself – implications for tumor metabolism and the genesis of cachexia. Crit Rev Oncol Hematol 26:77–100
Stehle G, Wunder A, Sinn H, Schrenk HH, Schutt S, Frei E, Hartung G, Maier-Borst W, Heene DL (1997b) Pharmacokinetics of methotrexate-albumin conjugates in tumor-bearing rats. Anticancer Drugs 8:835–844
Storm T, Tranebjaerg L, Frykholm C, Birn H, Verroust PJ, Neveus T, Sundelin B, Hertz JM, Holmstrom G, Ericson K, Christensen EI, Nielsen R (2013) Renal phenotypic investigations of megalin-deficient patients: novel insights into tubular proteinuria and albumin filtration. Nephrol Dial Transplant 28:585–591
Subramanian GM, Fiscella M, Lamouse-Smith A, Zeuzem S, McHutchison JG (2007) Albinterferon alpha-2b: a genetic fusion protein for the treatment of chronic hepatitis C. Nat Biotechnol 25:1411–1419

Sugio S, Kashima A, Mochizuki S, Noda M, Kobayashi K (1999) Crystal structure of human serum albumin at 2.5 angstrom resolution. Protein Eng 12:439–446

Sung C, Nardelli B, Lafleur DW, Blatter E, Corcoran M, Olsen HS, Birse CE, Pickeral OK, Zhang JL, Shah D, Moody G, Gentz S, Beebe L, Moore PA (2003) An IFN-beta-albumin fusion protein that displays improved pharmacokinetic and pharmacodynamic properties in nonhuman primates. J Interf Cytokine Res 23:25–36

Syed S, Schuyler PD, Kulczycky M, Sheffield WP (1997) Potent antithrombin activity and delayed clearance from the circulation characterize recombinant hirudin genetically fused to albumin. Blood 89:3243–3252

Tam SH, McCarthy SG, Brosnan K, Goldberg KM, Scallon BJ (2013) Correlations between pharmacokinetics of IgG antibodies in primates vs. FcRn-transgenic mice reveal a rodent model with predictive capabilities. MAbs 5:397–405

Tenten V, Menzel S, Kunter U, Sicking EM, van Roeyen CRC, Sanden SK, Kaldenbach M, Boor P, Fuss A, Uhlig S, Lanzmich R, Willemsen B, Dijkman H, Grepl M, Wild K, Kriz W, Smeets B, Floege J, Moeller MJ (2013) Albumin is recycled from the primary urine by tubular transcytosis. J Am Soc Nephrol 24:1966–1980

Tesar DB, Tiangco NE, Bjorkman PJ (2006) Ligand valency affects transcytosis, recycling and intracellular trafficking mediated by the neonatal Fc receptor. Traffic 7:1127–1142

Tijink BM, Laeremans T, Budde M, Stigter-van Walsum M, Dreier T, de Haard HJ, Leemans CR, van Dongen GA (2008) Improved tumor targeting of anti-epidermal growth factor receptor nanobodies through albumin binding: taking advantage of modular nanobody technology. Mol Cancer Ther 7:2288–2297

Tiwari B, Junghans RP (2005) Functional analysis of the mouse Fcgrt 5′ proximal promoter. Biochim Biophys Acta 1681:88–98

Tolmachev V, Orlova A, Pehrson R, Galli J, Baastrup B, Andersson K, Sandstrom M, Rosik D, Carlsson J, Lundqvist H, Wennborg A, Nilsson FY (2007) Radionuclide therapy of HER2-positive microxenografts using a Lu-177-labeled HER2-specific affibody molecule. Cancer Res 67:2773–2782

Tzaban S, Massol RH, Yen E, Hamman W, Frank SR, Lapierre LA, Hansen SH, Goldenring JR, Blumberg RS, Lencer WI (2009) The recycling and transcytotic pathways for IgG transport by FcRn are distinct and display an inherent polarity. J Cell Biol 185:673–684

Vidarsson G, Stemerding AM, Stapleton NM, Spliethoff SE, Janssen H, Rebers FE, de Haas M, van de Winkel JG (2006) FcRn: an IgG receptor on phagocytes with a novel role in phagocytosis. Blood 108:3573–3579

Viuff D, Antunes F, Evans L, Cameron J, Dyrnesli H, Thue Ravn B, Stougaard M, Thiam K, Andersen B, Kjaerulff S, Howard KA (2016) Generation of a double transgenic humanized neonatal Fc receptor (FcRn)/albumin mouse to study the pharmacokinetics of albumin-linked drugs. J Control Release: Off J Control Release Soc 223:22–30

Von Hoff DD, Ramanathan RK, Borad MJ, Laheru DA, Smith LS, Wood TE, Korn RL, Desai N, Trieu V, Iglesias JL, Zhang H, Soon-Shiong P, Shi T, Rajeshkumar NV, Maitra A, Hidalgo M (2011) Gemcitabine plus nab-paclitaxel is an active regimen in patients with advanced pancreatic cancer: a phase I/II trial. J Clin Oncol 29:4548–4554

Waldmann TA, Terry WD (1990) Familial hypercatabolic hypoproteinemia. A disorder of endogenous catabolism of albumin and immunoglobulin. J Clin Invest 86:2093–2098

Walker A, Dunlevy G, Rycroft D, Topley P, Holt LJ, Herbert T, Davies M, Cook F, Holmes S, Jespers L, Herring C (2010) Anti-serum albumin domain antibodies in the development of highly potent, efficacious and long-acting interferon. Protein Eng Des Sel 23:271–278

Wani MA, Haynes LD, Kim J, Bronson CL, Chaudhury C, Mohanty S, Waldmann TA, Robinson JM, Anderson CL (2006) Familial hypercatabolic hypoproteinemia caused by deficiency of the neonatal Fc receptor, FcRn, due to a mutant beta2-microglobulin gene. Proc Natl Acad Sci U S A 103:5084–5089

Ward ES, Gussow D, Griffiths AD, Jones PT, Winter G (1989) Binding activities of a repertoire of single immunoglobulin variable domains secreted from Escherichia coli. Nature 341:544–546

Ward ES, Zhou J, Ghetie V, Ober RJ (2003) Evidence to support the cellular mechanism involved in serum IgG homeostasis in humans. Int Immunol 15:187–195

Ward ES, Martinez C, Vaccaro C, Zhou J, Tang Q, Ober RJ (2005) From sorting endosomes to exocytosis: association of Rab4 and Rab11 GTPases with the Fc receptor, FcRn, during recycling. Mol Biol Cell 16:2028–2038

Wernick NL, Haucke V, Simister NE (2005) Recognition of the tryptophan-based endocytosis signal in the neonatal Fc receptor by the mu subunit of adaptor protein-2. J Biol Chem 280:7309–7316

West AP Jr, Bjorkman PJ (2000) Crystal structure and immunoglobulin G binding properties of the human major histocompatibility complex-related Fc receptor(,). Biochemistry 39:9698–9708

Wu Z, Simister NE (2001) Tryptophan- and dileucine-based endocytosis signals in the neonatal Fc receptor. J Biol Chem 276:5240–5247

Wunder A, Muller-Ladner U, Stelzer E, Neumann E, Sinn H, Gay S, Fiehn C (2003) Albumin-based drug delivery as novel therapeutic approach for rheumatoid arthritis. Arthritis Res Ther 5:S4

Yazaki PJ, Kassa T, Cheung CW, Crow DM, Sherman MA, Bading JR, Anderson ALJ, Colcher D, Raubitschek A (2008) Biodistribution and tumor imaging of an anti-CEA single-chain antibody-albumin fusion protein. Nucl Med Biol 35:151–158

Yeh P, Landais D, Lemaitre M, Maury I, Crenne JY, Becquart J et al (1992) Design of yeast-secreted albumin derivatives for human therapy: biological and antiviral properties of a serum albumin-CD4 genetic conjugate. Proc Natl Acad Sci U S A 89(5):1904–1908

Zalevsky J, Chamberlain AK, Horton HM, Karki S, Leung IWL, Sproule TJ, Lazar GA, Roopenian DC, Desjarlais JR (2010) Enhanced antibody half-life improves in vivo activity. Nat Biotechnol 28:157–159

Zhen Y, Stenmark H (2015) Cellular functions of Rab GTPases at a glance. J Cell Sci 128:3171–3176

Zhu XP, Meng G, Dickinson BL, Li XT, Mizoguchi E, Miao LL, Wang YS, Robert C, Wu BY, Smith PD, Lencer WI, Blumberg RS (2001) MHC class I-related neonatal Fc receptor for IgG is functionally expressed in monocytes, intestinal macrophages, and dendritic cells. J Immunol 166:3266–3276

Chapter 11
Human Serum Albumin in Blood Detoxification Treatment

Victor Tuan Giam Chuang, Toru Maruyama, and Masaki Otagiri

Abstract Human serum albumin (HSA) is known to bind a broad spectrum of endogenous and exogenous substances. The ligand-binding property of albumin has been utilized to remove endogenous toxins in extracorporeal blood detoxification methods such as single-pass albumin dialysis (SPAD), fractionated plasma separation and adsorption (FPSA), Prometheus®, and molecular adsorbent recirculating system (MARS). Production of recombinant HSA including individual domains has been successfully attempted by a number of researchers. The albumin domains retain similar structural characteristics of the HSA. The ligand-binding properties of albumin domains are identical to those of HSA but with lower binding affinity and percentage for most of the ligands studied. The albumin domains have an increased elimination profile compared to that of the HSA. Molecular modification of the albumin domains through site-directed mutagenesis for strengthening toxin binding is a feasible approach for improving the efficiency and effectiveness of blood detoxification treatment.

Keywords Human serum albumin • Recombinant technology • Domain • Detoxification

---

V.T.G. Chuang (✉)
School of Pharmacy, Faculty of Health Sciences, Curtin University,
GPO Box U1987, Perth 6845, WA, Australia
e-mail: V.Chuang@curtin.edu.au

T. Maruyama
Department of Biopharmaceutics, Graduate School of Pharmaceutical Sciences, Kumamoto University, 5-1, Oe-honmachi, Chuo-ku, Kumamoto 862-0973, Japan

Center for Clinical Pharmaceutical Sciences, School of Pharmacy, Kumamoto University, 5-1, Oe-honmachi, Chuo-ku, Kumamoto 862-0973, Japan

M. Otagiri
School of Pharmacy, Faculty of Health Sciences, Curtin University,
GPO Box U1987, Perth 6845, WA, Australia

School of Pharmaceutical Sciences, Kumamoto University,
1-5 Oe-honmachi, Kumamoto-shi, Kumamoto 862-0973, Japan

School of Pharmaceutical Sciences, Sojo University,
4-22-1 Ikeda, Kumamoto-shi, Kumamoto 860-0082, Japan

M. Otagiri, V.T.G. Chuang (eds.), *Albumin in Medicine*,
DOI 10.1007/978-981-10-2116-9_11

11.1 Introduction

Albumin is an important transport protein especially for transporting hydrophobic protein-bound endogenous substances, drugs, and uremic toxins (Table 11.1). Most of these endogenous substances are produced in the body and are excreted into urine and feces via the kidney and liver (bile), respectively, in healthy subjects. A decreased detoxification function associated with liver insufficiency is typically manifested by hyperbilirubinemia (jaundice), encephalopathy, and a reduction in synthetic function leading to the development of hypoalbuminemia. The accumulation of albumin-bound water-insoluble toxins, as a result of impaired hepatic detoxification and/or metabolism, includes, among others, aromatic amino acids, tryptophan, indoles, mercaptans, endogenous benzodiazepines, ammonia, prostanoids, and nitric oxide. The accumulation of albumin-bound toxins is responsible for the associated end-organ dysfunctions (the kidney, circulation, brain). These toxins accumulate in the body and cause severe and life-threatening complications including coagulopathy, encephalopathy, cerebral edema, the hepatorenal syndrome, aggravating hyperdynamic circulation, and an increased susceptibility to infections (Stockmann and IJzermans 2002; Hoofnagle et al. 1995; Jalan and Williams 2002; Losser and Payen 1996; Shakil et al. 2000).

On the other hand, in renal impairment or failure, a myriad of compounds with different water solubilities are retained and accumulate in the patients. Most of the water-insoluble compounds are protein-bound, mainly to albumin. The protein-bound uremic solutes identified so far that contribute to the uremic syndrome, or uremic toxins, include p-cresol, indoxyl sulfate, hippuric acid, 3-carboxy-4-methyl-5-propyl-2-furan-propionic acid (CMPF), and homocysteine. Uremic toxins at concentrations that are seen during uremia have been reported to cause a variety of negative effects on almost all organ systems, including the cardiovascular system and the development of immunological and neurological symptoms. They are also known to contribute to the increase of the rate of progression of renal failure.

Table 11.1 Protein-bound endogenous substances and toxins

Compound	Primary association constant, k_1	Secondary association constant, k_2	n	Reference
Bilirubin	5.50×10^7		1	(Jacobsen 1969)
CMPF	1.30×10^7		1	(Tsutsumi et al. 1999)
p-cresyl sulfate	1.00×10^5	1.96×10^4	2	(Watanabe et al. 2012)
Indoxyl sulfate	0.98×10^5	8.0×10^3	2	(Watanabe et al. 2012)
Tryptophan	1.14×10^4		1	(Mingrone et al. 1997)
Bile acids	$1–20 \times 10^4$	$3.0 \times 10^2–4.1 \times 10^4$	2	(Roda et al. 1982)

CMPF 3-Carboxy-4-methyl-5-propyl-2-furanpropionate

11.2 Albumin in Extracorporeal Blood Detoxification

Numerous procedures have been used to remove these toxins, including plasma and blood exchange (Kondrup et al. 1992; Larsen et al. 1994; Trey et al. 1966), hemodialysis (Keller et al. 1995; Kiley et al. 1958), hemoadsorption using activated charcoal (Gimson et al. 1982; O'Grady et al. 1988), and hemodiabsorption with charcoal or ion-exchange resins (Hughes et al. 1994). However, it has been reported that these procedures are not capable of effectively removing protein-binding toxins such as bilirubin in both liver and renal failure, where these protein-binding toxins accumulate to significant levels. In order to overcome this shortcoming, the toxin-binding property of albumin can be exploited for the removal of these protein-binding endogenous toxins in several types of albumin dialysis systems (Tsipotis et al. 2015). Three main albumin dialysis methods have recently emerged as the most promising extracorporeal blood detoxification method (Table 11.2).

11.2.1 Single-Pass Albumin Dialysis (SPAD)

SPAD is the simplest form of albumin dialysis (Fig. 11.1). The use of SPAD in the clinic was initially described by Seige et al. in 1999 (Seige et al. 1999). In this procedure, plasma flows from a reservoir through a standard albumin-impermeable high-flux dialyzer and is dialyzed against a low concentration of an albumin dialysate solution. Molecules that are small enough to pass through the membrane pores are retained in the dialysate as a result of binding to albumin. However, the albumin dialysate is not recyclable and is disposed of after a single pass.

Table 11.2 A comparison of the characteristics of albumin dialysis systems

	SPAD	Prometheus	MARS®
Circuit	Refer to Fig. 11.1	Refer to Fig. 11.2	Refer to Fig. 11.3
Membrane with cutoff	<50 kDa	<250 kDa	<50 kDa
Adsorbents	No adsorbent	Neutral resin Anion exchanger	Uncoated activated charcoal Anion exchanger
Albumin dialysate (concentration)	Commercial HSA without precleaning (4.4 %)	No albumin dialysate	Commercial HSA with precleaning (20 %)
Extracorporeal blood volume	Small amount of extracorporeal blood volume	Large amount of extracorporeal blood volume	Small amount of extracorporeal blood volume

Modified from Mitzner et al. (2006)

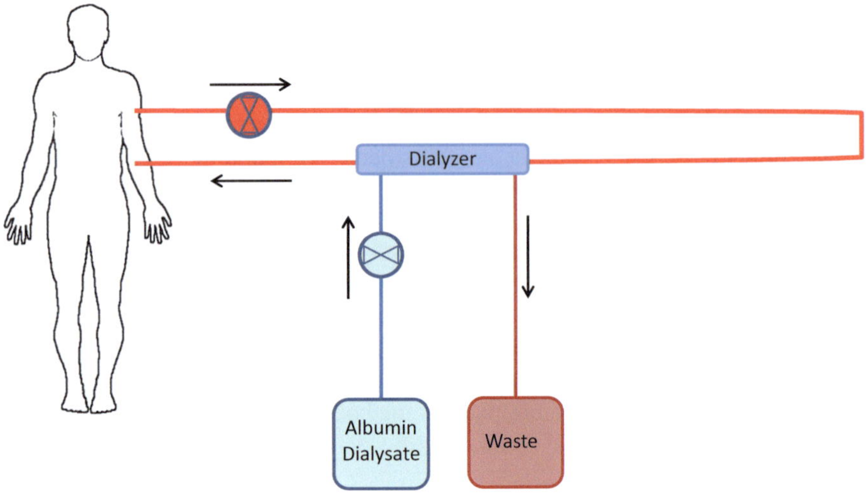

Fig. 11.1 Schematic configuration of the SPAD system

11.2.2 Fractionated Plasma Separation and Adsorption (FPSA): Prometheus®

The FPSA (Prometheus®, Fresenius Medical Care Allgemeine Gesellschaft, Bad Homburg, Germany) system was introduced by Falkenhagen in 1999, for the removal of both albumin-bound and water-soluble toxins (Falkenhagen et al. 1999) (Fig. 11.2). All toxin molecules with sizes that are equal to or lower than albumin are first separated from the blood by a capillary dialyser (AlbuFlow®, Fresenius Medical Care AG, Bad Homburg, Germany) made of polysulfone hollow fibers with a size-selection threshold of 250 kDa which is permeable to both albumin and albumin-bound substances. After passage through the AlbuFlow®, the patient's blood is dialyzed through a high-flux dialyser (FX50), to eliminate water-soluble toxins. The albumin filtrate is then perfused through a column containing a neutral resin (prometh01) and a second column containing an anion-exchange resin adsorber (prometh02). All of the bound or unbound toxin molecules are captured by direct contact with the high-affinity adsorbing material, thereby purifying the endogenous albumin in the albumin filtrate. The purified endogenous albumin is then returned to the patient. In some cases, it is necessary to perform an infusion of an albumin solution to make up for albumin that is lost after the Prometheus treatment (Rifai et al. 2003).

11.2.3 Molecular Adsorbent Recirculating System (MARS®)

The MARS® (Gambro, Stockholm, Sweden) system, based on the principles of dialysis, filtration, and adsorption, was developed by Stange and Mitzner in 1993 and was applied for the first time in humans in 1996 (Stange et al. 1993) (Fig. 11.3).

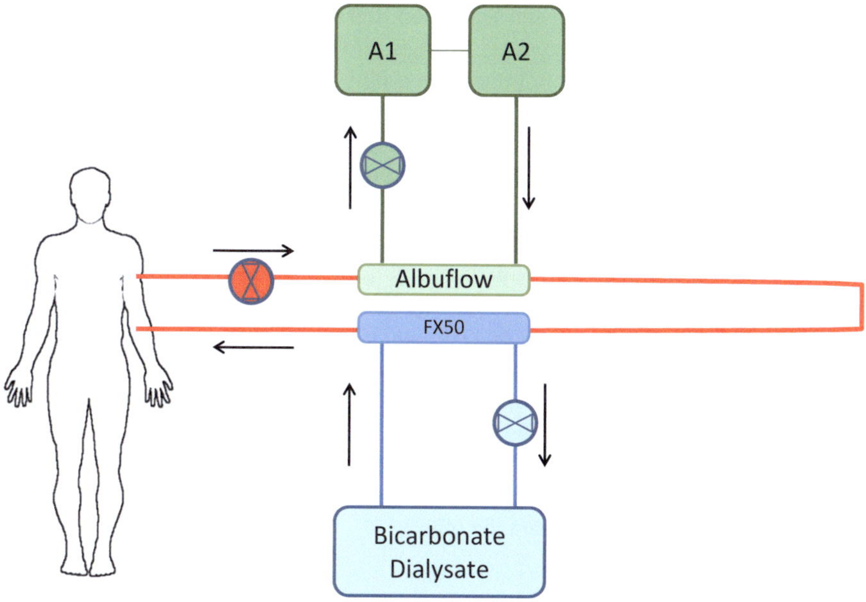

Fig. 11.2 Schematic configuration of the Prometheus® system (Fresenius, Germany). *A1* prometh01, *A2* prometh02

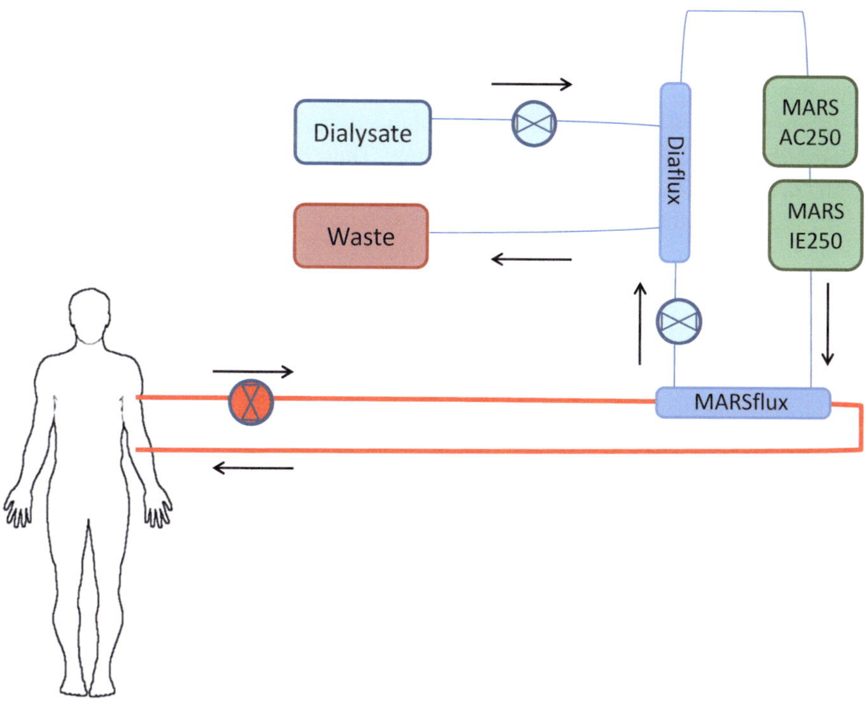

Fig. 11.3 Schematic configuration of the MARS® (Gambro, Germany)

MARS® does not require a plasma separation step or direct plasma perfusion over a sorbent. The patient's blood is passed through a hollow-fiber dialysis module where it is dialyzed across an albumin impregnated polysulfone membrane (MARS Flux dialyzer; membrane thickness: 100 nm; pore size: 50 kDa, surface area: 2.1 m^2), against an albumin-rich dialysate (600 ml of a 20 % human albumin solution). This enables water-soluble and protein-bound toxins, such as ammonia, bilirubin, bile acids, medium- and short-chain fatty acids, creatine, and urea, to be removed (Mitzner et al. 2001). The albumin-coated membrane transiently absorbs and holds the toxins upon making contact with the membrane, and the toxins are then released to the other side of the membrane by virtue of a concentration gradient where the toxins are retained as the result of dialysis against the albumin-containing dialysate. The toxin-enriched albumin dialysate is then passed through another dialyzer countercurrent to a standard buffered dialysis solution to remove water-soluble toxins by diffusion. Meanwhile, the protein-bound toxins are removed from the albumin by passing the albumin dialysate through two columns: an anion-exchanger column and an uncoated charcoal column to regenerate purified albumin in the dialysate for the further removal of toxins from the blood.

11.3 Recombinant HSA Domains

The intracorporeal blood detoxification methods that are in common use include forced diuresis, gastrointestinal adsorption, device-assisted or monitoring large intestinal purification, peritoneal dialysis, and intestinal lavage. Like the extracorporeal blood detoxification methods, most, if not all of the detoxification procedures, require that the patients be immobilized on a bed with the detoxifying equipment attached for a certain length of time. Thus, both intra- and extracorporeal blood detoxification methods are not without their problems.

In particular, the removal of highly protein-bound uremic toxins remains an issue. Furthermore, a lack of a convenient, flexible, and quick blood detoxification method for the immediate relief of acute intoxication symptoms warrants further development of a novel or improved blood detoxification method. In this regard, it would be logical to explore the use of a molecularly modified form of albumin for the production of a more robust biomaterial that could satisfy the various needs of the blood detoxification treatment procedures. One logical approach would be to increase the efficiency of the molecule in terms of capturing toxins, as well as the dialyzable volume via the use of a compacted albumin matrix in the cleansing column or dialysate. This concept could be achieved by developing a binding protein with a smaller size than albumin that would occupy a smaller volume in the column.

HSA is a heart-shaped protein comprised of three homologous domains, each of which is further divided into two subdomains A and B (Curry et al. 1998). The helical pattern is similar for all three domains (Fig. 11.4). In addition to a high level of structural similarity among these three domains, the physicochemical properties of

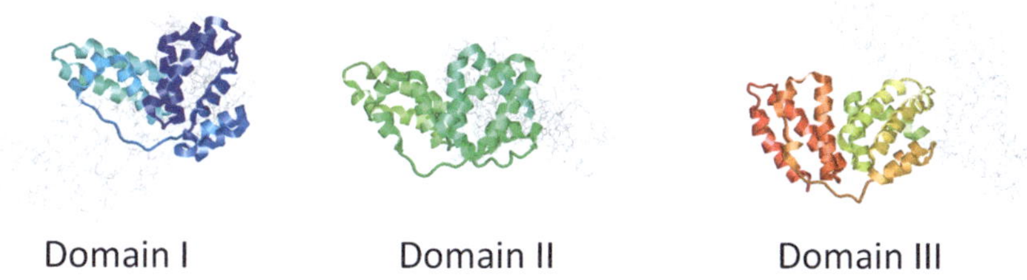

Fig. 11.4 Three HSA domains, namely, Domains *I*, *II*, and *III*, prepared with RASMOL using pdb file 1E7E. Each highlighted domain is depicted in a cartoon representation and colored in the default domain mode, while the others are depicted in *wireframe* representation and colored in *monochrome*

each domain are influenced by the presence of different reactive amino acids that are unique to each domain (Peters 1996). Hence, identification and production of the albumin domains that bind toxins could be explored for use as a detoxifying matrix.

11.3.1 Recombinant Production of Albumin Domains

The *in vitro* expression of albumin can be traced back to the early 1980s when attempts were made to achieve this in rat hepatoma cells (Cassio et al. 1981). Okabayashi et al. reported the secretary expression of HSA gene in yeast, *Saccharomyces cerevisiae* (Okabayashi et al. 1991). Shani et al. reported on the expression of HSA in milk of transgenic mice (Shani et al. 1992), while Fleer et al. expressed HSA using *Kluyveromyces* yeasts (Fleer et al. 1991). To date, *Pichia pastoris* is a popular yeast expression system for expressing HSA or for conducting site-directed mutagenesis for HSA studies (Kobayashi et al. 1998; Ohtani et al. 1997, 1998a, b; Ikegaya et al. 1997; Watanabe et al. 2001; Chuang and Otagiri 2007). For recombinant albumin domain production, domains I and III were reported to have been successfully expressed and secreted in *Saccharomyces cerevisiae* (Kjeldsen et al. 1998) and *Escherichia coli* (Mao et al. 2000) expression systems. However, neither of these expression systems were able to produce domain II satisfactorily. In contrast, the expression of all three human albumin domains was successfully accomplished using *Pichia pastoris* (Dockal et al. 1999; Matsushita et al. 2004; Park et al. 1999).

11.3.2 Structural Properties of Albumin Domains

HSA has a high α-helical content (~67 %), no β-sheet, 10 % β-turn, 23 % extended chain and an unusually large number of 17 disulfide bonds in total (Peters 1996). It is anticipated that all three individual albumin domains would have undergone a certain degree of structural change in comparison to HSA, as evidenced by studies conducted by Matsushita et al. (2004) and Dockal et al. (1999). In general, all stand-alone domains showed a decrease in alpha-helix content (to less than 66 %), an increase in beta-sheet content (to more than 6 %) except for domain III (0 %), a decrease in coil content (to less than 26 %) except for domain III (37 %), and an increase in turn content for all domains (more than 2 %) (Dockal et al. 1999). It is interesting to note that the ellipticities, calculated by combining three of the domain signals, approached that of a wild-type HSA (Matsushita et al. 2004). The structural changes of the domains as a consequence of genetic truncation or chemical cleavage can affect the physicochemical as well as functional properties of each of the domains, since the flexibility of each domain would be expected to increase.

Domain I of HSA comprises the amino acid sequence starting from 1 to 197. Domain I undergoes a structural rearrangement with only minor changes in its secondary structure as the pH is reduced to between pH 5.0 and 3.5 of the N-F transition. At the pH 7.0 and 9.0 region of the N-B transition of the HSA, domain I undergoes a tertiary structural isomerization (Peters 1996; Dockal et al. 2000a). In a functional analysis of the domain I protein expressed using a yeast system, Matsushita et al. reported the existence of weak esterase-like as well as enolase-like activity in domain I. Domain I has also been shown to exhibit antioxidant activity comparable to that of rHSA in the same study (Matsushita et al. 2004).

In the N-F transition (pH 5.0–3.5), domain II transforms to a molten globule-like state as the pH is reduced, but in the pH region of the N-B transition (pH 7.0–9.0), domain II experiences a tertiary structural isomerization (Dockal et al. 2000a). Domain II has enolase-like activity in a pH 9.2 carbonate buffer and a low activity in a pH 7.4 phosphate buffer, but esterase-like activity has not been detected (Matsushita et al. 2004).

The loosening of the HSA structure in the N-F transition takes place primarily in domain III; in contrast, no alterations in tertiary structure are observed in the pH region of the N-B transition of HSA (Dockal et al. 2000a). Independent domain III was found to have retained about 45 % of its esterase-like activity, in addition to showing a low enolase-like activity in a pH 7.4 phosphate buffer (Matsushita et al. 2004).

11.3.3 Ligand-Binding Properties of Albumin Domains

Domain I contains a cysteine residue with a free thiol unit at 34 that can interact covalently with a number of drugs such as bucillamine and ethacrynic acid (Narazaki et al. 1996; Narazaki and Otagiri 1997; Bertucci et al. 1998). In addition, Cys34 may account for more than 40 % of the total antioxidant effect of HSA in vivo (Anraku et al. 2011). S-Nitrosylation of Cys34 has been shown to provide a cytoprotective effect on liver cells in a rat ischemia/reperfusion model (Ishima et al. 2008). HSA is known to have a metal-binding site for copper at the N-terminal involving His3 (Dixon and Sarkar 1972). Shearer et al. reported an analogous

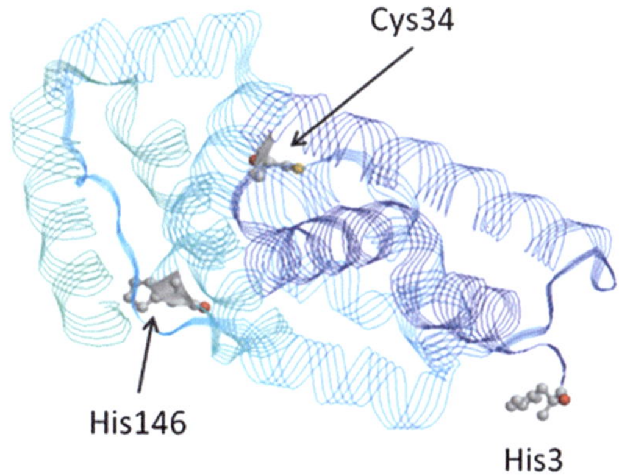

Fig. 11.5 Domain I showing the reactive amino acid residues prepared with RASMOL using pdb file 1E7E

copper-binding site (Asp-Thr-His) with peptide 1–24 of bovine serum albumin (Shearer et al. 1967). Although no study has been carried out using a single domain to evaluate the binding of copper thus far, a metal-binding site is still predicted to exist on the single domain protein, since it involves the N-terminal sequence of the protein (Fig. 11.5).

Hemin and long-chain fatty acids were found to bind subdomain IB of HSA (Zunszain et al. 2003). This hemin-binding cavity in subdomain IB appears to be preserved when domain I is a stand-alone protein (Dockal et al. 1999; Monzani et al. 2002). 4Z,15Z-bilirubin-IXalpha bilirubin is an insoluble yellow-orange pigment derived from the catabolism of heme. Although the binding region for 4Z,15Z-bilirubin-IXalpha bilirubin was proposed to be located in subdomain IIA (Sudlow site I), the more soluble and excretable isomer, 4Z,15E-bilirubin-IXalpha isomer, binds to an L-shaped pocket in subdomain IB (Zunszain et al. 2008) (Fig. 11.6).

One molecule of a long-chain fatty acid was found to bind at subdomain 1B, while another binds at the interface of subdomains IA–IIA of HSA (Curry et al. 1998, 1999) (Fig. 11.7). However, a fatty acid-binding site at the interface of subdomains IA–IIA may not exist on the independent domain I protein.

Domain II, a major drug-binding site (Sudlow site I), contains the only tryptophan residue of HSA and spans from residue 189 to residue 385. From studies conducted by Dockal et al. and Matsushita et al., it appears that the binding of ligands to site I of domain II was somehow compromised (Dockal et al. 1999, 2000b). Negligible binding of warfarin and DNSA (normally bound at subbinding regions Ia and Ic of site I, respectively) was found to domain II, but the domain retains a considerable portion of subregion Ic (n-butyl p-AB) ligand-binding property (9.4 % compared with 14.4 % of rHSA). The other two domains I and III showed comparatively low and insignificant binding percentages to all of these three site I binding subregion probes (Matsushita et al. 2004). This confirms the importance of domain I on the binding of site I ligands to HSA, since Tyr150 from subdomain IB has been shown to play a role in the binding of site I ligands such as warfarin in drug-HSA complex crystal structures (Petitpas et al. 2001; Ghuman et al. 2005) (Fig. 11.8).

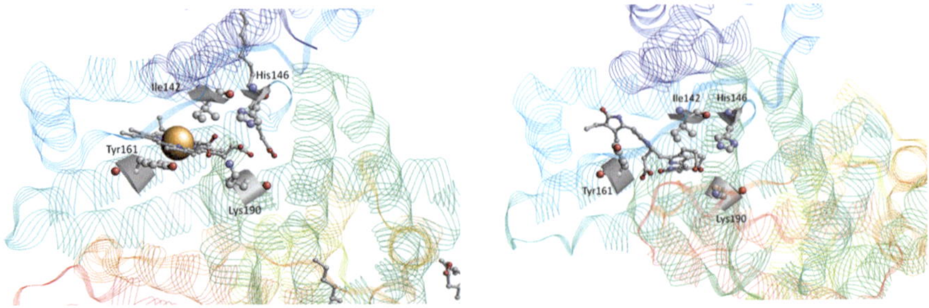

A. B.

Fig. 11.6 (**a**) Binding conformation of heme in the presence of myristates at subdomain 1B of HSA. (**b**) Binding conformation of 4Z,15E-bilirubin-IXalpha at subdomain IB of HSA. Both structures were prepared with RASMOL using pdb file 1N5U(A) and 2VUE(B)

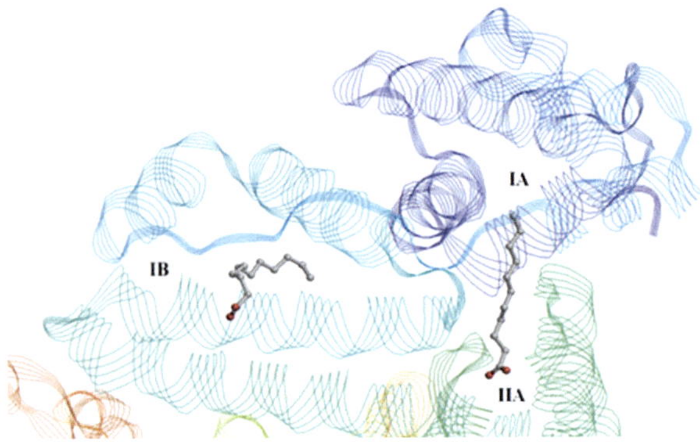

Fig. 11.7 Binding conformation of two myristate molecules at the interface of subdomains IA–IIA and subdomain 1B of HSA prepared with RASMOL using pdb file 1BJ5

In addition to drugs, HSA is a major transporter of endogenous hydrophobic substances such as fatty acids, thyroxine, uremic toxin CMPF, and bilirubin. Two myristate molecules can bind at the interfaces of subdomains IA–IIA and IIB–IIIA. The drug-binding site at subdomain IIA is considered to be an insignificant binding site for fatty acids (Fig. 11.9).

Petitpas et al. (based on crystal analyses) reported four binding sites for thyroxine that were distributed in subdomains IIA, IIIA, and IIIB (Fig. 11.10a). In addition, the mutation of residue R218 (R218H and R218P) within subdomain IIA greatly enhanced the affinity for thyroxine. A fatty acid was shown to replace thyroxine at all four sites and induced conformational changes that created a fifth hormone-binding site in the cleft between domains I and III, at least 9 Å from R218 (Petitpas et al. 2003). On the other hand, Park et al. produced an HSA fragment that was comprised of subdomains IA, IB, and IIA (amino acids 1–297) which had an affinity for thyroxine and several thyroxine analogs similar to that observed for

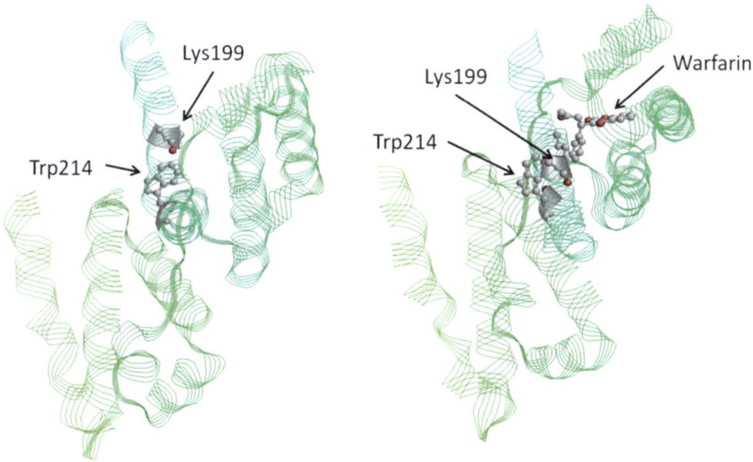

Fig. 11.8 HSA-DOM II, 189–385 prepared with RASMOL using pdb file

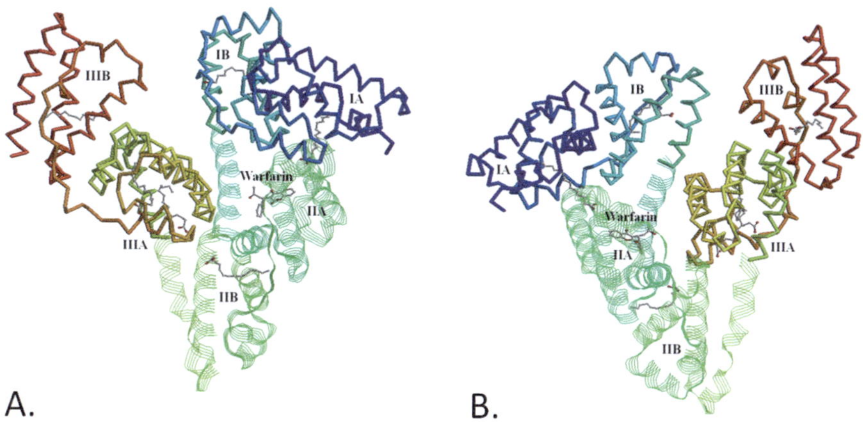

Fig. 11.9 (**a**) Binding conformation of warfarin at Sudlow drug-binding site I in the presence of myristates. (**b**) Back view of the same structure prepared with RASMOL using pdb file 2BXD

wild-type HSA (Park et al. 1999). The protein-bound uremic toxin, CMPF, was also found to bind to subdomain IIA (Fig. 11.10b).

Domain III spans from residues 381 to the last residue 585 and contains a binding pocket at subdomain IIIA (Sudlow site II) to which a number of drugs bind with high affinity. In contrast to domain II, domain III appears to have retained a good level of binding to site II ligands. Dockal et al. reported that, in independent domain III, the diazepam-binding site is preserved, in the absence of the other two domains (Dockal et al. 1999) (Fig. 11.11). This finding is in agreement with the findings reported by Matsushita et al. where the percentage of binding for ketoprofen was 64 % (80 % for rHSA) and 39 % for DNSS (62 % for rHSA), while the binding per-

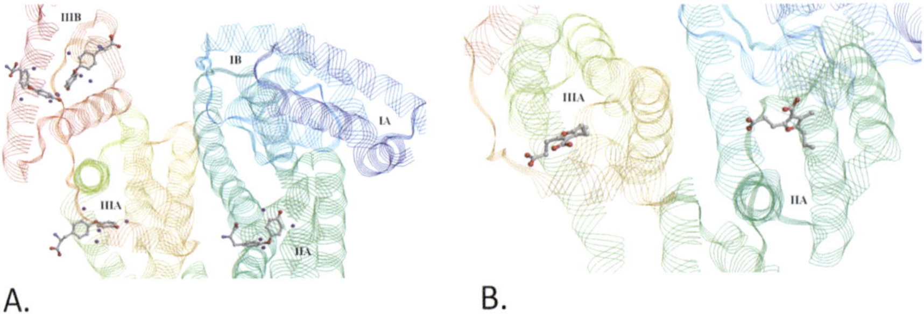

Fig. 11.10 (**a**) Binding conformation of thyroxine at the two major drug-binding sites as well as subdomain 1IIB of HSA. (**b**) Binding conformation of CMPF at the two major drug-binding sites of HSA. Both structures were prepared with RASMOL using pdb file 1HK1 (A) and 2BXA (B)

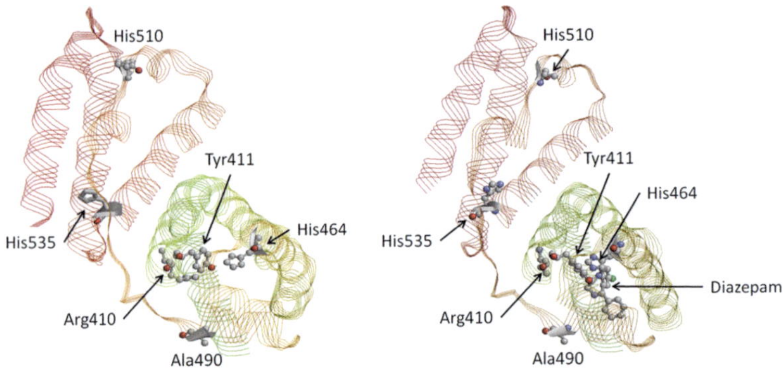

Fig. 11.11 HSA-DOM III, 381–585 prepared with RASMOL using pdb file

centage of these two ligands to the other two domains was insignificant (Matsushita et al. 2004). Liu et al. expressed the wild-type HSA domain III and a site-directed mutant of domain III (Y411W) in a yeast expression system. Both domain proteins retained their secondary structures and anesthetic-binding characteristics of an intact HSA molecule, but with fewer binding sites. The Y411W mutant domain III had a decreased binding affinity for propofol but not for 2-bromo-2-chloro-1,1,1-trifluoroethane (Liu et al. 2005).

11.3.4 Pharmacokinetics of Albumin Domains

Matsushita et al. genetically expressed each domain of HSA in *Pichia pastoris* and evaluated its total clearance as well as renal and hepatic uptake clearance in mice (Matsushita et al. 2004). All three domains exhibited a high level of renal clearance compared to rHSA, with about a 50-fold difference in total clearance between rHSA and the domains, but no significant difference among the three domains (rHSA;

3.65 ± 0.78 μl/min, domain I; 178.35 ± 9.72 μl/min, domain II; 181.37 ± 12.12 μl/min, domain III; 184.80 ± 15.21 μl/min) was found.

Domain I showed a slightly lower renal clearance than the other two domains (rHSA; 0.28 ± 0.10 μl/min, domain I; 147.33 ± 10.57 μl/min, domain II; 173.38 ± 12.35 μl/min, domain III; 171.14 ± 5.63 μl/min). Interestingly, domain III showed about a fourfold higher liver clearance than the other two domains (rHSA; 0.69 ± 0.21 μl/min, domain I; 5.12 ± 0.98 μl/min, domain II; 5.20 ± 2.52 μl/min, domain III; 19.09 ± 4.16 μl/min).

The results reported by Matsushita et al. were in agreement with the findings reported by Sheffield et al. where N-linked glycosylation and the truncation of rabbit albumin led to the shortening of the plasma half-lives of the proteins. The mean catabolic half-life was reduced to 2.87 days for the D494N glycosylated variant and to less than 0.071 days for all domains. These domains were found in the urine in tissue distribution experiments, suggesting a renal route of clearance. This indicates all three internally repeated albumin domains are required to maintain the slow in vivo clearance profile of albumin (Sheffield et al. 2000).

11.4 Potential Clinical Use of Domains

Bilirubin encephalopathy (kernicterus) is caused by very high levels of bilirubin, especially in the first few weeks of life in newborns (Allen et al. 2009). Hyperbilirubinemia may be due to the underdevelopment of the liver in newborns and especially in premature infants, in whom this organ is not able to convert 4*Z*, 15*Z*-BR to water-soluble forms, eg 4*Z*,15*E*-BR, that can be excreted into the bile. Bilirubin-albumin binding has been reported to be an important parameter in the evaluation of jaundiced newborns (Ahlfors and Wennberg 2004). A 5 % albumin infusion prior to exchange transfusion was found to significantly reduce post-exchange bilirubin levels and the phototherapy requirement (Mitra et al. 2011). The major preventative treatment for 4*Z*, 15*Z*-BR-induced encephalopathy is phototherapy, during which 4*Z* is converted to 15*Z*-BR, which primarily binds to its high-affinity HSA-binding site, into more soluble nontoxic structural isomers that are more easily eliminated from the circulatory system. The rate of formation of isomers and the isomeric composition of the reaction products were shown to be highly dependent on the conformation and electronic environment of 4*Z*, 15*Z*-BR bound to HSA (Khan et al. 2000). Onishi et al. reported that, as the serum bilirubin concentration increases to high and potentially dangerous levels, HSA helps to promote its photocyclization. During irradiation, HSA appears to act by retaining low, useful, concentrations of bilirubin while facilitating the irreversible photoisomerization of excess bilirubin (Onishi et al. 1989).

Tsutsumi et al. reported that the primary binding site for CMPF (Fig. 11.10b) and bilirubin was located at subdomain IIA (Tsutsumi et al. 1999). A site-directed mutagenesis study of HSA by Petersen et al. suggested the existence of a dynamic, unusually flexible high-affinity binding site for bilirubin on HSA (Petersen et al.

2000). We previously reported that both Lys195 and Lys199 in subdomain IIA of human serum albumin are important for the high-affinity binding of 4Z, 15Z-bilirubin-IXα (Minomo et al. 2011, 2013). Further screening of mutants that exhibit higher bilirubin binding affinity with phage display identified one such pan3_3-13 domain II mutant that has an increased binding affinity for bilirubin, but also increased urinary excretion in disease model mice as compared to treatment with the wild-type domain II. These results suggest that pan3_3-13 has great potential as a therapeutic agent that can be administered intravenously to promote urinary BR excretion in hyperbilirubinemia. This also indicates that an albumin domain II mutant could be more useful than human serum albumin in albumin dialysis as a matrix for the dialysis column or imprinting onto a membrane to remove toxins, including bilirubin, from blood to improve the efficiency of toxin removal.

11.5 Conclusion

Similar development approaches could be implemented on the other two albumin domains. For example, domain I, which is known to bind copper, could be developed further as a therapeutic agent for treating Wilson's disease. Further clinical and basic studies of the production of different albumin domain mutants including dimerization of albumin or albumin domains are expected to be of great use in the development of improved intra- as well as extracorporeal blood detoxification devices and beneficial treatment to prevent disease progression or alleviate toxicity symptoms for patients with liver or renal failure.

Acknowledgment Thank you

References

Ahlfors CE, Wennberg RP (2004) Bilirubin-albumin binding and neonatal jaundice. Semin Perinatol 28:334–339

Allen NM, Mohammad F, Foran A, Corcoran D, Clarke T (2009) Severe hyperbilirubinaemia and kernicterus: more caution is needed in newborn jaundice surveillance. Ir Med J 102:228–229

Anraku M, Takeuchi K, Watanabe H, Kadowaki D, Kitamura K, Tomita K, Kuniyasu A, Suenaga A, Maruyama T, Otagiri M (2011) Quantitative analysis of cysteine-34 on the antioxidative properties of human serum albumin in hemodialysis patients. J Pharm Sci 100:3968–3976

Bertucci C, Nanni B, Raffaelli A, Salvadori P (1998) Chemical modification of human albumin at cys34 by ethacrynic acid: structural characterisation and binding properties. J Pharm Biomed Anal 18:127–136

Cassio D, Weiss MC, Ott MO, Sala-Trepat JM, Fries J, Erdos T (1981) Expression of the albumin gene in rat hepatoma cells and their dedifferentiated variants. Cell 27:351–358

Chuang VT, Otagiri M (2007) Recombinant human serum albumin. Drugs Today (Barc) 43:547–561

Curry S, Mandelkow H, Brick P, Franks N (1998) Crystal structure of human serum albumin complexed with fatty acid reveals an asymmetric distribution of binding sites. Nat Struct Biol 5:827–835
Curry S, Brick P, Franks NP (1999) Fatty acid binding to human serum albumin: new insights from crystallographic studies. Biochim Biophys Acta 1441:131–140
Dixon JW, Sarkar B (1972) Absence of a specific copper(II) binding site in dog albumin is due to amino acid mutation in position 3. Biochem Biophys Res Commun 48:197–200
Dockal M, Carter DC, Ruker F (1999) The three recombinant domains of human serum albumin. Structural characterization and ligand binding properties. J Biol Chem 274:29303–29310
Dockal M, Carter DC, Ruker F (2000a) Conformational transitions of the three recombinant domains of human serum albumin depending on pH. J Biol Chem 275:3042–3050
Dockal M, Chang M, Carter DC, Ruker F (2000b) Five recombinant fragments of human serum albumin-tools for the characterization of the warfarin binding site. Protein Sci 9:1455–1465
Falkenhagen D, Strobl W, Vogt G, Schrefl A, Linsberger I, Gerner FJ, Schoenhofen M (1999) Fractionated plasma separation and adsorption system: a novel system for blood purification to remove albumin bound substances. Artif Organs 23:81–86
Fleer R, Yeh P, Amellal N, Maury I, Fournier A, Bacchetta F, Baduel P, Jung G, L'Hote H, Becquart J et al (1991) Stable multicopy vectors for high-level secretion of recombinant human serum albumin by Kluyveromyces yeasts. Biotechnology (N Y) 9:968–975
Ghuman J, Zunszain PA, Petitpas I, Bhattacharya AA, Otagiri M, Curry S (2005) Structural basis of the drug-binding specificity of human serum albumin. J Mol Biol 353:38–52
Gimson AE, Braude S, Mellon PJ, Canalese J, Williams R (1982) Earlier charcoal haemoperfusion in fulminant hepatic failure. Lancet 2:681–683
Hoofnagle JH, Carithers RL Jr, Shapiro C, Ascher N (1995) Fulminant hepatic failure: summary of a workshop. Hepatology 21:240–252
Hughes RD, Pucknell A, Routley D, Langley PG, Wendon JA, Williams R (1994) Evaluation of the BioLogic-DT sorbent-suspension dialyser in patients with fulminant hepatic failure. Int J Artif Organs 17:657–662
Ikegaya K, Hirose M, Ohmura T, Nokihara K (1997) Complete determination of disulfide forms of purified recombinant human serum albumin, secreted by the yeast Pichia pastoris. Anal Chem 69:1986–1991
Ishima Y, Akaike T, Kragh-Hansen U, Hiroyama S, Sawa T, Suenaga A, Maruyama T, Kai T, Otagiri M (2008) S-nitrosylated human serum albumin-mediated cytoprotective activity is enhanced by fatty acid binding. J Biol Chem 283:34966–34975
Jacobsen J (1969) Binding of bilirubin to human serum albumin – determination of the dissociation constants. FEBS Lett 5:112–114
Jalan R, Williams R (2002) Acute-on-chronic liver failure: pathophysiological basis of therapeutic options. Blood Purif 20:252–261
Keller F, Heinze H, Jochimsen F, Passfall J, Schuppan D, Buttner P (1995) Risk factors and outcome of 107 patients with decompensated liver disease and acute renal failure (including 26 patients with hepatorenal syndrome): the role of hemodialysis. Ren Fail 17:135–146
Khan MM, Muzammil S, Tayyab S (2000) Role of salt bridge(s) in the binding and photoconversion of bilirubin bound to high affinity site on human serum albumin. Biochim Biophys Acta – Protein Struct Mol Enzymol 1479:103–113
Kiley JE, Pender JC, Welch HF, Welch CS (1958) Ammonia intoxication treated by hemodialysis. N Engl J Med 259:1156–1161
Kjeldsen T, Pettersson AF, Drube L, Kurtzhals P, Jonassen I, Havelund S, Hansen PH, Markussen J (1998) Secretory expression of human albumin domains in Saccharomyces cerevisiae and their binding of myristic acid and an acylated insulin analogue. Protein Expr Purif 13:163–169
Kobayashi K, Nakamura N, Sumi A, Ohmura T, Yokoyama K (1998) The development of recombinant human serum albumin. Ther Apher 2:257–262

Kondrup J, Almdal T, Vilstrup H, Tygstrup N (1992) High volume plasma exchange in fulminant hepatic failure. Int J Artif Organs 15:669–676
Larsen FS, Hansen BA, Jorgensen LG, Secher NH, Kirkegaard P, Tygstrup N (1994) High-volume plasmapheresis and acute liver transplantation in fulminant hepatic failure. Transplant Proc 26:1788
Liu R, Yang J, Ha CE, Bhagavan NV, Eckenhoff RG (2005) Truncated human serum albumin retains general anaesthetic binding activity. Biochem J 388:39–45
Losser MR, Payen D (1996) Mechanisms of liver damage. Semin Liver Dis 16:357–367
Mao H, Gunasekera AH, Fesik SW (2000) Expression, refolding, and isotopic labeling of human serum albumin domains for NMR spectroscopy. Protein Expr Purif 20:492–499
Matsushita S, Isima Y, Chuang VT, Watanabe H, Tanase S, Maruyama T, Otagiri M (2004) Functional analysis of recombinant human serum albumin domains for pharmaceutical applications. Pharm Res 21:1924–1932
Mingrone G, De Smet R, Greco AV, Bertuzzi A, Gandolfi A, Ringoir S, Vanholder R (1997) Serum uremic toxins from patients with chronic renal failure displace the binding of L-tryptophan to human serum albumin. Clin Chim Acta 260:27–34
Minomo A, Ishima Y, Kragh-Hansen U, Chuang VT, Uchida M, Taguchi K, Watanabe H, Maruyama T, Morioka H, Otagiri M (2011) Biological characteristics of two lysines on human serum albumin in the high-affinity binding of 4Z,15Z-bilirubin-IXalpha revealed by phage display. FEBS J 278:4100–4111
Minomo A, Ishima Y, Chuang VT, Suwa Y, Kragh-Hansen U, Narisoko T, Morioka H, Maruyama T, Otagiri M (2013) Albumin domain II mutant with high bilirubin binding affinity has a great potential as serum bilirubin excretion enhancer for hyperbilirubinemia treatment. Biochim Biophys Acta 1830:2917–2923
Mitra S, Samanta M, Sarkar M, Kumar De A, Chatterjee S (2011) Pre-exchange 5% albumin infusion in low birth weight neonates with intensive phototherapy failure – a randomized controlled trial. J Trop Pediatr 57: 217–221
Mitzner SR, Stange J, Klammt S, Peszynski P, Schmidt R (2001) Albumin dialysis using the molecular adsorbent recirculating system. Curr Opin Nephrol Hypertens 10:777–783
Mitzner S, Klammt S, Stange J, Schmidt R (2006) Albumin regeneration in liver support-comparison of different methods. Ther Apher Dial 10:108–117
Monzani E, Curto M, Galliano M, Minchiotti L, Aime S, Baroni S, Fasano M, Amoresano A, Salzano AM, Pucci P, Casella L (2002) Binding and relaxometric properties of heme complexes with cyanogen bromide fragments of human serum albumin. Biophys J 83:2248–2258
Narazaki R, Otagiri M (1997) Covalent binding of a bucillamine derivative with albumin in sera from healthy subjects and patients with various diseases. Pharm Res 14:351–353
Narazaki R, Hamada M, Harada K, Otagiri M (1996) Covalent binding between bucillamine derivatives and human serum albumin. Pharm Res 13:1317–1321
O'Grady JG, Gimson AE, O'Brien CJ, Pucknell A, Hughes RD, Williams R (1988) Controlled trials of charcoal hemoperfusion and prognostic factors in fulminant hepatic failure. Gastroenterology 94:1186–1192
Ohtani W, Masaki A, Ikeda Y, Hirose M, Chuganji M, Takeshima K, Kondo M, Sumi A, Ohmura T (1997) Structure of recombinant human serum albumin from Pichia pastoris. Yakugaku Zasshi 117:220–232
Ohtani W, Nawa Y, Takeshima K, Kamuro H, Kobayashi K, Ohmura T (1998a) Physicochemical and immunochemical properties of recombinant human serum albumin from Pichia pastoris. Anal Biochem 256:56–62
Ohtani W, Ohda T, Sumi A, Kobayashi K, Ohmura T (1998b) Analysis of Pichia pastoris components in recombinant human serum albumin by immunological assays and by HPLC with pulsed amperometric detection. Anal Chem 70:425–429
Okabayashi K, Nakagawa Y, Hayasuke N, Ohi H, Miura M, Ishida Y, Shimizu M, Murakami K, Hirabayashi K, Minamino H et al (1991) Secretory expression of the human serum albumin gene in the yeast, Saccharomyces cerevisiae. J Biochem 110:103–110

Onishi S, Itoh S, Isobe K, Ochi M, Kunikata T, Imai T (1989) Effect of the binding of bilirubin to either the first class or the second class of binding sites of the human serum albumin molecule on its photochemical reaction. Biochem J 257:711–714
Park DS, Petersen CE, Ha C, Harohalli K, Feix JB, Bhagavan NV (1999) Expression of a human serum albumin fragment (consisting of subdomains IA, IB, and IIA) and a study of its properties. IUBMB Life 48:169–174
Peters T (1996) All about albumin: biochemistry, genetics, and medical applications. Academic, San Diego
Petersen CE, Ha CE, Harohalli K, Feix JB, Bhagavan NV (2000) A dynamic model for bilirubin binding to human serum albumin. J Biol Chem 275:20985–20995
Petitpas I, Bhattacharya AA, Twine S, East M, Curry S (2001) Crystal structure analysis of warfarin binding to human serum albumin: anatomy of drug site I. J Biol Chem 276:22804–22809
Petitpas I, Petersen CE, Ha CE, Bhattacharya AA, Zunszain PA, Ghuman J, Bhagavan NV, Curry S (2003) Structural basis of albumin-thyroxine interactions and familial dysalbuminemic hyperthyroxinemia. Proc Natl Acad Sci U S A 100:6440–6445
Rifai K, Ernst T, Kretschmer U, Bahr MJ, Schneider A, Hafer C, Haller H, Manns MP, Fliser D (2003) Prometheus – a new extracorporeal system for the treatment of liver failure. J Hepatol 39:984–990
Roda A, Cappelleri G, Aldini R, Roda E, Barbara L (1982) Quantitative aspects of the interaction of bile acids with human serum albumin. J Lipid Res 23:490–495
Seige M, Kreymann B, Jeschke B, Schweigart U, Kopp KF, Classen M (1999) Long-term treatment of patients with acute exacerbation of chronic liver failure by albumin dialysis. Transplant Proc 31:1371–1375
Shakil AO, Kramer D, Mazariegos GV, Fung JJ, Rakela J (2000) Acute liver failure: clinical features, outcome analysis, and applicability of prognostic criteria. Liver Transpl 6:163–169
Shani M, Barash I, Nathan M, Ricca G, Searfoss GH, Dekel I, Faerman A, Givol D, Hurwitz DR (1992) Expression of human serum albumin in the milk of transgenic mice. Transgenic Res 1:195–208
Shearer WT, Bradshaw RA, Gurd FR, Peters T Jr (1967) The amino acid sequence and copper(II)-binding properties of peptide (1–24) of bovine serum albumin. J Biol Chem 242:5451–5459
Sheffield WP, Marques JA, Bhakta V, Smith IJ (2000) Modulation of clearance of recombinant serum albumin by either glycosylation or truncation. Thromb Res 99:613–621
Stange J, Ramlow W, Mitzner S, Schmidt R, Klinkmann H (1993) Dialysis against a recycled albumin solution enables the removal of albumin-bound toxins. Artif Organs 17:809–813
Stockmann HB, IJzermans JN (2002) Prospects for the temporary treatment of acute liver failure. Eur J Gastroenterol Hepatol 14:195–203
Trey C, Burns DG, Saunders SJ (1966) Treatment of hepatic coma by exchange blood transfusion. N Engl J Med 274:473–481
Tsipotis E, Shuja A, Jaber BL (2015) Albumin dialysis for liver failure: a systematic review. Adv Chronic Kidney Dis 22:382–390
Tsutsumi Y, Maruyama T, Takadate A, Goto M, Matsunaga H, Otagiri M (1999) Interaction between two dicarboxylate endogenous substances, bilirubin and an uremic toxin, 3-carboxy-4-methyl-5-propyl-2-furanpropanoic acid, on human serum albumin. Pharm Res 16:916–923
Watanabe H, Yamasaki K, Kragh-Hansen U, Tanase S, Harada K, Suenaga A, Otagiri M (2001) In vitro and in vivo properties of recombinant human serum albumin from Pichia pastoris purified by a method of short processing time. Pharm Res 18:1775–1781
Watanabe H, Noguchi T, Miyamoto Y, Kadowaki D, Kotani S, Nakajima M, Miyamura S, Ishima Y, Otagiri M, Maruyama T (2012) Interaction between two sulfate-conjugated uremic toxins, p-cresyl sulfate and indoxyl sulfate, during binding with human serum albumin. Drug Metab Dispos 40:1423–1428
Zunszain PA, Ghuman J, Komatsu T, Tsuchida E, Curry S (2003) Crystal structural analysis of human serum albumin complexed with hemin and fatty acid. BMC Struct Biol 3:6
Zunszain PA, Ghuman J, McDonagh AF, Curry S (2008) Crystallographic analysis of human serum albumin complexed with 4Z,15E-bilirubin-IXalpha. J Mol Biol 381:394–406

Bibliography

Aguilera A, Selgas R, Diez JJ, Bajo MA, Codoceo R, Alvarez V (2001) Anorexia in end-stage renal disease: pathophysiology and treatment. Expert Opin Pharmacother 2:1825–1838 (**Chapter 2**)

Ahlfors CE, Wennberg RP (2004) Bilirubin-albumin binding and neonatal jaundice. Semin Perinatol 28:334–339 (**Chapter 11**)

Ahmad B, Ahmed MZ, Haq SK, Khan RH (2005) Guanidine hydrochloride denaturation of human serum albumin originates by local unfolding of some stable loops in domain III. Biochim Biophys Acta 1750:93–102 (**Chapter 2**)

Ahouse JJ, Hagerman CL, Mittal P, Gilbert DJ, Copeland NG, Jenkins NA, Simister NE (1993) Mouse MHC class I-like Fc receptor encoded outside the MHC. J Immunol 151:6076–6088 (**Chapter 10**)

Akilesh S, Christianson GJ, Roopenian DC, Shaw AS (2007) Neonatal FcR expression in bone marrow-derived cells functions to protect serum IgG from catabolism. J Immunol 179:4580–4588 (**Chapter 10**)

Akilesh S, Huber TB, Wu H, Wang G, Hartleben B, Kopp JB, Miner JH, Roopenian DC, Unanue ER, Shaw AS (2008) Podocytes use FcRn to clear IgG from the glomerular basement membrane. Proc Natl Acad Sci U S A 105:967–972 (**Chapter 10**)

Alagic A, Koprianiuk A, Kluger R (2005) Hemoglobin-superoxide dismutase-chemical linkages that create a dual-function protein. J Am Chem Soc 127:8036–8043 (**Chapter 9**)

Allen NM, Mohammad F, Foran A, Corcoran D, Clarke T (2009) Severe hyperbilirubinaemia and kernicterus: more caution is needed in newborn jaundice surveillance. Ir Med J 102:228–229 (**Chapter 11**)

Alter J, Sennoga CA, Lopes DM, Eckersley RJ, Wells DJ (2009) Microbubble stability is a major determinant of the efficiency of ultrasound and microbubble mediated in vivo gene transfer. Ultrasound Med Biol 35:976–984 (**Chapter 7**)

Alvarez R, Musteanu M, Garcia-Garcia E, Lopez-Casas PP, Megias D, Guerra C, Munoz M, Quijano Y, Cubillo A, Rodriguez-Pascual J, Plaza C, de Vicente E, Prados S, Tabernero S, Barbacid M, Lopez-Rios F, Hidalgo M (2013) Stromal disrupting effects of nab-paclitaxel in pancreatic cancer. Br J Cancer 109:926–933 (**Chapter 6**)

American Diabetes Association (2013) Diagnosis and classification of diabetes mellitus. Diabetes Care 36(Suppl 1):S67–S74 (**Chapter 3**)

Amsellem S, Gburek J, Hamard G, Nielsen R, Willnow TE, Devuyst O, Nexo E, Verroust PJ, Christensen EI, Kozyraki R (2010) Cubilin is essential for albumin reabsorption in the renal proximal tubule. J Am Soc Nephrol 21:1859–1867 (**Chapter 10**)

Anand U, Mukherjee S (2013) Reversibility in protein folding: effect of beta-cyclodextrin on bovine serum albumin unfolded by sodium dodecyl sulphate. Phys Chem Chem Phys 15:9375–9383 (**Chapter 2**)

M. Otagiri, V.T.G. Chuang (eds.), *Albumin in Medicine*,
DOI 10.1007/978-981-10-2116-9

Anand U, Ray S, Ghosh S, Banerjee R, Mukherjee S (2015) Structural aspects of a protein-surfactant assembly: native and reduced States of human serum albumin. Protein J 34:147–157 (**Chapter 2**)

Andersen JT, Dee Qian J, Sandlie I (2006) The conserved histidine 166 residue of the human neonatal Fc receptor heavy chain is critical for the pH-dependent binding to albumin. Eur J Immunol 36:3044–3051 (**Chapter 10**)

Andersen JT, Daba MB, Berntzen G, Michaelsen TE, Sandlie I (2010a) Cross-species binding analyses of mouse and human neonatal Fc receptor show dramatic differences in immunoglobulin G and albumin binding. J Biol Chem 285:4826–4836 (**Chapter 10**)

Andersen JT, Daba MB, Sandlie I (2010b) FcRn binding properties of an abnormal truncated analbuminemic albumin variant. Clin Biochem 43:367–372 (**Chapter 10**)

Andersen JT, Pehrson R, Tolmachev V, Daba MB, Abrahmsen L, Ekblad C (2011) Extending half-life by indirect targeting of the neonatal Fc receptor (FcRn) using a minimal albumin binding domain. J Biol Chem 286:5234–5241 (**Chapter 10**)

Andersen JT, Dalhus B, Cameron J, Daba MB, Plumridge A, Evans L, Brennan SO, Gunnarsen KS, Bjoras M, Sleep D, Sandlie I (2012a) Structure-based mutagenesis reveals the albumin-binding site of the neonatal Fc receptor. Nat Commun 3:610 (**Chapter 10**)

Andersen JT, Foss S, Kenanova VE, Olafsen T, Leikfoss IS, Roopenian DC, Wu AM, Sandlie I (2012b) Anti-carcinoembryonic antigen single-chain variable fragment antibody variants bind mouse and human neonatal Fc receptor with different affinities that reveal distinct cross-species differences in serum half-life. J Biol Chem 287:22927–22937 (**Chapter 10**)

Andersen JT, Cameron J, Plumridge A, Evans L, Sleep D, Sandlie I (2013) Single-chain variable fragment albumin fusions bind the neonatal Fc receptor (FcRn) in a species-dependent manner: implications for in vivo half-life evaluation of albumin fusion therapeutics. J Biol Chem 288:24277–24285 (**Chapter 10**)

Andersen JT, Dalhus B, Viuff D, Ravn BT, Gunnarsen KS, Plumridge A, Bunting K, Antunes F, Williamson R, Athwal S, Allan E, Evans L, Bjørås M, Kjærulff S, Sleep D, Sandlie I, Cameron J (2014) Extending serum half-life of albumin by engineering neonatal Fc receptor (FcRn) binding. J Biol Chem 289:13492–13502 (**Chapters 4 and 10**)

Androulakis N, Kourousis C, Dimopoulos MA, Samelis G, Kakolyris S, Tsavaris N, Genatas K, Aravantinos G, Papadimitriou C, Karabekios S, Stathopoulos GP, Georgoulias V (1999) Treatment of pancreatic cancer with docetaxel and granulocyte colony-stimulating factor: a multicenter phase II study. J Clin Oncol 17:1779–1785 (**Chapter 6**)

Anhorn MG, Mahler HC, Langer K (2008) Freeze drying of human serum albumin (HSA) nanoparticles with different excipients. Int J Pharm 363:162–169 (**Chapter 2**)

Anraku M, Kitamura K, Shinohara A, Adachi M, Suenga A, Maruyama T, Miyanaka K, Miyoshi T, Shiraishi N, Nonoguchi H, Otagiri M, Tomita K, Suenaga A (2004a) Intravenous iron administration induces oxidation of serum albumin in hemodialysis patients. Kidney Int 66:841–848 (**Chapter 3**)

Anraku M, Tsurusaki Y, Watanabe H, Maruyama T, Kragh-Hansen U, Otagiri M (2004b) Stabilizing mechanisms in commercial albumin preparations: octanoate and N-acetyl-L-tryptophanate protect human serum albumin against heat and oxidative stress. Biochim Biophys Acta 1702:9–17 (**Chapters 1, 2 and 8**)

Anraku M, Kouno Y, Kai T, Tsurusaki Y, Yamasaki K, Otagiri M (2007) The role of N-acetyl-methioninate as a new stabilizer for albumin products. Int J Pharm 329:19–24 (**Chapter 2**)

Anraku M, Kitamura K, Shintomo R, Takeuchi K, Ikeda H, Nagano J, Ko T, Mera K, Tomita K, Otagiri M (2008) Effect of intravenous iron administration frequency on AOPP and inflammatory biomarkers in chronic hemodialysis patients: a pilot study. Clin Biochem 41:1168–1174 (**Chapter 3**)

Anraku M, Takeuchi K, Watanabe H, Kadowaki D, Kitamura K, Tomita K, Kuniyasu A, Suenaga A, Maruyama T, Otagiri M (2011) Quantitative analysis of cysteine-34 on the antioxidative properties of human serum albumin in hemodialysis patients. J Pharm Sci 100:3968–3976 (**Chapters 3 and 11**)

Anraku M, Chuang VT, Maruyama T, Otagiri M (2013) Redox properties of serum albumin. Biochim Biophys Acta 1830:5465–5472 (**Chapters 1 and 3**)

Anraku M, Shintomo R, Taguchi K, Kragh-Hansen U, Kai T, Maruyama T, Otagiri M (2015) Amino acids of importance for the antioxidant activity of human serum albumin as revealed by recombinant mutants and genetic variants. Life Sci 134:36–41 (**Chapters 1 and 2**)

Antohe F, Radulescu L, Gafencu A, Ghetie V, Simionescu M (2001) Expression of functionally active FcRn and the differentiated bidirectional transport of IgG in human placental endothelial cells. Hum Immunol 62:93–105 (**Chapter 10**)

Antonini E, Brunori M (1971) Hemoglobin and myoglobin in their reactions with ligands. In: Neuberger A, Tatum EL (eds) North-Holland research monographs. Frontiers of Biology, vol 21. North-Holland, Amsterdam, pp 13–39 (**Chapter 9**)

Aoi A, Watanabe Y, Mori S, Takahashi M, Vassaux G, Kodama T (2008) Herpes simplex virus thymidine kinase-mediated suicide gene therapy using nano/microbubbles and ultrasound. Ultrasound Med Biol 34:425–434 (**Chapter 7**)

Arakawa T, Kita Y (2000a) Stabilizing effects of caprylate and acetyltryptophanate on heat-induced aggregation of bovine serum albumin. Biochim Biophys Acta 1479:32–36 (**Chapter 2**)

Arakawa T, Kita Y (2000b) Protection of bovine serum albumin from aggregation by Tween 80. J Pharm Sci 89:646–651 (**Chapter 2**)

Arnedo A, Espuelas S, Irache JM (2002) Albumin nanoparticles as carriers for a phosphodiester oligonucleotide. Int J Pharm 244:59–72 (**Chapter 5**)

Arner ES, Holmgren A (2000) Physiological functions of thioredoxin and thioredoxin reductase. Eur J Biochem 267:6102–6109 (**Chapter 4**)

Aseem O, Smith BT, Cooley MA, Wilkerson BA, Argraves KM, Remaley AT, Argraves WS (2014) Cubilin maintains blood levels of HDL and albumin. J Am Soc Nephrol 25:1028–1036 (**Chapter 10**)

Association AD (2013) Diagnosis and classification of diabetes mellitus. Diabetes Care 36(Suppl 1):S67–S74 (**Chapter 3**)

Averkiou M, Bruce M, Jensen S, Rafter P, Brock-Fisher T, Powers J (eds) (2004) Pulsing schemes for the detection of nonlinear echoes from contrast microbubbles. 9th European Symposium on Ultrasound Contrast Imaging, Rotterdam (**Chapter 7**)

Awasthi N, Zhang C, Schwarz AM, Hinz S, Wang C, Williams NS, Schwarz MA, Schwarz RE (2013) Comparative benefits of Nab-paclitaxel over gemcitabine or polysorbate-based docetaxel in experimental pancreatic cancer. Carcinogenesis 34:2361–2369 (**Chapter 6**)

Azuma H, Tomita N, Sakamoto T, Kiyama S, Inamoto T, Takahara K, Kotake Y, Segawa N, Morishita R, Takahara S, Hayasaki H, Otsuki Y, Horie S, Tanigawa N, Katsuoka Y (2008) Marked regression of liver metastasis by combined therapy of ultrasound-mediated NF kappaB-decoy transfer and transportal injection of paclitaxel, in mouse. Int J Cancer 122:1645–1656 (**Chapter 7**)

Baek S, Hrabie J, Keefer L, Hou D, Fineberg N, Rhoades R, March K (2002) Augmentation of intrapericardial nitric oxide level by a prolonged-release nitric oxide donor reduces luminal narrowing after porcine coronary angioplasty. Circulation 105:2779–2784 (**Chapter 8**)

Baggio LL, Huang QL, Brown TJ, Drucker DJ (2004) A recombinant human glucagon-like peptide (GLP)-1-albumin protein (Albugon) mimics peptidergic activation of GLP-1 receptor-dependent pathways coupled with satiety, gastrointestinal motility, and glucose homeostasis. Diabetes 53:2492–2500 (**Chapter 10**)

Baggio LL, Huang Q, Cao X, Drucker DJ (2008) An albumin-exendin-4 conjugate engages central and peripheral circuits regulating murine energy and glucose homeostasis. Gastroenterology 134:1137–1147 (**Chapter 10**)

Bain VG, Kaita KD, Yoshida EM, Swain MG, Heathcote EJ, Neumann AU, Fiscella M, Yu R, Osborn BL, Cronin PW, Freimuth WW, McHutchison JG, Subramanian GM (2006) A phase 2 study to evaluate the antiviral activity, safety, and pharmacokinetics of recombinant human albumin-interferon alfa fusion protein in genotype 1 chronic hepatitis C patients. J Hepatol 44:671–678 (**Chapters 2 and 10**)

Bal W, Sokolowska M, Kurowska E, Faller P (2013) Binding of transition metal ions to albumin: sites, affinities and rates. Biochim Biophys Acta 1830:5444–5455 (**Chapter 1**)

Bao S, Thrall BD, Miller DL (1997) Transfection of a reporter plasmid into cultured cells by sonoporation in vitro. Ultrasound Med Biol 23:953–959 (**Chapter 7**)

Bar-Or D, Bar-Or R, Rael LT, Gardner DK, Slone DS, Craun ML (2005a) Heterogeneity and oxidation status of commercial human albumin preparations in clinical use. Crit Care Med 33:1638–1641 (**Chapter 3**)

Bar-Or D, Heyborne KD, Bar-Or R, Rael LT, Winkler JV, Navot D (2005b) Cysteinylation of maternal plasma albumin and its association with intrauterine growth restriction. Prenat Diagn 25:245–249 (**Chapter 3**)

Becher H, Burns PN (2000) Handbook of contrast echocardiography: left ventricular function and myocardial perfusion. Springer, Berlin/Heidelberg (**Chapter 7**)

BeneFIX. BeneFIX (coagulation Factor IX Recombinant) (2015) Electronic medicines compendium. https://www.medicines.org.uk/emc/medicine/20376/SPC/BeneFIX/#PHARMACODINAMIC_PROPS. Accessed 16 Feb 2016 (**Chapter 10**)

Bennhold H, Kallee E (1959) Comparative studies on the half-life of I 131-labeled albumins and nonradioactive human serum albumin in a case of analbuminemia. J Clin Invest 38:863–872 (**Chapter 10**)

Berg AH, Drechsler C, Wenger J, Buccafusca R, Hod T, Kalim S, Ramma W, Parikh SM, Steen H, Friedman DJ, Danziger J, Wanner C, Thadhani R, Karumanchi SA (2013) Carbamylation of serum albumin as a risk factor for mortality in patients with kidney failure. Sci Transl Med 5:175ra129 (**Chapter 3**)

Bern M, Knudsen Sand KM, Nilsen J, Sandlie I, Andersen JT (2015a) The role of albumin receptors in regulation of albumin homeostasis: implications for drug delivery. J Control Release 211:144–162 (**Chapter 1**)

Bern M, Sand KM, Nilsen J, Sandlie I, Andersen JT (2015b) The role of albumin receptors in regulation of albumin homeostasis: implications for drug delivery. J Control Release Off J Control Release Soc 211:144–162 (**Chapter 10**)

Bertolini J, Goss N, Curling J (2012) Production of plasma proteins for therapeutic use. Wiley, Hoboken (**Chapter 2**)

Bertucci C, Nanni B, Raffaelli A, Salvadori P (1998) Chemical modification of human albumin at cys34 by ethacrynic acid: structural characterisation and binding properties. J Pharm Biomed Anal 18:127–136 (**Chapter 11**)

Bhattacharya AA, Grüne T, Curry S (2000) Crystallographic analysis reveals common modes of binding of medium and long-chain fatty acids to human serum albumin. J Mol Biol 303:721–732 (**Chapter 1**)

Bienk K, Dagnæs-Hansen F, Wengel J, Kragh-Hansen U, Howard KA (2015) Albumin-mediated protection, reduced immunogenicity and extended circulatory half-life of cholesterol modified small interfering RNA. Presented at the 2015 Annual Meeting of Controlled Release Society, Edinburgh, July 26–29. Abstract 844 (**Chapter 1**)

Birn H, Christensen EI (2006) Renal albumin absorption in physiology and pathology. Kidney Int 69:440–449 (**Chapter 10**)

Birn H, Fyfe JC, Jacobsen C, Mounier F, Verroust PJ, Orskov H, Willnow TE, Moestrup SK, Christensen EI (2000) Cubilin is an albumin binding protein important for renal tubular albumin reabsorption. J Clin Investig 105:1353–1361 (**Chapter 10**)

Blindauer CA, Harvey I, Bunyan KE, Stewart AJ, Sleep D, Harrison DJ, Berezenko S, Sadler PJ (2009) Structure, properties and engineering of the major zinc binding site on human albumin. J Biol Chem 284:23116–23124 (**Chapter 1**)

Blum JL, Savin MA, Edelman G, Pippen JE, Robert NJ, Geister BV, Kirby RL, Clawson A, O'Shaughnessy JA (2007) Phase II study of weekly albumin-bound paclitaxel for patients with metastatic breast cancer heavily pretreated with taxanes. Clin Breast Cancer 7:850–856 (**Chapter 6**)

Boisvert MR, Koski KG, Skinner CD (2010) Increased oxidative modifications of amniotic fluid albumin in pregnancies associated with gestational diabetes mellitus. Anal Chem 82:1133–1137 (**Chapter 3**)

Bolling C, Graefe T, Lubbing C, Jankevicius F, Uktveris S, Cesas A, Meyer-Moldenhauer WH, Starkmann H, Weigel M, Burk K, Hanauske AR (2006) Phase II study of MTX-HSA in combination with cisplatin as first line treatment in patients with advanced or metastatic transitional cell carcinoma. Invest New Drugs 24:521–527 (**Chapter 10**)

Borvak J, Richardson J, Medesan C, Antohe F, Radu C, Simionescu M, Ghetie V, Ward ES (1998) Functional expression of the MHC class I-related receptor, FcRn, in endothelial cells of mice. Int Immunol 10:1289–1298 (**Chapter 10**)

Bourdon E, Loreau N, Lagrost L, Blache D (2005) Differential effects of cysteine and methionine residues in the antioxidant activity of human serum albumin. Free Radic Res 39:15–20 (**Chapter 2**)

Brambell FW, Hemmings WA, Morris IG (1964) A theoretical model of gamma-globulin catabolism. Nature 203:1352–1354 (**Chapter 10**)

Brouwer E, Verweij J, De Bruijn P, Loos WJ, Pillay M, Buijs D, Sparreboom A (2000) Measurement of fraction unbound paclitaxel in human plasma. Drug Metab Dispos 28:1141–1145 (**Chapter 6**)

Buehler PW, Boyskins RA, Jia Y, Norris S, Freedberg DI, Alayash AI (2005) Structural and functional characterization of glutaraldehyde-polymerized bovine hemoglobin and its isolated fractions. Anal Chem 77:3466–3478 (**Chapter 9**)

Buehler PW, Boykins RA, Norris S, Alayash AL (2006) Chemical characterization of diaspirin cross-linked hemoglobin polymerized with poly(ethylene glycol). Anal Chem 78:4634–4641 (**Chapter 9**)

Bujacz A (2012) Structures of bovine, equine and leporine serum albumin. Acta Cryst D68:1278–1289 (**Chapter 1**)

Burger AM, Hartung G, Stehle G, Sinn H, Fiebig HH (2001) Pre-clinical evaluation of a methotrexate-albumin conjugate (MTX-HSA) in human tumor xenografts in vivo. Int J Cancer 92:718–724 (**Chapter 2**)

Burmeister WP, Gastinel LN, Simister NE, Blum ML, Bjorkman PJ (1994a) Crystal structure at 2.2 A resolution of the MHC-related neonatal Fc receptor. Nature 372:336–343 (**Chapter 10**)

Burmeister WP, Huber AH, Bjorkman PJ (1994b) Crystal structure of the complex of rat neonatal Fc receptor with Fc. Nature 372:379–383 (**Chapter 10**)

Cai C, Zhou K, Wu Y, Wu L (2006) Enhanced liver targeting of 5-fluorouracil using galactosylated human serum albumin as a carrier molecule. J Drug Target 14:55–61 (**Chapter 1**)

Carballal S, Radi R, Kirk MC, Barnes S, Freeman BA, Alvarez B (2003) Sulfenic acid formation in human serum albumin by hydrogen peroxide and peroxynitrite. Biochemistry 42:9906–9914 (**Chapter 2**)

Carter DC, Ho JX (1994) Structure of serum albumin. Adv Protein Chem 45:153–203 (**Chapter 2**)

Cassio D, Weiss MC, Ott MO, Sala-Trepat JM, Fries J, Erdos T (1981) Expression of the albumin gene in rat hepatoma cells and their dedifferentiated variants. Cell 27:351–358 (**Chapter 11**)

Castle J, Feinstein SB (2014) Ultrasound-directed, site-specific gene delivery. Methods Mol Biol (Clifton, NJ) 1141:67–76 (**Chapter 7**)

Castle J, Butts M, Healey A, Kent K, Marino M, Feinstein SB (2013) Ultrasound-mediated targeted drug delivery: recent success and remaining challenges. Am J Physiol Heart Circ Physiol 304:H350–H357 (**Chapter 7**)

Castle JW, Kent KP, Fan Y, Wallace KD, Davis CE, Roberts JC, Marino ME, Thomenius KE, Lim HW, Coles E, Davidson MH, Feinstein SB, DeMaria A (2015) Therapeutic ultrasound: increased HDL-cholesterol following infusions of acoustic microspheres and apolipoprotein A-I plasmids. Atherosclerosis 241:92–99 (**Chapter 7**)

Cauza K, Hinterhuber G, Dingelmaier-Hovorka R, Brugger K, Klosner G, Horvat R, Wolff K, Foedinger D (2005) Expression of FcRn, the MHC class I-related receptor for IgG, in human keratinocytes. J Invest Dermatol 124:132–139 (**Chapter 10**)

Chan B, Dodsworth N, Woodrow J, Tucker A, Harris R (1995) Site-specific N-terminal autodegradation of human serum albumin. Eur J Biochem 227:524–528 (**Chapter 3**)

Chapman AP (2002) PEGylated antibodies and antibody fragments for improved therapy: a review. Adv Drug Deliv Rev 54:531–545 (**Chapter 4**)

Chaudhury C, Mehnaz S, Robinson JM, Hayton WL, Pearl DK, Roopenian DC, Anderson CL (2003) The major histocompatibility complex-related Fc receptor for IgG (FcRn) binds albumin and prolongs its lifespan. J Exp Med 197:315–322 (**Chapter 10**)

Chaudhury C, Brooks CL, Carter DC, Robinson JM, Anderson CL (2006) Albumin binding to FcRn: distinct from the FcRn-IgG interaction. Biochemistry 45:4983–4990 (**Chapter 10**)

Chen JH, Zhang XG, Jiang YT, Yan LY, Tang L, Yin YW, Cheng DS, Chen J, Wang M (2010) Bioactivity and pharmacokinetics of two human serum albumin-thymosin alpha1-fusion proteins, rHSA-talpha1 and rHSA-L-talpha1, expressed in recombinant Pichia pastoris. Cancer Immunol Immunother 59:1335–1345 (**Chapter 4**)

Chen Y-C, Jiang L-P, Liu N-X, Ding L, Liu X-L, Wang Z-H, Hong K, Zhang Q-P (2011) Enhanced gene transduction into skeletal muscle of mice in vivo with pluronic block copolymers and ultrasound exposure. Cell Biochem Biophys 60:267–273 (**Chapter 7**)

Chen Z, He Y, Shi B, Yang D (2013) Human serum albumin from recombinant DNA technology: challenges and strategies. Biochim Biophys Acta 1830:5515–5525 (**Chapter 1**)

Chen N, Li Y, Ye Y, Palmisano M, Chopra R, Zhou S (2014) Pharmacokinetics and pharmacodynamics of nab-paclitaxel in patients with solid tumors: disposition kinetics and pharmacology distinct from solvent-based paclitaxel. J Clin Pharmacol 54:1097–1107 (**Chapter 6**)

Chen N, Brachmann C, Liu X, Pierce DW, Dey J, Kerwin WS, Li Y, Zhou S, Hou S, Carleton M, Klinghoffer RA, Palmisano M, Chopra R (2015) Albumin-bound nanoparticle (nab) paclitaxel exhibits enhanced paclitaxel tissue distribution and tumor penetration. Cancer Chemother Pharmacol 76:699–712 (**Chapter 6**)

Choi JJ, Pernot M, Brown TR, Small SA, Konofagou EE (2007a) Spatio-temporal analysis of molecular delivery through the blood-brain barrier using focused ultrasound. Phys Med Biol 52:5509–5530 (**Chapter 7**)

Choi JJ, Pernot M, Small SA, Konofagou EE (2007b) Noninvasive, transcranial and localized opening of the blood-brain barrier using focused ultrasound in mice. Ultrasound Med Biol 33:95–104 (**Chapter 7**)

Choi N, Kim SM, Hong KS, Cho G, Cho JH, Lee C, Ryu EK (2011) The use of the fusion protein RGD-HSA-TIMP2 as a tumor targeting imaging probe for SPECT and PET. Biomaterials 32:7151–7158 (**Chapter 4**)

Choi S, Park S, Kim S, Lim C, Kim J, Cha DR, Oh J (2012) Recombinant fusion protein of albumin-retinol binding protein inactivates stellate cells. Biochem Biophys Res Commun 418:191–197 (**Chapter 4**)

Chou DK, Krishnamurthy R, Randolph TW, Carpenter JF, Manning MC (2005) Effects of Tween 20 and Tween 80 on the stability of albutropin during agitation. J Pharm Sci 94:1368–1381 (**Chapter 2**)

Christensen M, Knop FK (2010) Once-weekly GLP-1 agonists: how do they differ from exenatide and liraglutide? Curr Diabetes Rep 10:124–132 (**Chapter 2**)

Chuang VT, Otagiri M (2007) Recombinant human serum albumin. Drugs Today (Barc) 43:547–561 (**Chapters 4 and 11**)

Chuang VT, Kragh-Hansen U, Otagiri M (2002) Pharmaceutical strategies utilizing recombinant human serum albumin. Pharm Res 19:569–577 (**Chapters 1 and 4**)

Cianga P, Cianga C, Cozma L, Ward ES, Carasevici E (2003) The MHC class I related Fc receptor, FcRn, is expressed in the epithelial cells of the human mammary gland. Hum Immunol 64:1152–1159 (**Chapter 10**)

Cianga P, Cianga C, Plamadeala P, Branisteanu D, Carasevici E (2007) The neonatal Fc receptor (FcRn) expression in the human skin. Virchows Arch 451:859–860 (**Chapter 10**)

Clancy R, Abramson S (1995) Nitric oxide: a novel mediator of inflammation. Proc Soc Exp Biol Med 210:93–101 (**Chapter 8**)

Clancy R, Levartovsky D, Leszczynska-Piziak J, Yegudin J, Abramson S (1994) Nitric oxide reacts with intracellular glutathione and activates the hexose monophosphate shunt in human neutrophils: evidence for S-nitrosoglutathione as a bioactive intermediary. Proc Natl Acad Sci U S A 91:3680–3684 (**Chapter 8**)

Claypool SM, Dickinson BL, Wagner JS, Johansen FE, Venu N, Borawski JA, Lencer WI, Blumberg RS (2004) Bidirectional transepithelial IgG transport by a strongly polarized basolateral membrane Fc gamma-receptor. Mol Biol Cell 15:1746–1759 (**Chapter 10**)

Cochran M, Wheatley MA (2013) In vitro gene delivery with ultrasound-triggered polymer microbubbles. Ultrasound Med Biol 39:1102–1119 (**Chapter 7**)

Cohen MP (2013) Clinical, pathophysiological and structure/function consequences of modification of albumin by Amadori-glucose adducts. Biochim Biophys Acta 1830:5480–5485 (**Chapter 1**)

Cohen JL, Cheirif J, Segar DS, Gillam LD, Gottdiener JS, Hausnerova E, Bruns DE (1998) Improved left ventricular endocardial border delineation and opacification with OPTISON (FS069), a new echocardiographic contrast agent. Results of a phase III multicenter trial. J Am Coll Cardiol 32:746–752 (**Chapter 7**)

Cohn EJ, Strong LE, Hughes WL, Mulford DJ, Ashworth JN, Melin M, Taylor HL (1946) Preparation and properties of serum and plasma proteins. IV. A system for the separation into fractions of the protein and lipoprotein components of biological tissues and fluids1a, b, c, d. J Am Chem Soc 68:459–475 (**Chapter 2**)

Colombo G, Clerici M, Giustarini D, Rossi R, Milzani A, Dalle-Donne I (2012) Redox albuminomics: oxidized albumin in human diseases. Antioxid Redox Signal 17:1515–1527 (**Chapter 3**)

Commisso C, Davidson SM, Soydaner-Azeloglu RG, Parker SJ, Kamphorst JJ, Hackett S, Grabocka E, Nofal M, Drebin JA, Thompson CB, Rabinowitz JD, Metallo CM, Vander Heiden MG, Bar-Sagi D (2013) Macropinocytosis of protein is an amino acid supply route in Ras-transformed cells. Nature 497:633–637 (**Chapter 6**)

Cordes AA, Carpenter JF, Randolph TW (2012a) Selective domain stabilization as a strategy to reduce human serum albumin-human granulocyte colony stimulating factor aggregation rate. J Pharm Sci 101:2009–2016 (**Chapter 2**)

Cordes AA, Platt CW, Carpenter JF, Randolph TW (2012b) Selective domain stabilization as a strategy to reduce fusion protein aggregation. J Pharm Sci 101:1400–1409 (**Chapter 2**)

Cormode EJ, Lyster DM, Israels S (1975) Analbuminemia in a neonate. J Pediatr 86:862–867 (**Chapter 10**)

Cortes J, Saura C (2010) Nanoparticle albumin-bound (nab™)-paclitaxel: improving efficacy and tolerability by targeted drug delivery in metastatic breast cancer. Eur J Cancer Suppl 8:1–10 (**Chapter 5**)

Crane M, Ollosson R, Moore K, Rossi A, Megson I (2002) Novel role for low molecular weight plasma thiols in nitric oxide-mediated control of platelet function. J Biol Chem 277:46858–46863 (**Chapter 8**)

Crommelin DJ, Florence AT (2013) Towards more effective advanced drug delivery systems. Int J Pharm 454:496–511 (**Chapter 6**)

Cui SY, Verroust PJ, Moestrup SK, Christensen EI (1996) Megalin/gp330 mediates uptake of albumin in renal proximal tubule. Am J Physiol-Ren 271:F900–F907 (**Chapter 10**)

Curry S, Mandelkow H, Brick P, Franks N (1998) Crystal structure of human serum albumin complexed with fatty acid reveals an asymmetric distribution of binding sites. Nat Struct Biol 5:827–835 (**Chapters 2, 9 and 11**)

Curry S, Brick P, Franks NP (1999) Fatty acid binding to human serum albumin: new insights from crystallographic studies. Biochim Biophys Acta 1441:131–140 (**Chapter 11**)

Czajkowsky DM, Hu J, Shao Z, Pleass RJ (2012) Fc-fusion proteins: new developments and future perspectives. EMBO Mol Med 4:1015–1028 (**Chapter 10**)

D'Agnilloo F, Chang TMS (1998) Polyhemoglobin-superoxide dismutase-catalase as a blood substitute with antioxidant properties. Nat Biotechnol 16:667–671 (**Chapter 9**)

Daijima Y, Komatsu T (2014) Haemoglobin wrapped covalently by human serum albumin mutants containing Mn(III) protoporphyrin IX: an O_2 complex stable in H_2O_2 solution. Chem Commun 50:14716–14719 (**Chapter 9**)

Dall'Acqua WF, Kiener PA, Wu H (2006) Properties of human IgG1s engineered for enhanced binding to the neonatal Fc receptor (FcRn). J Biol Chem 281:23514–23524 (**Chapter 10**)

Dancey J (2010) mTOR signaling and drug development in cancer. Nat Rev Clin Oncol 7:209–219 (**Chapter 6**)

Davies MJ (2005) The oxidative environment and protein damage. Biochim Biophys Acta 1703:93–109 (**Chapter 2**)

de Bold MK, Sheffield WP, Martinuk A, Bhakta V, Eltringham-Smith L, de Bold AJ (2012) Characterization of a long-acting recombinant human serum albumin-atrial natriuretic factor (ANF) expressed in Pichia pastoris. Regul Pept 175:7–10 (**Chapter 4**)

Delalande A, Kotopoulis S, Postema M, Midoux P, Pichon C (2013) Sonoporation: mechanistic insights and ongoing challenges for gene transfer. Gene 6:00364–00368 (**Chapter 7**)

Dennis MS, Zhang M, Meng YG, Kadkhodayan M, Kirchhofer D, Combs D, Damico LA (2002) Albumin binding as a general strategy for improving the pharmacokinetics of proteins. J Biol Chem 277:35035–35043 (**Chapter 10**)

Dennis MS, Jin HK, Dugger D, Yang RH, McFarland L, Ogasawara A, Williams S, Cole MJ, Ross S, Schwall R (2007) Imaging tumors with an albumin-binding Fab, a novel tumor-targeting agent. Cancer Res 67:254–261 (**Chapter 10**)

Desai N (2007) Nanoparticle albumin bound (nab) technology: targeting tumors through the endothelial gp60 receptor and SPARC. Nanomedicine 3:337–346 (**Chapter 5**)

Desai N (2012a) Albumin drug nanoparticles. In: Kratz F, Senter P, Steinhagen H (eds) Drug delivery in oncology: from basic research to cancer therapy, 1st edn. Wiley-VCH Verlag GmbH & Co. KGaA, Weinheim, pp 1133–1161 (**Chapter 6**)

Desai N (2012b) Challenges in development of nanoparticle-based therapeutics. AAPS J 14:282–295 (**Chapter 6**)

Desai N (2013) Integration of nab-technology into clinical drug development. In: Bischoff J (ed) Nanotechnologie beim Mammakarzinom- Grundlagen und aktuelle Perspektiven. UNI-MED Verlag AG, Bremen, pp 22–31 (**Chapter 6**)

Desai N, Trieu V, Yao Z, Louie L, Ci S, Yang A, Tao C, De T, Beals B, Dykes D, Noker P, Yao R, Labao E, Hawkins M, Soon-Shiong P (2006) Increased antitumor activity, intratumor paclitaxel concentrations, and endothelial cell transport of cremophor-free, albumin-bound paclitaxel, ABI-007, compared with cremophor-based paclitaxel. Clin Cancer Res 12:1317–1324 (**Chapters 2 and 6**)

Desai NP, Trieu V, Hwang LY, Wu R, Soon-Shiong P, Gradishar WJ (2008) Improved effectiveness of nanoparticle albumin-bound (nab) paclitaxel versus polysorbate-based docetaxel in multiple xenografts as a function of HER2 and SPARC status. Anticancer Drugs 19:899–909 (**Chapter 6**)

Desai N, Trieu V, Damascelli B, Soon-Shiong P (2009) SPARC expression correlates with tumor response to albumin-bound paclitaxel in head and neck cancer patients. Transl Oncol 2:59–64 (**Chapter 1**)

Descamps-Latscha B, Witko-Sarsat V (2001) Importance of oxidatively modified proteins in chronic renal failure. Kidney Int Suppl 78:S108–S113 (**Chapter 3**)

Dickinson BL, Badizadegan K, Wu Z, Ahouse JC, Zhu XP, Simister NE, Blumberg RS, Lencer WI (1999) Bidirectional FcRn-dependent IgG transport in a polarized human intestinal epithelial cell line. J Clin Investig 104:903–911 (**Chapter 10**)

Dickinson BL, Claypool SM, D'Angelo JA, Aiken ML, Venu N, Yen EH, Wagner JS, Borawski JA, Pierce AT, Hershberg R, Blumberg RS, Lencer WI (2008) Ca2+-dependent calmodulin binding to FcRn affects immunoglobulin G transport in the transcytotic pathway. Mol Biol Cell 19:414–423 (**Chapter 10**)

Dimcevski G Ultrasound-assisted treatment of inoperable pancreatic cancer. ClinicalTrials.gov. https://clinicaltrials.gov/show/NCT01674556 (**Chapter 7**)

Ding Y, Fan J, Li W, Yang R, Peng Y, Deng L, Wu Y, Fu Q (2013) The effect of albumin fusion patterns on the production and bioactivity of the somatostatin-14 fusion protein in Pichia pastoris. Appl Biochem Biotechnol 170:1637–1648 (**Chapter 4**)

Ding Y, Fan J, Li W, Peng Y, Yang R, Deng L, Fu Q (2014a) The effect of albumin fusion structure on the production and bioactivity of the somatostatin-28 fusion protein in Pichia pastoris. J Ind Microbiol Biotechnol 41:997–1006 (**Chapter 4**)

Ding Y, Peng Y, Deng L, Wu Y, Fu Q, Jin J (2014b) The effects of fusion structure on the expression and bioactivity of human brain natriuretic peptide (BNP) albumin fusion proteins. Curr Pharm Biotechnol 15:856–863 (**Chapter 4**)

Dixon JW, Sarkar B (1972) Absence of a specific copper(II) binding site in dog albumin is due to amino acid mutation in position 3. Biochem Biophys Res Commun 48:197–200 (**Chapter 11**)

Dockal M, Carter DC, Ruker F (1999) The three recombinant domains of human serum albumin – structural characterization and ligand binding properties. J Biol Chem 274:29303–29310 (**Chapters 10 and 11**)

Dockal M, Carter DC, Ruker F (2000a) Conformational transitions of the three recombinant domains of human serum albumin depending on pH. J Biol Chem 275:3042–3050 (**Chapter 11**)

Dockal M, Chang M, Carter DC, Ruker F (2000b) Five recombinant fragments of human serum albumin-tools for the characterization of the warfarin binding site. Protein Sci 9:1455–1465 (**Chapter 11**)

Doherty DH, Doyle MP, Curry SR, Vali RJ, Fattor TJ, Olson JS, Lemon DD (1998) Rate of reaction with nitric oxide determines the hypertensive effect of cell-free hemoglobin. Nat Biotechnol 16:672–676 (**Chapter 9**)

Doinikov AA, Bouakaz A (2010) Acoustic microstreaming around an encapsulated particle. J Acoust Soc Am 127:1218–1227 (**Chapter 7**)

Domenicali M, Baldassarre M, Giannone FA, Naldi M, Mastroroberto M, Biselli M, Laggetta M, Patrono D, Bertucci C, Bernardi M, Caraceni P (2014) Posttranscriptional changes of serum albumin: clinical and prognostic significance in hospitalized patients with cirrhosis. Hepatology 60:1851–1860 (**Chapter 3**)

Drapier J, Pellat C, Henry Y (1991) Generation of EPR-detectable nitrosyl-iron complexes in tumor target cells cocultured with activated macrophages. J Biol Chem 266:10162–10167 (**Chapter 8**)

Drechsler C, Kalim S, Wenger JB, Suntharalingam P, Hod T, Thadhani RI, Karumanchi SA, Wanner C, Berg AH (2015) Protein carbamylation is associated with heart failure and mortality in diabetic patients with end-stage renal disease. Kidney Int 87:1201–1208 (**Chapter 3**)

Ducharme E, Weinberg JM (2008) Etanercept. Expert Opin Biol Ther 8:491–502 (**Chapter 10**)

Duggan EL, Luck JM (1948) The combination of organic anions with serum albumin; stabilization against urea denaturation. J Biol Chem 172:205–220 (**Chapter 2**)

Duttaroy A, Kanakaraj P, Osborn BL, Schneider H, Pickeral OK, Chen C, Zhang GY, Kaithamana S, Singh M, Schulingkamp R, Crossan D, Bock J, Kaufman TE, Reavey P, Carey-Barber M, Krishnan SR, Garcia A, Murphy K, Siskind JK, McLean MA, Cheng S, Ruben S, Birse CE, Blondel O (2005) Development of a long-acting insulin analog using albumin fusion technology. Diabetes 54:251–258 (**Chapters 4 and 10**)

Duvshani-Eshet M, Machluf M (2007) Efficient transfection of tumors facilitated by long-term therapeutic ultrasound in combination with contrast agent: from in vitro to in vivo setting. Cancer Gene Ther 14:306–315 (**Chapter 7**)

Duvshani-Eshet M, Baruch L, Kesselman E, Shimoni E, Machluf M (2006) Therapeutic ultrasound-mediated DNA to cell and nucleus: bioeffects revealed by confocal and atomic force microscopy. Gene Ther 13:163–172 (**Chapter 7**)

Duvshani-Eshet M, Benny O, Morgenstern A, Machluf M (2007) Therapeutic ultrasound facilitates antiangiogenic gene delivery and inhibits prostate tumor growth. Mol Cancer Ther 6:2371–2382 (**Chapter 7**)

Dye D, Watkins J (1980) Suspected anaphylactic reaction to Cremophor EL. Br Med J 280:1353 (**Chapter 6**)

Edeling MA, Smith C, Owen D (2006) Life of a clathrin coat: insights from clathrin and AP structures. Nat Rev Mol Cell Biol 7:32–44 (**Chapter 10**)

Elmer J, Zorc K, Rameez S, Zhou Y, Cabrales P, Palmer AF (2012) Hypervolemic infusion of *Lumbricus terrestris* erythrocruorin purified by tangential-flow filtration. Transfusion 52:1729–1740 (**Chapter 9**)

Elsadek B, Kratz F (2012) Impact of albumin on drug delivery – new applications on the horizon. J Control Release 157:4–28 (**Chapters 1 and 10**)

Elzoghby AO, Samy WM, Elgindy NA (2012) Albumin-based nanoparticles as potential controlled release drug delivery systems. J Control Release 157:168–182 (**Chapter 1**)

Endoh M, Koibuchi N, Sato M, Morishita R, Kanzaki T, Murata Y, Kaneda Y (2002) Fetal gene transfer by intrauterine injection with microbubble-enhanced ultrasound. Mol Ther J Am Soc Gene Ther 5:501–508 (**Chapter 7**)

Era S, Kuwata K, Imai H, Nakamura K, Hayashi T, Sogami M (1995) Age-related change in redox state of human serum albumin. Biochim Biophys Acta 1247:12–16 (**Chapter 3**)

Escoffre JM, Zeghimi A, Novell A, Bouakaz A (2013) In-vivo gene delivery by sonoporation: recent progress and prospects. Curr Gene Ther 13:2–14 (**Chapter 7**)

Evans TW (2002) Review article: albumin as a drug – biological effects of albumin unrelated to oncotic pressure. Aliment Pharmacol Ther 16:6–11 (**Chapter 1**)

Evans L, Hughes M, Waters J, Cameron J, Dodsworth N, Tooth D, Greenfield A, Sleep D (2010) The production, characterisation and enhanced pharmacokinetics of scFv-albumin fusions expressed in Saccharomyces cerevisiae. Protein Express Purif 73:113–124 (**Chapter 10**)

Ewing J, Young D, Janero D, Garvey D, Grinnell T (1997) Nitrosylated bovine serum albumin derivatives as pharmacologically active nitric oxide congeners. J Pharmacol Exp Ther 283:947–954 (**Chapter 8**)

Fahey JL, Robinson AG (1963) Factors controlling serum gamma-globulin concentration. J Exp Med 118:845–868 (**Chapter 10**)

Falkenhagen D, Strobl W, Vogt G, Schrefl A, Linsberger I, Gerner FJ, Schoenhofen M (1999) Fractionated plasma separation and adsorption system: a novel system for blood purification to remove albumin bound substances. Artif Organs 23:81–86 (**Chapter 11**)

Fanali G, Trezza V, Marino M, Fasano M, Ascenzi P (2012) Human serum albumin: from bench to bedside. Mol Asp Med 33:209–290 (**Chapter 1**)

Ferrer ML, Duchowicz R, Carrasco B, de la Torre JG, Acuna AU (2001) The conformation of serum albumin in solution: a combined phosphorescence depolarization-hydrodynamic modelling study. Biophys J 80:2422–2430 (**Chapter 1**)

Fersht A (1999) Structure and mechanism in protein science: a guide to enzyme catalysis and protein folding. W. H. Freeman, New York (**Chapter 2**)

Fiume L, Di Stefano G (2010) Lactosaminated human albumin, a hepatotropic carrier of drugs. Eur J Pharm Sci 40:253–262 (**Chapter 1**)

Fleer R, Yeh P, Amellal N, Maury I, Fournier A, Bacchetta F, Baduel P, Jung G, L'Hote H, Becquart J et al (1991) Stable multicopy vectors for high-level secretion of recombinant human serum albumin by Kluyveromyces yeasts. Biotechnology (N Y) 9:968–975 (**Chapter 11**)

Flora K, Brennan JD, Baker GA, Doody MA, Bright FV (1998) Unfolding of acrylodan-labeled human serum albumin probed by steady-state and time-resolved fluorescence methods. Biophys J 75:1084–1096 (**Chapter 2**)

Forbes MM, Steinberg RL, O'Brien WD Jr (2008) Examination of inertial cavitation of optison in producing sonoporation of Chinese hamster ovary cells. Ultrasound Med Biol 34:2009–2018 (**Chapter 7**)

Frahm GE, Smith DGS, Kane A, Lorbetskie B, Cyr TD, Girard M, Johnston MJW (2014) Determination of supplier-to-supplier and lot-to-lot variability in glycation of recombinant human serum albumin expressed in Oryza sativa. PLoS ONE 9, e109893 (**Chapter 1**)

Frank PG, Pavlides S, Lisanti MP (2009) Caveolae and transcytosis in endothelial cells: role in atherosclerosis. Cell Tissue Res 335:41–47 (**Chapter 6**)

Frenkel PA, Chen S, Thai T, Shohet RV, Grayburn PA (2002) DNA-loaded albumin microbubbles enhance ultrasound-mediated transfection in vitro. Ultrasound Med Biol 28:817–822 (**Chapter 7**)

Frese KK, Neesse A, Cook N, Bapiro TE, Lolkema MP, Jodrell DI, Tuveson DA (2012) nab-Paclitaxel potentiates gemcitabine activity by reducing cytidine deaminase levels in a mouse model of pancreatic cancer. Cancer Discov 2:260–269 (**Chapter 6**)

Frodin JE, Lefvert AK, Mellstedt H (1990) Pharmacokinetics of the mouse monoclonal antibody 17-1A in cancer patients receiving various treatment schedules. Cancer Res 50:4866–4871 (**Chapter 10**)

Furchgott R, Zawadzki J (1980) The obligatory role of endothelial cells in the relaxation of arterial smooth muscle by acetylcholine. Nature 288:373–376 (**Chapter 8**)

Furukawa M, Tanaka R, Chuang VT, Ishima Y, Taguchi K, Watanabe H, Maruyama T, Otagiri M (2011) Human serum albumin-thioredoxin fusion protein with long blood retention property is effective in suppressing lung injury. J Control Release 154:189–195 (**Chapter 4**)

Furusyo N, Hayashi J (2013) Glycated albumin and diabetes mellitus. Biochim Biophys Acta 1830:5509–5514 (**Chapters 1 and 3**)

Galantini L, Leggio C, Pavel NV (2008) Human serum albumin unfolding: a small-angle X-ray scattering and light scattering study. J Phys Chem B 112:15460–15469 (**Chapter 2**)

Galantini L, Leggio C, Konarev PV, Pavel NV (2010) Human serum albumin binding ibuprofen: a 3D description of the unfolding pathway in urea. Biophys Chem 147:111–122 (**Chapter 2**)

Galis Z, Ghitescu L, Simionescu M (1988) Fatty acids binding to albumin increases its uptake and transcytosis by the lung capillary endothelium. Eur J Cell Biol 47:358–365 (**Chapter 1**)

Gan Z, Ram S, Ober RJ, Ward ES (2013) Using multifocal plane microscopy to reveal novel trafficking processes in the recycling pathway. J Cell Sci 126:1176–1188 (**Chapter 10**)

Gao Y, LaFleur D, Shah R, Zhao Q, Singh M, Brimijoin S (2008) An albumin-butyrylcholinesterase for cocaine toxicity and addiction: catalytic and pharmacokinetic properties. Chem Biol Interact 175:83–87 (**Chapter 4**)

Garcia-Martinez R, Noiret L, Sen S, Mookerjee R, Jalan R (2015) Albumin infusion improves renal blood flow autoregulation in patients with acute decompensation of cirrhosis and acute kidney injury. Liver Int 35:335–343 (**Chapter 3**)

Gardner ER, Dahut WL, Scripture CD, Jones J, Aragon-Ching JB, Desai N, Hawkins MJ, Sparreboom A, Figg WD (2008) Randomized crossover pharmacokinetic study of solvent-based paclitaxel and nab-paclitaxel. Clin Cancer Res 14:4200–4205 (**Chapter 6**)

Gaston B, Reilly J, Drazen J, Fackler J, Ramdev P, Arnelle D, Mullins M, Sugarbaker D, Chee C, Singel D (1993) Endogenous nitrogen oxides and bronchodilator S-nitrosothiols in human airways. Proc Natl Acad Sci U S A 90:10957–10961 (**Chapter8**)

Gaze DC (2009) Ischemia modified albumin: a novel biomarker for the detection of cardiac ischemia. Drug Metab Pharmacokinet 24:333–341 (**Chapter 1**)

Gelderblom H, Verweij J, Nooter K, Sparreboom A (2001) Cremophor EL: the drawbacks and advantages of vehicle selection for drug formulation. Eur J Cancer 37:1590–1598 (**Chapter 6**)

Geng Y, Petersson A, Wennmalm A, Hansson G (1994) Cytokine-induced expression of nitric oxide synthase results in nitrosylation of heme and nonheme iron proteins in vascular smooth muscle cells. Exp Cell Res 214:418–428 (**Chapter 8**)

Gerlza T, Winkler S, Atlic A, Zankl C, Konya V, Kitic N, Strutzmann E, Knebl K, Adage T, Heinemann A, Weis R, Kungl AJ (2015) Designing a mutant CCL2-HSA chimera with high glycosaminoglycan-binding affinity and selectivity. Protein Eng Des Sel 28:231–240 (**Chapter 4**)

Ghetie V, Popov S, Borvak J, Radu C, Matesoi D, Medesan C, Ober RJ, Ward ES (1997) Increasing the serum persistence of an IgG fragment by random mutagenesis. Nat Biotechnol 15:637–640 (**Chapter 10**)

Ghuman J, Zunszain PA, Petitpas I, Bhattacharya AA, Otagiri M, Curry S (2005) Structural basis of the drug-binding specificity of human serum albumin. J Mol Biol 353:38–52 (**Chapters 1 and 11**)

Gianni L, Kearns CM, Giani A, Capri G, Vigano L, Locatelli A, Bonadonna G, Egorin MJ (1995) Nonlinear pharmacokinetics and metabolism of paclitaxel and its pharmacokinetic/pharmacodynamic relationships in humans. J Clin Oncol 13:180–190 (**Chapter 6**)

Giannoukakis N (2003) CJC-1131. ConjuChem. Curr Opin Investig Drugs 4:1245–1249 (**Chapter 10**)

Gilje OH (ed) (2015) Recent Norwegian experiences with sonazoid. The 30th anniversary bubble conference, September 10–12, 2015. 10–12 Sept 2015. Chicago (**Chapter 7**)

Gimson AE, Braude S, Mellon PJ, Canalese J, Williams R (1982) Earlier charcoal haemoperfusion in fulminant hepatic failure. Lancet 2:681–683 (**Chapter 11**)

Goldberg BB, Liu J-B, Forsberg F (1994) Ultrasound contrast agents: a review. Ultrasound Med Biol 20:319–333 (**Chapter 7**)

Golor G, Bensen-Kennedy D, Haffner S, Easton R, Jung K, Moises T, Lawo JP, Joch C, Veldman A (2013) Safety and pharmacokinetics of a recombinant fusion protein linking coagulation factor VIIa with albumin in healthy volunteers. J Thromb Haemost 11:1977–1985 (**Chapter 4**)

Goncharova EA (2013) mTOR and vascular remodeling in lung diseases: current challenges and therapeutic prospects. FASEB J 27:1796–1807 (**Chapter 6**)

Gong J, Huo M, Zhou J, Zhang Y, Peng X, Yu D, Zhang H, Li J (2009) Synthesis, characterization, drug-loading capacity and safety of novel octyl modified serum albumin micelles. Int J Pharm 376:161–168 (**Chapter 5**)

Gonzalez-Angulo AM, Meric-Bernstam F, Chawla S, Falchook G, Hong D, Akcakanat A, Chen H, Naing A, Fu S, Wheler J, Moulder S, Helgason T, Li S, Elias I, Desai N, Kurzrock R (2013) Weekly nab-rapamycin in patients with advanced nonhematologic malignancies: final results of a phase I trial. Clin Cancer Res 19:5474–5484 (**Chapter 6**)

Gonzalez-Jimenez J, Cortijo M (2002) Urea-induced denaturation of human serum albumin labeled with acrylodan. J Protein Chem 21:75–79 (**Chapter 2**)

Gordon RS Jr, Bartter FC, Waldmann T (1959) Idiopathic hypoalbuminemias: clinical staff conference at the National Institutes of Health. Ann Intern Med 51:553–576 (**Chapter 10**)

Gradishar WJ (2006) Albumin-bound paclitaxel: a next-generation taxane. Expert Opin Pharmacother 7:1041–1053 (**Chapter 6**)

Gradishar WJ, Tjulandin S, Davidson N, Shaw H, Desai N, Bhar P, Hawkins M, O'Shaughnessy J (2005) Phase III trial of nanoparticle albumin-bound paclitaxel compared with polyethylated castor oil-based paclitaxel in women with breast cancer. J Clin Oncol 23:7794–7803 (**Chapters 6 and 10**)

Greenleaf WJ, Bolander ME, Sarkar G, Goldring MB, Greenleaf JF (1998) Artificial cavitation nuclei significantly enhance acoustically induced cell transfection. Ultrasound Med Biol 24:587–595 (**Chapter 7**)

Gryglewski R, Palmer R, Moncada S (1986) Superoxide anion is involved in the breakdown of endothelium-derived vascular relaxing factor. Nature 320:454–456 (**Chapter 8**)

Gryzunov Y, Arroyo A, Vigne J, Zhao Q, Tyurin V, Hubel C, Gandley R, Vladimirov Y, Taylor R, Kagan V (2003) Binding of fatty acids facilitates oxidation of cysteine-34 and converts copper-albumin complexes from antioxidants to prooxidants. Arch Biochem Biophys 413:53–66 (**Chapter 8**)

Gupta N, Hatoum H, Dy GK (2014) First line treatment of advanced non-small-cell lung cancer – specific focus on albumin bound paclitaxel. Int J Nanomedicine 9:209–221 (**Chapter 10**)

Guzman HR, Nguyen DX, Khan S, Prausnitz MR (2001) Ultrasound-mediated disruption of cell membranes. II. Heterogeneous effects on cells. J Acoust Soc Am 110:597–606 (**Chapter 7**)

Guzman HR, McNamara AJ, Nguyen DX, Prausnitz MR (2003) Bioeffects caused by changes in acoustic cavitation bubble density and cell concentration: a unified explanation based on cell-to-bubble ratio and blast radius. Ultrasound Med Biol 29:1211–1222 (**Chapter 7**)

Haley B, Frenkel E (2008) Nanoparticles for drug delivery in cancer treatment. Urol Oncol 26:57–64 (**Chapter 6**)

Hallow DM, Mahajan AD, McCutchen TE, Prausnitz MR (2006) Measurement and correlation of acoustic cavitation with cellular bioeffects. Ultrasound Med Biol 32:1111–1122 (**Chapter 7**)

Halpern W, Riccobene TA, Agostini H, Baker K, Stolow D, Gu ML, Hirsch J, Mahoney A, Carrell J, Boyd E, Grzegorzewski KJ (2002) Albugranin (TM), a recombinant human granulocyte colony stimulating factor (G-CSF) genetically fused to recombinant human albumin induces prolonged myelopoietic effects in mice and monkeys. Pharm Res 19:1720–1729 (**Chapters 4 and 10**)

Hamasaki T, Kashiwagi T, Imada T, Nakamichi N, Aramaki S, Toh K, Morisawa S, Shimakoshi H, Hisaeda Y, Shirahata S (2008) Kinetic analysis of superoxide radical-scavenging and hydroxyl radical-scavenging activities of platinum nanoparticles. Langmuir 24:7354–7364 (**Chapter 9**)

Han Y, Jin BS, Lee SB, Sohn Y, Joung JW, Lee JH (2007) Effects of sugar additives on protein stability of recombinant human serum albumin during lyophilization and storage. Arch Pharm Res 30:1124–1131 (**Chapter 2**)

Hansen RE, Roth D, Winther JR (2009) Quantifying the global cellular thiol-disulfide status. Proc Natl Acad Sci U S A 106:422–427 (**Chapter 3**)

Haraldsson B, Nyström J, Deen WM (2008) Properties of the glomerular barrier and mechanisms of proteinuria. Physiol Rev 88:451–487 (**Chapter 9**)

Hart DL, Heidkamp MC, Iyengar R, Vijayan K, Szotek EL, Barakat JA, Leya M, Henze M, Scrogin K, Henderson KK, Samarel AM (2008) CRNK gene transfer improves function and reverses the myosin heavy chain isoenzyme switch during post-myocardial infarction left ventricular remodeling. J Mol Cell Cardiol 45:93–105 (**Chapter 7**)

Haruki R, Kimura T, Iwsaki H, Yamada K, Kamiyama I, Kohno M, Taguchi K, Nagao S, Maruyama T, Otagiri M, Komatsu T (2015) Safety evaluation of hemoglobin-albumin cluster "HemoAct" as a red blood cell substitute. Sci Rep 5:12778, 1–9 (**Chapter 9**)

Hashiya N, Aoki M, Tachibana K, Taniyama Y, Yamasaki K, Hiraoka K, Makino H, Yasufumi K, Ogihara T, Morishita R (2004) Local delivery of E2F decoy oligodeoxynucleotides using ultrasound with microbubble agent (Optison) inhibits intimal hyperplasia after balloon injury in rat carotid artery model. Biochem Biophys Res Commun 317:508–514 (**Chapter 7**)

Hawkins MJ, Desai N, Soon-Shiong P (ed) (2003) Rationale, preclinical support, and clinical proof-of-concept for formulating waterinsoluble therapeutics as albumin-stabilized nanoparticles: experience with paclitaxel. [abstr 442]. American Association for Cancer Research (AACR) Annual Meeting. Anaheim (**Chapter 6**)

Hawkins MJ, Soon-Shiong P, Desai N (2008) Protein nanoparticles as drug carriers in clinical medicine. Adv Drug Deliv Rev 60:876–885 (**Chapters 2 and 6**)

Haymann JP, Levraud JP, Bouet S, Kappes V, Hagege J, Nguyen G, Xu Y, Rondeau E, Sraer JD (2000) Characterization and localization of the neonatal Fc receptor in adult human kidney. J Am Soc Nephrol 11:632–639 (**Chapter 10**)

He XM, Carter DC (1992) Atomic structure and chemistry of human serum albumin. Nature 358:209–215 (**Chapter 1**)

Hein KL, Kragh-Hansen U, Morth JP, Jeppesen MD, Otzen D, Møller JV, Nissen P (2010) Crystallographic analysis reveals a unique lidocaine binding site on human serum albumin. J Struct Biol 171:353–360 (**Chapter 1**)

Hernot S, Klibanov AL (2008) Microbubbles in ultrasound-triggered drug and gene delivery. Adv Drug Deliv Rev 60:1153–1166 (**Chapter 7**)

Herring C, Schon O (2012) AlbudAb™ technology platform–versatile albumin binding domains for the development of therapeutics with tunable half-lives. In: Kontermann RE (ed) Therapeutic proteins: strategies to modulate their plasma half-lives. Wiley-VCH Verlag GmbH & Co., Weinheim (**Chapter 10**)

Hersh EM, O'Day SJ, Ribas A, Samlowski WE, Gordon MS, Shechter DE, Clawson AA, Gonzalez R (2010) A phase 2 clinical trial of nab-paclitaxel in previously treated and chemotherapy-naive patients with metastatic melanoma. Cancer Am Cancer Soc 116:155–163 (**Chapter 10**)

Hibbs JJ, Taintor R, Vavrin Z, Rachlin E (1988) Nitric oxide: a cytotoxic activated macrophage effector molecule. Biochem Biophys Res Commun 157:87–94 (**Chapter 8**)

Hinton PR, Johlfs MG, Xiong JM, Hanestad K, Ong KC, Bullock C, Keller S, Tang MT, Tso JY, Vasquez M, Tsurushita N (2004) Engineered human IgG antibodies with longer serum half-lives in primates. J Biol Chem 279:6213–6216 (**Chapter 10**)

Hinton PR, Xiong JM, Johlfs MG, Tang MT, Keller S, Tsurushita N (2006) An engineered human IgG1 antibody with longer serum half-life. J Immunol 176:346–356 (**Chapter 10**)

Hirata K, Maruyama T, Watanabe H, Maeda H, Nakajou K, Iwao Y, Ishima Y, Katsumi H, Hashida M, Otagiri M (2010) Genetically engineered mannosylated-human serum albumin as a versatile carrier for liver-selective therapeutics. J Control Release 145:9–16 (**Chapter 2**)

Hobbs SK, Monsky WL, Yuan F, Roberts WG, Griffith L, Torchilin VP, Jain RK (1998) Regulation of transport pathways in tumor vessels: role of tumor type and microenvironment. Proc Natl Acad Sci U S A 95:4607–4612 (**Chapter 6**)

Holt LJ, Basran A, Jones K, Chorlton J, Jespers LS, Brewis ND, Tomlinson IM (2008) Anti-serum albumin domain antibodies for extending the half-lives of short lived drugs. Protein Eng Des Sel 21:283–288 (**Chapter 10**)

Hoofnagle JH, Carithers RL Jr, Shapiro C, Ascher N (1995) Fulminant hepatic failure: summary of a workshop. Hepatology 21:240–252 (**Chapter 11**)

Hosaka H, Haruki R, Yamada K, Böttcher C, Komatsu T (2014) Hemoglobin–albumin cluster incorporating a Pt nanoparticle: artificial O_2 carrier with antioxidant activities. PLoS ONE 9, e110541:1–9 (**Chapter 9**)

Hou S, Li C, Huan Y, Liu S, Liu Q, Sun S, Jiang Q, Jia C, Shen Z (2015) Effets of E2HSA, a long-acting glucagon like peptide-1 receptor agonist, on glycemic control and beta cell function in spontaneous diabetic db/db mice. J Diabetes Res 2015:817839 (**Chapter 4**)

Houghton PJ, Kurmasheva RT, Kolb EA, Gorlick R, Maris JM, Wu J, Tong Z, Arnold MA, Chatterjee M, Williams TM, Smith MA (2015) Initial testing (stage 1) of the tubulin binding agent nanoparticle albumin-bound (nab) paclitaxel (Abraxane((R))) by the Pediatric Preclinical Testing Program (PPTP). Pediatr Blood Cancer 62:1214–1221 (**Chapter 6**)

Howard CM, Forsberg F, Minimo C, Liu JB, Merton DA, Claudio PP (2006) Ultrasound guided site specific gene delivery system using adenoviral vectors and commercial ultrasound contrast agents. J Cell Physiol 209:413–421 (**Chapter 7**)

Hu D, Kluger R (2008) Functional cross-linked hemoglobin bis-tetramers: geometry and cooperativity. Biochemistry 47:12551–12561 (**Chapter 9**)

Hu X, Kheirolomoom A, Mahakian LM, Beegle JR, Kruse DE, Lam KS, Ferrara KW (2012) Insonation of targeted microbubbles produces regions of reduced blood flow within tumor vasculature. Investig Radiol 47:398 (**Chapter 7**)

Huang YS, Chen Z, Chen YQ, Ma GC, Shan JF, Liu W, Zhou LF (2008) Preparation and characterization of a novel exendin-4 human serum albumin fusion protein expressed in Pichia pastoris. J Pept Sci 14:588–595 (**Chapter 4**)

Huang YS, Wen XF, Yang ZY, Wu YL, Lu Y, Zhou LF (2014) Development and characterization of a novel fusion protein of a mutated granulocyte colony-stimulating factor and hman serum albumin in Pichia pastoris. PLoS ONE 9, e115840 (**Chapter 4**)

Hughes RD, Pucknell A, Routley D, Langley PG, Wendon JA, Williams R (1994) Evaluation of the BioLogic-DT sorbent-suspension dialyser in patients with fulminant hepatic failure. Int J Artif Organs 17:657–662 (**Chapter 11**)

Ibrahim NK, Desai N, Legha S, Soon-Shiong P, Theriault RL, Rivera E, Esmaeli B, Ring SE, Bedikian A, Hortobagyi GN, Ellerhorst JA (2002) Phase I and pharmacokinetic study of ABI-007, a cremophor-free, protein-stabilized, nanoparticle formulation of paclitaxel. Clin Cancer Res 8:1038–1044 (**Chapter 6**)

Ignarro L (2000) The unique role of nitric oxide as a signaling molecule in the cardiovascular system. Ital Heart J 1:28–29 (**Chapter 8**)

Ignarro L, Wei L (2002) Visiting professorial lecture: nitric oxide in the regulation of vascular function: an historical overview. J Card Surg 17:301–306 (**Chapter 8**)

Ikebe N, Akaike T, Miyamoto Y, Hayashida K, Yoshitake J, Ogawa M, Maeda H (2000) Protective effect of S-nitrosylated alpha(1)-protease inhibitor on hepatic ischemia-reperfusion injury. J Pharmacol Exp Ther 295:904–911 (**Chapter 8**)

Ikegaya K, Hirose M, Ohmura T, Nokihara K (1997) Complete determination of disulfide forms of purified recombinant human serum albumin, secreted by the yeast Pichia pastoris. Anal Chem 69:1986–1991 (**Chapter 11**)

Ikuta S, Chuang VT, Ishima Y, Nakajou K, Furukawa M, Watanabe H, Maruyama T, Otagiri M (2010) Albumin fusion of thioredoxin – the production and evaluation of its biological activity for potential therapeutic applications. J Control Release 147:17–23 (**Chapters 2 and 4**)

Intaglietta M (2004) Microvascular transport factors in the design of effective blood substitutes. In: Messmer K, Burhop KE, Hutter J (eds) Microcirculatory effects of hemoglobin solutions. Karger AG, Basel, pp 8–15 (**Chapter 9**)

Irache JM, Merodio M, Arnedo A, Camapanero MA, Mirshahi M, Espuelas S (2005) Albumin nanoparticles for the intravitreal delivery of anticytomegaloviral drugs. Mini Rev Med Chem 5:293–305 (**Chapter 5**)

Irizarry L, Luu T, McKoy J, Samaras A, Fisher M, Carias E, Raisch D, Calhoun E, Bennett C (2009) Cremophor EL-containing paclitaxel-induced anaphylaxis: a call to action. Commun Oncol 6:132–134 (**Chapter 6**)

Ishima Y, Maruyama T (2016) Human serum albumin as carrier in drug delivery systems. Yakugaku Zasshi 136:39–47 (**Chapter 8**)

Ishima Y, Akaike T, Kragh-Hansen U, Hiroyama S, Sawa T, Maruyama T, Kai T, Otagiri M (2007a) Effects of endogenous ligands on the biological role of human serum albumin in S-nitrosylation. Biochem Biophys Res Commun 364:790–795 (**Chapter 8**)

Ishima Y, Sawa T, Kragh-Hansen U, Miyamoto Y, Matsushita S, Akaike T, Otagiri M (2007b) S-nitrosylation of human variant albumin liprizzi (R410C) confers potent antibacterial and cytoprotective properties. J Pharmacol Exp Ther 320:969–977 (**Chapter 8**)

Ishima Y, Akaike T, Kragh-Hansen U, Hiroyama S, Sawa T, Suenaga A, Maruyama T, Kai T, Otagiri M (2008) S-nitrosylated human serum albumin-mediated cytoprotective activity is enhanced by fatty acid binding. J Biol Chem 283:34966–34975 (**Chapters 8 and 11**)

Ishima Y, Hiroyama S, Kragh-Hansen U, Maruyama T, Sawa T, Akaike T, Kai T, Otagiri M (2010) One-step preparation of S-nitrosated human serum albumin with high biological activities. Nitric Oxide 23:121–127 (**Chapters 2 and 8**)

Ishima Y, Yoshida F, Kragh-Hansen U, Watanabe K, Katayama N, Nakajou K, Akaike T, Kai T, Maruyama T, Otagiri M (2011) Cellular uptake mechanisms and responses to NO transferred from mono- and poly-S-nitrosated human serum albumin. Free Radic Res 45:1196–1206 (**Chapter 8**)

Ishima Y, Chen D, Fang J, Maeda H, Minomo A, Kragh-Hansen U, Kai T, Maruyama T, Otagiri M (2012a) S-nitrosated human serum albumin dimer is not only a novel anti-tumor drug but also a potentiator for anti-tumor drugs with augmented EPR effects. Bioconjug Chem 23:264–271 (**Chapters 1 and 8**)

Ishima Y, Hoshino H, Shinagawa T, Watanabe K, Akaike T, Sawa T, Kragh-Hansen U, Kai T, Watanabe H, Maruyama T, Otagiri M (2012b) *S*-Guanylation of human serum albumin is a unique posttranslational modification and results in a novel class of antibacterial agents. J Pharm Sci 101:3222–3229 (**Chapter 1**)

Ishima Y, Hara M, Kragh-Hansen U, Inoue A, Suenaga A, Kai T, Watanabe H, Otagiri M, Maruyama T (2012c) Elucidation of the therapeutic enhancer mechanism of poly-S-nitrosated human serum albumin against multidrug-resistant tumor in animal models. J Control Release 164:1–7 (**Chapter 8**)

Ishima Y, Fang J, Kragh-Hansen U, Yin H, Liao L, Katayama N, Watanabe H, Kai T, Suenaga A, Maeda H, Otagiri M, Maruyama T (2014) Tuning of poly-S-nitrosated human serum albumin as superior antitumor nanomedicine. J Pharm Sci 103:2184–2188 (**Chapter 1**)

Ishima Y, Inoue A, Fang J, Kinoshita R, Ikeda M, Watanabe H, Maeda H, Otagiri M, Maruyama T (2015) Poly-S-nitrosated human albumin enhances the antitumor and antimetastasis effect of bevacizumab, partly by inhibiting autophagy through the generation of nitric oxide. Cancer Sci 106:194–200 (**Chapter 8**)

Israel EJ, Patel VK, Taylor SF, Marshakrothstein A, Simister N (1995) Requirement for a beta(2)-microglobulin-associated Fc receptor for acquisition of maternal Igg by fetal and neonatal mice. J Immunol 154:6246–6251 (**Chapter 10**)

Israel EJ, Wilsker DF, Hayes KC, Schoenfeld D, Simister NE (1996) Increased clearance of IgG in mice that lack beta(2)-microglobulin: possible protective role of FcRn. Immunology 89:573–578 (**Chapter 10**)

Israel EJ, Taylor S, Wu Z, Mizoguchi E, Blumberg RS, Bhan A, Simister NE (1997) Expression of the neonatal Fc receptor, FcRn, on human intestinal epithelial cells. Immunology 92:69–74 (**Chapter 10**)

Iwanaga K, Tominaga K, Yamamoto K, Habu M, Maeda H, Akifusa S, Tsujisawa T, Okinaga T, Fukuda J, Nishihara T (2007) Local delivery system of cytotoxic agents to tumors by focused sonoporation. Cancer Gene Ther 14:354–363 (**Chapter 7**)

Iwao Y, Anraku M, Yamasaki K, Kragh-Hansen U, Kawai K, Maruyama T, Otagiri M (2006) Oxidation of Arg-410 promotes the elimination of human serum albumin. Biochim Biophys Acta 1764:743–749 (**Chapter 1**)

Iwao Y, Hiraike M, Kragh-Hansen U, Mera K, Noguchi T, Anraku M, Kawai K, Maruyama T, Otagiri M (2007) Changes of net charge and α-helical content affect the pharmacokinetic properties of human serum albumin. Biochim Biophys Acta 1774:1582–1590 (**Chapter 1**)

Iwao Y, Hiraike M, Kragh-Hansen U, Kawai K, Suenaga A, Maruyama T, Otagiri M (2009) Altered chain-length and glycosylation modify the pharmacokinetics of human serum albumin. Biochim Biophys Acta 1794:634–641 (**Chapter 1**)

Iwao Y, Ishima Y, Yamada J, Noguchi T, Kragh-Hansen U, Mera K, Honda D, Suenaga A, Maruyama T, Otagiri M (2012) Quantitative evaluation of the role of cysteine and methionine residues in the antioxidant activity of human serum albumin using recombinant mutants. IUBMB Life 64:450–454 (**Chapters 1, 2 and 3**)

Iyengar R, Stuehr D, Marletta M (1987) Macrophage synthesis of nitrite, nitrate, and N-nitrosamines: precursors and role of the respiratory burst. Proc Natl Acad Sci U S A 84:6369–6373 (**Chapter 8**)

Jacobs AD, Otero H, Picozzi V (1999) Gemcitabine (G) and taxotere® (T) in patients with unresectable pancreatic carcinoma. Proc Am Soc Clin Oncol 18:1103A (**Chapter 6**)

Jacobs SA, Gibbs AC, Conk M, Yi F, Maguire D, Kane C, O'Neil KT (2015) Fusion to a highly stable consensus albumin binding domain allows for tunable pharmacokinetics. Protein Eng Des Sel 28:385–393 (**Chapter 10**)

Jacobsen J (1969) Binding of bilirubin to human serum albumin – determination of the dissociation constants. FEBS Lett 5:112–114 (**Chapter 11**)

Jahanshahi M, Babaei Z (2008) Protein nanoparticle: a unique system as drug delivery vehicles. Afr J Biotechnol 7:4926–4934 (**Chapter 5**)

Jahanshahi M, Zhang Z, Lyddiatt A (2005) Subtractive chromatography for purification and recovery of nano-bioproducts. J IET Nanobiol Technol 152:121–126 (**Chapter 5**)

Jahr JS, Sadighi A, Doherty L, Li A, Kim HW (2011) Hemoglobin-based oxygen carriers: history, limits, brief summary of the state of the art, including clinical trials. In: Bettati S, Mozzarelli A (eds) Chemistry and biochemistry of oxygen therapeutics: from transfusion to artificial blood. Wiley, West Sussex, pp 301–316 (**Chapter 9**)

Jalan R, Kapoor D (2004) Reversal of diuretic-induced hepatic encephalopathy with infusion of albumin but not colloid. Clin Sci (Lond) 106:467–474 (**Chapter 3**)

Jalan R, Williams R (2002) Acute-on-chronic liver failure: pathophysiological basis of therapeutic options. Blood Purif 20:252–261 (**Chapter 11**)

Jalan R, Schnurr K, Mookerjee RP, Sen S, Cheshire L, Hodges S, Muravsky V, Williams R, Matthes G, Davies NA (2009) Alterations in the functional capacity of albumin in patients with decompensated cirrhosis is associated with increased mortality. Hepatology 50:555–564 (**Chapter 3**)

Jaroslav Turánek ADM, Zuzana Kauerová, Róbert Lukáč, Josef Mašek ŠKaMRL-BNaMML-BBPfivIaT, Advances in Bioengineering, Prof. Pier Andrea Serra (ed). ISBN: 978-953-51-2141-1, InTech, doi:10.5772/59870. Available from: http://www.intechopen.com/books/advances-in-bioengineering/lipid-based-nanoparticles-and-microbubbles-multifunctional-lipid-based-biocompatible-particles-for-i. Lipid-based nanoparticles and microbubbles – multifunctional lipid-based biocompatible particles for in vivo imaging and theranostics, Advances in Bioengineering (**Chapter 7**)

Jelenc J, Jelenc J, Miklavčič D, Lebar AM (2012) Low-frequency sonoporation in vitro: experimental system evaluation. Strojniški Vestn J Mech Eng 58:319–326 (**Chapter 7**)

Jespers L, Schon O, Famm K, Winter G (2004) Aggregation-resistant domain antibodies selected on phage by heat denaturation. Nat Biotechnol 22:1161–1165 (**Chapter 10**)

Jiang L, Wang J, Solorzano-Vargas RS, Tsai HV, Gutierrez EM, Ontiveros LO, Kiela PR, Wu SV, Martin MG (2004) Characterization of the rat intestinal Fc receptor (FcRn) promoter: transcriptional regulation of FcRn gene by the Sp family of transcription factors. Am J Physiol Gastrointest Liver Physiol 286:G922–G931 (**Chapter 10**)

Joerger M, Huitema AD, van den Bongard DH, Schellens JH, Beijnen JH (2006) Quantitative effect of gender, age, liver function, and body size on the population pharmacokinetics of paclitaxel in patients with solid tumors. Clin Cancer Res 12:2150–2157 (**Chapter 6**)

Joerger M, Huitema AD, Richel DJ, Dittrich C, Pavlidis N, Briasoulis E, Vermorken JB, Strocchi E, Martoni A, Sorio R, Sleeboom HP, Izquierdo MA, Jodrell DI, Calvert H, Boddy AV, Hollema H, Fety R, Van der Vijgh WJ, Hempel G, Chatelut E, Karlsson M, Wilkins J, Tranchand B, Schrijvers AH, Twelves C, Beijnen JH, Schellens JH (2007) Population pharmacokinetics and pharmacodynamics of paclitaxel and carboplatin in ovarian cancer patients: a study by the European organization for research and treatment of cancer-pharmacology and molecular mechanisms group and new drug development group. Clin Cancer Res 13:6410–6418 (**Chapter 6**)

Johansson MU, Frick IM, Nilsson H, Kraulis PJ, Hober S, Jonasson P, Linhult M, Nygren PA, Uhlen M, Bjorck L, Drakenberg T, Forsen S, Wikstrom M (2002) Structure, specificity, and mode of interaction for bacterial albumin-binding modules. J Biol Chem 277:8114–8120 (**Chapter 10**)

John TA, Vogel SM, Tiruppathi C, Malik AB, Minshall RD (2003) Quantitative analysis of albumin uptake and transport in the rat microvessel endothelial monolayer. Am J Physiol Lung Cell Mol Physiol 284:L187–L196 (**Chapter 6**)

Jonsson A, Dogan J, Herne N, Abrahmsen L, Nygren PA (2008) Engineering of a femtomolar affinity binding protein to human serum albumin. Protein Eng Des Sel 21:515–527 (**Chapter 10**)

Joshi MR, Yao N, Myers KA, Zn L (2013) Human serum albumin and p53-activating peptide fusion protein is able to promote apoptosis and deliver fatty acid-modified molecules. PLoS ONE 8, e80926 (**Chapter 4**)

Joung CH, Shin JY, Koo JK, Lim JJ, Wang JS, Lee SJ, Tan HK, Kim SL, Lim SM (2009) Production and characterization of long-acting recombinant human serum albumin-EPO fusion protein expressed in CHO cell. Protein Expr Purif 68:137–145 (**Chapter 4**)

Junghans RP, Anderson CL (1996) The protection receptor for IgG catabolism is the beta(2)-microglobulin-containing neonatal intestinal transport receptor. Proc Natl Acad Sci U S A 93:5512–5516 (**Chapter 10**)

Ka SM, Huang XR, Lan HY, Tsai PY, Yang SM, Shui HA, Chen A (2007) Smad7 gene therapy ameliorates an autoimmune crescentic glomerulonephritis in mice. J Am Soc Nephrol JASN 18:1777–1788 (**Chapter 7**)

Kadowaki D, Anraku M, Tasaki Y, Kitamura K, Wakamatsu S, Tomita K, Gebicki JM, Maruyama T, Otagiri M (2007) Effect of olmesartan on oxidative stress in hemodialysis patients. Hypertens Res 30:395–402 (**Chapter 3**)

Kajita M, Hikosaka K, Iitsuka M, Kanayama A, Toshima N, Miyamoto Y (2007) Platinum nanoparticle is a useful scavenger of superoxide anion and hydrogen peroxide. Free Radic Res 41:615–626 (**Chapter 9**)

Kalli C, Teoh WC, Leen E (2014) Introduction of genes via sonoporation and electroporation. Adv Exp Med Biol 818:231–254 (**Chapter 7**)

Kamp O (ed) (2015) Reduction of microvasculature injury using a theranostic imaging strategy, an an unexpected finding. The 30th anniversary bubble conference, September 10–12, 2015. 10–12 Sept 2015. Chicago (**Chapter 7**)

Kamp O, Dilkmas PA. Ultrasound contrast agents to facilitate sonothrombolysis in patients with acute myocardial infarction. http://www.trialregister.nl/trialreg/admin/rctview.asp?TC=161 (**Chapter 7**)

Kandil E, Egashira M, Miyoshi O, Niikawa N, Ishibashi T, Kasahara M (1996) The human gene encoding the heavy chain of the major histocompatibility complex class I-like Fc receptor (FCGRT) maps to 19q13.3. Cytogenet Cell Genet 73:97–98 (**Chapter 10**)

Karshafian R, Bevan PD, Burns PN, Karshafian R, Samac S, Banerjee M, Bevan PD (ed) (2005) Ultrasound-induced uptake of different size markers in mammalian cells. Ultrasonics Symposium, 2005 IEEE. 18–21 Sept 2005 (**Chapter 7**)

Katsumi H, Nishikawa M, Yamashita F, Hashida M (2005) Development of polyethylene glycol-conjugated poly-S-nitrosated serum albumin, a novel S-Nitrosothiol for prolonged delivery of nitric oxide in the blood circulation in vivo. J Pharmacol Exp Ther 314:1117–1124 (**Chapter 8**)

Katsumi H, Nishikawa M, Hashida M (2007) Development of nitric oxide donors for the treatment of cardiovascular diseases. Cardiovasc Hematol Agents Med Chem 5:204–208 (**Chapter 8**)

Kawakami A, Kubota K, Yamada N, Tagami U, Takehana K, Sonaka I, Suzuki E, Hirayama K (2006) Identification and characterization of oxidized human serum albumin. A slight structural change impairs its ligand-binding and antioxidant functions. FEBS J 273:3346–3357 (**Chapter 3**)

Keller F, Heinze H, Jochimsen F, Passfall J, Schuppan D, Buttner P (1995) Risk factors and outcome of 107 patients with decompensated liver disease and acute renal failure (including 26 patients with hepatorenal syndrome): the role of hemodialysis. Ren Fail 17:135–146 (**Chapter 11**)

Kessel D (1992) Properties of cremophor EL micelles probed by fluorescence. Photochem Photobiol 56:447–451 (**Chapter 6**)

Khan MM, Muzammil S, Tayyab S (2000) Role of salt bridge(s) in the binding and photoconversion of bilirubin bound to high affinity site on human serum albumin. Biochim Biophys Acta Protein Struct Mol Enzymol 1479:103–113 (**Chapter 11**)

Kiley JE, Pender JC, Welch HF, Welch CS (1958) Ammonia intoxication treated by hemodialysis. N Engl J Med 259:1156–1161 (**Chapter 11**)

Kim HJ, Greenleaf JF, Kinnick RR, Bronk JT, Bolander ME (1996) Ultrasound-mediated transfection of mammalian cells. Hum Gene Ther 7:1339–1346 (**Chapter 7**)

Kim JG, Baggio LL, Bridon DP, Castaigne JP, Robitaille MF, Jette L, Benquet C, Drucker DJ (2003) Development and characterization of a glucagon-like peptide 1-albumin conjugate: the ability to activate the glucagon-like peptide 1 receptor in vivo. Diabetes 52:751–759 (**Chapter 10**)

Kim H, Fariss RN, Zhang C, Robinson SB, Thill M, Csaky KG (2008) Mapping of the neonatal Fc receptor in the rodent eye. Invest Ophthalmol Vis Sci 49:2025–2029 (**Chapter 10**)

Kim TH, Jiang HH, Youn YS, Park CW, Tak KK, Lee S, Kim H, Jon S, Chen X, Lee KC (2011) Preparation and characterization of water-soluble albumin-bound curcumin nanoparticles with improved antitumor activity. Int J Pharm 403:285–291 (**Chapter 5**)

Kimura T, Shinohara R, Böttcher C, Komatsu T (2015) Core-shell clusters of human haemoglobin A and human serum albumin: artificial O_2-carriers having various O_2-affinicites. J Mater Chem B 3:6157–6164 (**Chapter 9**)

Kinoshita M, Hynynen K (2005) Intracellular delivery of Bak BH3 peptide by microbubble-enhanced ultrasound. Pharm Res 22:716–720 (**Chapter 7**)

Kinoshita M, Hynynen K (2006) Intracellular delivery of peptides and siRNAs using microbubble enhanced focused ultrasound. AIP Conf Proc 829:538–542 (**Chapter 7**)

Kinoshita J, Fushida S, Tsukada T, Oyama K, Watanabe T, Shoji M, Okamoto K, Nakanuma S, Sakai S, Makino I, Furukawa H, Hayashi H, Nakamura K, Inokuchi M, Nakagawara H, Miyashita T, Tajima H, Takamura H, Ninomiya I, Fujimura T, Masakazu Y, Hirakawa K, Ohta T (2014) Comparative study of the antitumor activity of Nab-paclitaxel and intraperitoneal solvent-based paclitaxel regarding peritoneal metastasis in gastric cancer. Oncol Rep 32:89–96 (**Chapter 6**)

Kinoshita R, Ishima Y, Ikeda M, Kragh-Hansen U, Fang J, Nakamura H, Chuang VT, Tanaka R, Maeda H, Kodama A, Watanabe H, Maeda H, Otagiri M, Maruyama T (2015) S-Nitrosated human serum albumin dimer as novel nano-EPR enhancer applied to macromolecular anti-tumor drugs such as micelles and liposomes. J Control Release 217:1–9 (**Chapters 1, 2 and 8**)

Kjeldsen T, Pettersson AF, Drube L, Kurtzhals P, Jonassen I, Havelund S, Hansen PH, Markussen J (1998) Secretory expression of human albumin domains in Saccharomyces cerevisiae and their binding of myristic acid and an acylated insulin analogue. Protein Expr Purif 13:163–169 (**Chapter 11**)

Klammt S, Mitzner S, Stange J, Brinkmann B, Drewelow B, Emmrich J, Liebe S, Schmidt R (2007) Albumin-binding function is reduced in patients with decompensated cirrhosis and correlates inversely with severity of liver disease assessed by model for end-stage liver disease. Eur J Gastroenterol Hepatol 19:257–263 (**Chapter 3**)

Klibanov AL, Maruyama K, Torichilin VP, Huang L (1900) Amphipatic polyethyleneglycols effectively prolong the circulation time of liposomes. FEBS Lett 268:235–238 (**Chapter 5**)

Kluger R (2010) Red cell substitutes from hemoglobin – do we start all over again? Curr Opin Chem Biol 14:538–543 (**Chapter 9**)

Kluger R, Lui FE (2013) HBOCs from chemical modification of Hb. In: Kim HW, Greenburg AG (eds) Hemoglobin-based oxygen carriers as red cell substitutes and oxygen therapeutics. Springer, Berlin, pp 159–183 (**Chapter 9**)

Kluger R, Zhang J (2003) Hemoglobin dendrimers: functional protein clusters. J Am Chem Soc 125:6070–6071 (**Chapter 7**)

Ko SY, Pegu A, Rudicell RS, Yang ZY, Joyce MG, Chen XJ, Wang KY, Bao S, Kraemer TD, Rath T, Zeng M, Schmidt SD, Todd JP, Penzak SR, Saunders KO, Nason MC, Haase AT, Rao SS, Blumberg RS, Mascola JR, Nabel GJ (2014) Enhanced neonatal Fc receptor function improves protection against primate SHIV infection. Nature 514:642 (**Chapter 10**)

Kobayashi K (2006) Summary of recombinant human serum albumin development. Biologicals 34:55–59 (**Chapter 2**)

Kobayashi K, Nakamura N, Sumi A, Ohmura T, Yokoyama K (1998) The development of recombinant human serum albumin. Ther Apher 2:257–262 (**Chapter 11**)

Kobayashi K, Qiao SW, Yoshida M, Baker K, Lencer WI, Blumberg RS (2009) An FcRn-dependent role for anti-flagellin immunoglobulin G in pathogenesis of colitis in mice. Gastroenterology 137:1746–1756 e1741 (**Chapter 10**)

Kodama A, Watanabe H, Tanaka R, Tanaka H, Chuang VT, Miyamoto Y, Wu Q, Endo M, Hamasaki K, Ishima Y, Fukagawa M, Otagiri M, Maruyama T (2013) A human serum albumin-thioredoxin fusion protein prevents experimental contrast-induced nephropathy. Kidney Int 83:446–454 (**Chapter 4**)

Kodama A, Watanabe H, Tanaka R, Kondo M, Chuang VT, Wu Q, Endo M, Ishima Y, Fukagawa M, Otagiri M, Maruyama T (2014) Albumin fusion renders thioredoxin an effective anti-oxidative and anti-inflammatory agent for preventing cisplatin-induced nephrotoxicity. Biochim Biophys Acta 1840:1152–1162 (**Chapter 4**)

Koike H, Tomita N, Azuma H, Taniyama Y, Yamasaki K, Kunugiza Y, Tachibana K, Ogihara T, Morishita R (2005) An efficient gene transfer method mediated by ultrasound and microbubbles into the kidney. J Gene Med 7:108–116 (**Chapter 7**)

Kondo I, Ohmori K, Oshita A, Takeuchi H, Fuke S, Shinomiya K, Noma T, Namba T, Kohno M (2004) Treatment of acute myocardial infarction by hepatocyte growth factor gene transfer: the first demonstration of myocardial transfer of a "functional" gene using ultrasonic microbubble destruction. J Am Coll Cardiol 44:644–653 (**Chapter 7**)

Kondrup J, Almdal T, Vilstrup H, Tygstrup N (1992) High volume plasma exchange in fulminant hepatic failure. Int J Artif Organs 15:669–676 (**Chapter 11**)

Kontermann RE, Brinkmann U (2015) Bispecific antibodies. Drug Discov Today 20:838–847 (**Chapter 4**)

Kosa T, Maruyama T, Otagiri M (1998) Species differences of serum albumins: II. Chemical and thermal stability. Pharm Res 15:449–454 (**Chapter 2**)

Kotopoulis S, Dimcevski G, Gilja OH, Hoem D, Postema M (2013) Treatment of human pancreatic cancer using combined ultrasound, microbubbles, and gemcitabine: a clinical case study. Med Phys 40:072902 (**Chapter 7**)

Kottschade LA, Suman VJ, Amatruda T, McWilliams RR, Mattar BI, Nikcevich DA, Behrens R, Fitch TR, Jaslowski AJ, Markovic SN (2011) A phase II trial of nab-paclitaxel (ABI-007) and carboplatin in patients with unresectable stage IV melanoma a North Central cancer treatment group study, N057E. Cancer Am Cancer Soc 117:1704–1710 (**Chapter 10**)

Kottschade LA, Suman VJ, Perez DG, McWilliams RR, Kaur JS, Amatruda TT, Geoffroy FJ, Gross HM, Cohen PA, Jaslowski AJ, Kosel ML, Markovic SN (2013) A randomized phase 2 study of temozolomide and bevacizumab or nab-paclitaxel, carboplatin, and bevacizumab in patients with unresectable stage IV melanoma a North Central cancer treatment group study, N0775. Cancer Am Cancer Soc 119:586–592 (**Chapter 10**)

Kouchakzadeh H, Shojaosadati SA, Maghsoudi A, Farahani EV (2010) Optimization of PEGylation conditions for BSA nanoparticles using response surface methodology. AAPS PharmSciTech 11:1206–1211 (**Chapter 5**)

Kouno Y, Anraku M, Yamasaki K, Okayama Y, Iohara D, Ishima Y, Maruyama T, Kragh-Hansen U, Hirayama F, Otagiri M (2014) N-acetyl-L-methionine is a superior protectant of human serum albumin against photo-oxidation and reactive oxygen species compared to N-acetyl-L-tryptophan. Biochim Biophys Acta 1840:2806–2812 (**Chapters 1 and 2**)

Kozyraki R, Fyfe J, Verroust PJ, Jacobsen C, Dautry-Varsat A, Gburek J, Willnow TE, Christensen EI, Moestrup SK (2001) Megalin-dependent cubilin-mediated endocytosis is a major pathway for the apical uptake of transferrin in polarized epithelia. Proc Natl Acad Sci U S A 98:12491–12496 (**Chapter 10**)

Kragh-Hansen U (1981) Molecular aspects of ligand binding to serum albumin. Pharmacol Rev 33:17–53 (**Chapters 1 and 8**)

Kragh-Hansen U (2013) Molecular and practical aspects of the enzymatic properties of human serum albumin and of albumin-ligand complexes. Biochim Biophys Acta 1830:5535–5544 (**Chapter 1**)

Kragh-Hansen U, Chuang VT, Otagiri M (2002) Practical aspects of the ligand-binding and enzymatic properties of human serum albumin. Biol Pharm Bull 25:695–704 (**Chapters 1, 2 and 8**)

Kragh-Hansen U, Saito S, Nishi K, Anraku M, Otagiri M (2005) Effect of genetic variation on the thermal stability of human serum albumin. Biochim Biophys Acta 1747:81–88 (**Chapter 2**)

Kragh-Hansen U, Watanabe H, Nakajou K, Iwao Y, Otagiri M (2006) Chain length-dependent binding of fatty acid anions to human serum albumin studied by site-directed mutagenesis. J Mol Biol 363:702–712 (**Chapter 1**)

Kragh-Hansen U, Minchiotti L, Galliano M, Peters T Jr (2013) Human serum albumin isoforms: genetic and molecular aspects and functional consequences. Biochim Biophys Acta 1830:5405–5417 (**Chapter 1**)

Kragh-Hansen U, Minchiotti L, Coletta A, Bienk K, Galliano M, Schiøtt B, Iwao Y, Ishima Y, Otagiri M (2016) Mutants and molecular dockings reveal that the primary L-thyroxine binding site in human serum albumin is not the one which can cause familial dysalbuminemic hyperthyroxinemia. Biochim Biophys Acta 1860:648–660 (**Chapter 1**)

Kratz F (2008) Albumin as a drug carrier: design of prodrugs, drug conjugates and nanoparticles. J Control Release 132:171–183 (**Chapters 2, 5 and 6**)

Kratz F (2014) A clinical update of using albumin as a drug vehicle – a commentary. J Control Release 190:331–336 (**Chapter 10**)

Kratz F, Muller-Driver R, Hofmann I, Drevs J, Unger C (2000) A novel macromolecular prodrug concept exploiting endogenous serum albumin as a drug carrier for cancer chemotherapy. J Med Chem 43:1253–1256 (**Chapter 10**)

Kratz F, Abu Ajaj K, Warnecke A (2007) Anticancer carrier-linked prodrugs in clinical trials. Expert Opin Investig Drug 16:1037–1058 (**Chapter 10**)

Kremer P, Hartung G, Bauder-Wust U, Schrenk HH, Wunder A, Heckl S, Zillmann U, Sinn H (2002) Efficacy and tolerability of an aminopterin-albumin conjugate in tumor-bearing rats. Anticancer Drugs 13:615–623 (**Chapter 6**)

Kubota K, Nakayama A, Takehana K, Kawakami A, Yamada N, Suzuki E (2009) A simple stabilization method of reduced albumin in blood and plasma for the reduced/oxidized albumin ratio measurement. Int J Biomed Sci 5:293–301 (**Chapter 3**)

Kuliszewski MA, Kobulnik J, Lindner JR, Stewart DJ, Leong-Poi H (2011) Vascular gene transfer of SDF-1 promotes endothelial progenitor cell engraftment and enhances angiogenesis in ischemic muscle. Mol Ther 19:895–902, doi: 810.1038/mt.2011.1018. Epub 2011 Mar 1031 (**Chapter 7**)

Lan HY, Mu W, Tomita N, Huang XR, Li JH, Zhu HJ, Morishita R, Johnson RJ (2003) Inhibition of renal fibrosis by gene transfer of inducible Smad7 using ultrasound-microbubble system in rat UUO model. J Am Soc Nephrol JASN 14:1535–1548 (**Chapter 7**)

Lancaster JJ (1994) Simulation of the diffusion and reaction of endogenously produced nitric oxide. Proc Natl Acad Sci U S A 91:8137–8141 (**Chapter 8**)

Lancaster JJ, Hibbs JJ (1990) EPR demonstration of iron-nitrosyl complex formation by cytotoxic activated macrophages. Proc Natl Acad Sci U S A 87:1223–1227 (**Chapter 8**)

Langer K, Balthasar S, Vogel V, Dinauer N, von Briesen H, Schubert D (2003) Optimization of the preparation process for human serum albumin (HSA) nanoparticles. Int J Pharm 257:169–180 (**Chapter 5**)

Larina IV, Evers BM, Esenaliev RO (2005) Optimal drug and gene delivery in cancer cells by ultrasound-induced cavitation. Anticancer Res 25:149–156 (**Chapter 7**)

Larsen FS, Hansen BA, Jorgensen LG, Secher NH, Kirkegaard P, Tygstrup N (1994) High-volume plasmapheresis and acute liver transplantation in fulminant hepatic failure. Transplant Proc 26:1788 (**Chapter 11**)

Lawrie A, Brisken AF, Francis SE, Cumberland DC, Crossman DC, Newman CM (2000) Microbubble-enhanced ultrasound for vascular gene delivery. Gene Ther 7:2023–2027 (**Chapter 7**)

Lawrie A, Brisken AF, Francis SE, Wyllie D, Kiss-Toth E, Qwarnstrom EE, Dower SK, Crossman DC, Newman CM (2003) Ultrasound-enhanced transgene expression in vascular cells is not dependent upon cavitation-induced free radicals. Ultrasound Med Biol 29:1453–1461 (**Chapter 7**)

Lee MS, Kim YH, Kim YJ, Kwon SH, Bang JK, Lee SM, Song YS, Hahm DH, Shim I, Han D, Her S (2011a) Pharmacokinetics and biodistribution of human serum albumin-TIMP-2 fusion protein using near-infrared optical imaging. J Pharm Sci 14:368–377 (**Chapter 4**)

Lee SH, Heng D, Ng WK, Chan HK, Tan RB (2011b) Nano spray drying: a novel method for preparing protein nanoparticles for protein therapy. Int J Pharm 403:192–200 (**Chapter 5**)

Lee MS, Jung JI, Kwon SH, Lee SM, Morita K, Her S (2012) TIMP-2 fusion protein with human serum albumin potentiates anti-angiogenesis-mediated inhibition of tumor growth by suppressing MMP-2 expression. PLoS ONE 7, e35710 (**Chapter 4**)

Lee H, Jeong H, Park S, Yoo W, Choi S, Choi K, Lee MG, Lee M, Cha D, Kim YS, Han J, Kim W, Park SH, Oh J (2015) Fusion protein of retinol-binding protein and albumin domain III reduces liver fibrosis. EMBO Mol Med 7:819–830 (**Chapter 4**)

Leger R, Thibaudeau K, Robitaille M, Quraishi O, van Wyk P, Bousquet-Gagnon N, Carette J, Castaigne JP, Bridon DP (2004) Identification of CJC-1131-albumin bioconjugate as a stable and bioactive GLP-1(7–36) analog. Bioorg Med Chem Lett 14:4395–4398 (**Chapter 10**)

Leggio C, Galantini L, Konarev PV, Pavel NV (2009) Urea-induced denaturation process on defatted human serum albumin and in the presence of palmitic acid. J Phys Chem B 113:12590–12602 (**Chapter 2**)

Leung K (2008) ^{125}I-T84.66 scFv-human serum albumin. Molecular Imaging and Contrast Agent Database (MICAD) [Internet] (**Chapter 4**)

Li D, Hu T, Manjula BN, Acharya SA (2008) Non-conservative surface decoration of hemoglobin: influence of neutralization of positive charges at PEGylation sites on molecular and functional properties of PEGylated hemoglobin. Biochim Biophys Acta 1784:1395–1401 (**Chapter 9**)

Li D, Hu T, Manjula BN, Acharya SA (2009) Extension arms facilitated pegylation of αα-hemoglobin with modifications targeted exclusively to amino groups: functional and structural advantages of free Cys-93(β) in the PEG-Hb adduct. Bioconjug Chem 20:2062–2070 (**Chapter 9**)

Li ZL, Palaniyandi S, Zeng RY, Tuo WB, Roopenian DC, Zhu XP (2011) Transfer of IgG in the female genital tract by MHC class I-related neonatal Fc receptor (FcRn) confers protective immunity to vaginal infection. Proc Natl Acad Sci U S A 108:4388–4393 (**Chapter 10**)

Li F, Meng F, Jin Q, Sun C, Li Y, Li H, Jin S (2014a) Fusion protein of single-chain variable domain fragments for treatment of myasthenia gravis. Neural Regen Res 9:851–856 (**Chapter 4**)

Li R, Zheng K, Hu P, Chen Z, Zhou S, Chen J, Yuan C, Chen S, Zheng W, Ma E, Zhang F, Xue J, Chen X, Huang M (2014b) A novel tumor targeting drug carrier for optical imaging and therapy. Theranostics 4:642–659 (**Chapter 4**)

Liang HD, Tang J, Halliwell M (2010) Sonoporation, drug delivery, and gene therapy. Proc Inst Mech Eng H J Eng Med 224:343–361 (**Chapter 7**)

Lim PS, Chen HP, Chen CH, Wu MY, Wu CY, Wu TK (2015) Association between redox status of serum albumin and peritoneal membrane transport properties in patients on peritoneal dialysis. Blood Purif 40:243–249 (**Chapter 3**)

Lin W, Coombes AG, Garnett MC, Davies MC, Schacht E, Davis SS, Illum L (1994) Preparation of sterically stabilized human serum albumin nanospheres using a novel dextranox-MPEG crosslinking agent. Pharm Res 11:1588–1592 (**Chapter 5**)

Lipsman N, Mainprize TG, Schwartz ML, Hynynen K, Lozano AM (2014) Intracranial applications of magnetic resonance-guided focused ultrasound. Neurother J Am Soc Exp Neurother 11:593–605 (**Chapter 7**)

Liu R, Yang J, Ha CE, Bhagavan NV, Eckenhoff RG (2005) Truncated human serum albumin retains general anaesthetic binding activity. Biochem J 388:39–45 (**Chapter 11**)

Liu X, Ye L, Christianson GJ, Yang JQ, Roopenian DC, Zhu X (2007) NF-kappaB signaling regulates functional expression of the MHC class I-related neonatal Fc receptor for IgG via intronic binding sequences. J Immunol 179:2999–3011 (**Chapter 10**)

Liu XD, Ye LL, Bai Y, Mojidi H, Simister NE, Zhu XP (2008) Activation of the JAK/STAT-1 signaling pathway by IFN-gamma can down-regulate functional expression of the MHC class I-related neonatal Fc receptor for IgG. J Immunol 181:449–463 (**Chapter 10**)

Liu M, Huang Y, Hu L, Liu G, Hu X, Liu D, Yang X (2012) Selective delivery of interleukine-1 receptor antagonist to inflamed joint by albumin fusion. BMC Biotechnol 12:68 (**Chapter 4**)

Losser MR, Payen D (1996) Mechanisms of liver damage. Semin Liver Dis 16:357–367 (**Chapter 11**)

Lu QL, Liang HD, Partridge T, Blomley MJ (2003) Microbubble ultrasound improves the efficiency of gene transduction in skeletal muscle in vivo with reduced tissue damage. Gene Ther 10:396–405 (**Chapter 7**)

Lum AFH, Borden MA, Dayton PA, Kruse DE, Simon SI, Ferrara KW (2006) Ultrasound radiation force enables targeted deposition of model drug carriers loaded on microbubbles. J Control Release 111:128–134 (**Chapter 7**)

Maeda H, Sawa T, Konno T (2001) Mechanism of tumor-targeted delivery of macromolecular drugs, including the EPR effect in solid tumor and clinical overview of the prototype polymeric drug SMANCS. J Control Release 74:47–61 (**Chapter 6**)

Majorek KA, Porebski PJ, Dayal A, Zimmerman MD, Jablonska K, Stewart AJ, Chruszcz M, Minor W (2012) Structural and immunologic characterization of bovine, horse, and rabbit serum albumins. Mol Immunol 52:174–182 (**Chapter 1**)

Manjula BN, Tsai A, Upadhya R, Perumalsamy K, Smith K, Malavalli A, Vandegriff K, Winslow RM, Intaglietta M, Prabhakaran M, Friedman JM, Acharya AS (2003) Site-specific PEGylation of hemoglobin at Cys-93(β): correlation between the colligative properties of the PEGylated protein and the length of the conjugated PEG chain. Bioconjug Chem 14:464–472 (**Chapter 9**)

Manoharan P, Wong YH, Tayyab S (2015) Stabilization of human serum albumin against urea denaturation by diazepam and ketoprofen. Protein Pept Lett 22:611–617 (**Chapter 2**)

Mao H, Gunasekera AH, Fesik SW (2000) Expression, refolding, and isotopic labeling of human serum albumin domains for NMR spectroscopy. Protein Expr Purif 20:492–499 (**Chapter 11**)

Marks D, Vita J, Folts J, Keaney JJ, Welch G, Loscalzo J (1995) Inhibition of neointimal proliferation in rabbits after vascular injury by a single treatment with a protein adduct of nitric oxide. J Clin Invest 96:2630–2638 (**Chapter 8**)

Marley R, Feelisch M, Holt S, Moore KA (2000) Chemiluminescense-based assay for S-nitrosoalbumin and other plasma S-nitrosothiols. Free Radic Res 32:1–9 (**Chapter 8**)

Marques JA, George JK, Smith IJ, Bhakta V, Sheffield WP (2001) A barbourin-albumin fusion protein that is slowly cleared in vivo retains the ability to inhibit platelet aggregation in vitro. Thromb Haemost 86:902–908 (**Chapter 4**)

Martin MG, Wu SV, Walsh JH (1993) Hormonal control of intestinal Fc receptor gene expression and immunoglobulin transport in suckling rats. J Clin Invest 91:2844–2849 (**Chapter 10**)

Martin MG, Wu SV, Walsh JH (1997) Ontogenetic development and distribution of antibody transport and Fc receptor mRNA expression in rat intestine. Dig Dis Sci 42:1062–1069 (**Chapter 10**)

Maruyama K, Yuda T, Okamoto A, Ishikura C, Kojima S, Iwatsuru M (1991) Effect of molecular weight in amphipathic polyethyleneglyco on prolonging the circulation time of large unilamellar liposomes. Chem Pharm Bull 39:1620–1622 (**Chapter 5**)

Mathias W, Kamp O, Porter T (eds) (2015) Safety and feasibility of diagnostic ultrasound high mechanical index impulses in restoring epicardial flow in acute ST segment elevation myocardial infarction in humans. The 30th anniversary bubble conference, September 10–12, 2015; Sept 10–12 2015. Chicago (**Chapter 7**)

Matsushita S, Isima Y, Chuang VT, Watanabe H, Tanase S, Maruyama T, Otagiri M (2004) Functional analysis of recombinant human serum albumin domains for pharmaceutical applications. Pharm Res 21:1924–1932 (**Chapter 11**)

Maunsbach AB (1966) Albumin absorption by renal proximal tubule cells. Nature 212:546–547 (**Chapter 10**)

McDannold N, Vykhodtseva N, Hynynen K (2008) Effects of acoustic parameters and ultrasound contrast agent dose on focused-ultrasound induced blood-brain barrier disruption. Ultrasound Med Biol 34:930–937 (**Chapter 7**)

McDonagh CF, Huhalov A, Harms BD, Adams S, Paragas V, Oyama S, Zhang B, Luus L, Overland R, Nguyen S, Gu JM, Kohli N, Wallace M, Feldhaus MJ, Kudla AJ, Schoeberl B, Nielsen UB (2012) Antitumor activity of a novel bispecific antibody that targets the ErbB2/ErbB3 oncogenic unit and inhibits heregulin-induced activation of ErbB3. Mol Cancer Ther 11:582–593 (**Chapters 4 and 10**)

McMahon HT, Gallop JL (2005) Membrane curvature and mechanisms of dynamic cell membrane remodelling. Nature 438:590–596 (**Chapter 10**)

Mehier-Humbert S, Guy RH (2005) Physical methods for gene transfer: improving the kinetics of gene delivery into cells. Adv Drug Deliv Rev 57:733–753 (**Chapter 7**)

Mehier-Humbert S, Yan F, Frinking P, Schneider M, Guy RH, Bettinger T (2007) Ultrasound-mediated gene delivery: influence of contrast agent on transfection. Bioconjug Chem 18:652–662 (**Chapter 7**)

Melder RJ, Osborn BL, Riccobene T, Kanakaraj P, Wei P, Chen G, Stolow D, Halpern WG, Migone TS, Wang Q, Grzegorzewski KJ, Gallant G (2005) Pharmacokinetics and in vitro and in vivo anti-tumor response of an interleukin-2-human serum albumin fusion protein in mice. Cancer Immunol Immunother 54:535–547 (**Chapters 2 and 4**)

Mendez CM, McClain CJ, Marsano LS (2005) Albumin therapy in clinical practice. Nutr Clin Pract 20:314–320 (**Chapters 1 and 2**)

Merisko-Liversidge E, Sarpotdar P, Bruno J, Hajj S, Wei L, Peltier N, Rake J, Shaw JM, Pugh S, Polin L, Jones J, Corbett T, Cooper E, Liversidge GG (1996) Formulation and antitumor activity evaluation of nanocrystalline suspensions of poorly soluble anticancer drugs. Pharm Res 13:272–278 (**Chapter 6**)

Merlot AM, Kalinowski DS, Richardson DR (2014) Unraveling the mysteries of serum albumin – more than just a serum protein. Front Physiol 5:299 (**Chapter 1**)

Merodio M, Arnedo A, Renedo MJ, Irache JM (2001) Ganciclovir-loaded albumin nanoparticles: characterization and in vitro release properties. Eur J Pharm Sci 12:251–259 (**Chapter 5**)

Metzner HJ, Weimer T, Kronthaler U, Lang W, Schulte S (2009) Genetic fusion to albumin improves the pharmacokinetic properties of factor IX. Thromb Haemost 102:634–644 (**Chapter 10**)

Mezo AR, Sridhar V, Badger J, Sakorafas P, Nienaber V (2010) X-ray crystal structures of monomeric and dimeric peptide inhibitors in complex with the human neonatal Fc receptor, FcRn. J Biol Chem 285:27694–27701 (**Chapter 10**)

Mi WT, Wanjie S, Lo ST, Gan Z, Pick-Herk B, Ober RJ, Ward ES (2008) Targeting the neonatal Fc receptor for antigen delivery using engineered Fc fragments. J Immunol 181:7550–7561 (**Chapter 10**)

Miao CH, Brayman AA, Loeb KR, Ye P, Zhou L, Mourad P, Crum LA (2005) Ultrasound enhances gene delivery of human factor IX plasmid. Hum Gene Ther 16:893–905 (**Chapter 7**)

Miele E, Spinelli GP, Miele E, Tomao F, Tomao S (2009) Albumin-bound formulation of paclitaxel (Abraxane (R) ABI-007) in the treatment of breast cancer. Int J Nanomedicine 4:99–105 (**Chapter 10**)

Mielke S, Sparreboom A, Mross K (2006) Peripheral neuropathy: a persisting challenge in paclitaxel-based regimes. Eur J Cancer 42:24–30 (**Chapter 6**)

Mikulska JE (2015) Correction: analysis of response elements involved in the regulation of the human neonatal Fc receptor gene (FCGRT). PLoS ONE 10, e0139744 (**Chapter 10**)

Mikulska JE, Simister NE (2000) Analysis of the promoter region of the human FcRn gene. Biochim Biophys Acta 1492:180–184 (**Chapter 10**)

Mikulska JE, Pablo L, Canel J, Simister NE (2000) Cloning and analysis of the gene encoding the human neonatal Fc receptor. Eur J Immunogenet 27:231–240 (**Chapter 10**)

Miller DL, Dou C, Song J (2003) DNA transfer and cell killing in epidermoid cells by diagnostic ultrasound activation of contrast agent gas bodies in vitro. Ultrasound Med Biol 29:601–607 (**Chapter 7**)

Miller DL, Averkiou MA, Brayman AA, Everbach EC, Holland CK, Wible JH Jr, Wu J (2008) Bioeffects considerations for diagnostic ultrasound contrast agents. J Ultrasound Med Off J Am Inst Ultrasound Med 27:611–632, quiz 633–616 (**Chapter 7**)

Minchiotti L, Galliano M, Caridi G, Kragh-Hansen U, Peters T Jr (2013) Congenital analbuminaemia: molecular defects and biochemical and clinical aspects. Biochim Biophys Acta 1830:5494–5502 (**Chapter 1**)

Mingrone G, De Smet R, Greco AV, Bertuzzi A, Gandolfi A, Ringoir S, Vanholder R (1997) Serum uremic toxins from patients with chronic renal failure displace the binding of L-tryptophan to human serum albumin. Clin Chim Acta 260:27–34 (**Chapter 11**)

Ministry of Health, Labor and Welfare, Japan (2014) Proceedings of blood donation promotion committee, pharmaceutical affairs and food sanitation council on 2 December, 2014. http://www.mhlw.go.jp/file/05-Shingikai-11121000-Iyakushokuhinkyoku-Soumuka/0000067177.pdf. Accessed 30 Dec 2015 (**Chapter 9**)

Minomo A, Ishima Y, Kragh-Hansen U, Chuang VT, Uchida M, Taguchi K, Watanabe H, Maruyama T, Morioka H, Otagiri M (2011) Biological characteristics of two lysines on human serum albumin in the high-affinity binding of 4Z,15Z-bilirubin-IXalpha revealed by phage display. FEBS J 278:4100–4111 (**Chapter 11**)

Minomo A, Ishima Y, Chuang VT, Suwa Y, Kragh-Hansen U, Narisoko T, Morioka H, Maruyama T, Otagiri M (2013) Albumin domain II mutant with high bilirubin binding affinity has a great potential as serum bilirubin excretion enhancer for hyperbilirubinemia treatment. Biochim Biophys Acta 1830:2917–2923 (**Chapters 1 and 11**)

Minshall RD, Tiruppathi C, Vogel SM, Niles WD, Gilchrist A, Hamm HE, Malik AB (2000) Endothelial cell-surface gp60 activates vesicle formation and trafficking via G(i)-coupled Src kinase signaling pathway. J Cell Biol 150:1057–1070 (**Chapter 6**)

Minshall R, Sessa W, Stan R, Anderson R, Malik A (2003) Caveolin regulation of endothelial function. Am J Physiol Lung Cell Mol Physiol 285:1179–1183 (**Chapter 8**)

Mitra S, Samanta M, Sarkar M, Kumar De A, Chatterjee S (2011) Pre-exchange 5 % albumin infusion in low birth weight neonates with intensive phototherapy failure – a randomized controlled trial. J Trop Pediatr 57:217–221 (**Chapter 11**)

Mitzner SR (2011) Extracorporeal liver support-albumin dialysis with the molecular adsorbent recirculating system (MARS). Ann Hepatol 10:S21–S28 (**Chapter 1**)

Mitzner SR, Stange J, Klammt S, Peszynski P, Schmidt R (2001) Albumin dialysis using the molecular adsorbent recirculating system. Curr Opin Nephrol Hypertens 10:777–783 (**Chapter 11**)

Mitzner S, Klammt S, Stange J, Schmidt R (2006) Albumin regeneration in liver support-comparison of different methods. Ther Apher Dial 10:108–117 (**Chapter 11**)

Miyakawa N, Nishikawa M, Takahashi Y, Ando M, Misaka M, Watanabe Y, Takakura Y (2011) Prolonged circulation half-life of interferon γ activity by gene delivery of interferon γ-serum albumin fusion protein in mice. J Pharm Sci 100:2350–2357 (**Chapter 4**)

Miyamura S, Imafuku T, Anraku M, Taguchi K, Yamasaki K, Tominaga Y, Maeda H, Ishima Y, Watanabe H, Otagiri M, Maruyama T (2016) Comparison of post-translational modification and the functional impairment of human serum albumin in commercial preparations. J Pharm Sci 105:1043–1049 (**Chapters 2 and 3**)

Mizutani T, Layon A (1996) Clinical applications of nitric oxide. Chest 110:506–524 (**Chapter 8**)

Moestrup SK, Kozyraki R, Kristiansen M, Kaysen JH, Rasmussen HH, Brault D, Pontillon F, Goda FO, Christensen EI, Hammond TG, Verroust PJ (1998) The intrinsic factor-vitamin B12 receptor and target of teratogenic antibodies is a megalin-binding peripheral membrane protein with homology to developmental proteins. J Biol Chem 273:5235–5242 (**Chapter 10**)

Moncada S, Palmer R, Higgs E (1991) Nitric oxide: physiology, pathophysiology, and pharmacology. Pharmacol Rev 43:109–142 (**Chapter 8**)

Mondery-Pawlowski CL, Tian LL, Pan V, Gupta AS (2013) Synthetic approaches to RBC mimicry and oxygen carrier systems. Biomacromolecules 14:939–948 (**Chapter 9**)

Montoyo HP, Vaccaro C, Hafner M, Ober RJ, Mueller W, Ward ES (2009) Conditional deletion of the MHC class I-related receptor FcRn reveals the sites of IgG homeostasis in mice. Proc Natl Acad Sci U S A 106:2788–2793 (**Chapter 10**)

Monzani E, Curto M, Galliano M, Minchiotti L, Aime S, Baroni S, Fasano M, Amoresano A, Salzano AM, Pucci P, Casella L (2002) Binding and relaxometric properties of heme complexes with cyanogen bromide fragments of human serum albumin. Biophys J 83:2248–2258 (**Chapter 11**)

Moriyama Y, Takeda K (2005) Protective effects of small amounts of bis(2-ethylhexyl) sulfosuccinate on the helical structures of human and bovine serum albumins in their thermal denaturations. Langmuir 21:5524–5528 (**Chapter 2**)

Mu X, Hu K, Shen M, Kong N, Fu C, Yan W, Wei A (2016) Protection against influenza A virus by vaccination with a recombinant fusion protein linking influenza M2e to human serum albumin (HSA). J Virol Methods 228:84–90 (**Chapter 4**)

Mueser TC, Rogers PH, Arnone A (2000) Interface sliding as illustrated by the multiple quaternary structures of liganded hemoglobin. Biochemistry 39:15353–15364 (**Chapter 9**)

Muller RH, Wallis KH (1993) Surface modification of i.v. injectable biodegradable nanoparticles with poloxamer polymers and poloxamine 908. Int J Pharm 89:25–31 (**Chapter 5**)

Muller D, Karle A, Meibburger B, Hofig I, Stork R, Kontermann RE (2007) Improved pharmacokinetics of recombinant bispecific antibody molecules by fusion to human serum albumin. J Biol Chem 282:12650–12660 (**Chapters 4 and 10**)

Muller MR, Saunders K, Grace C, Jin M, Piche-Nicholas N, Steven J, O'Dwyer R, Wu L, Khetemenee L, Vugmeyster Y, Hickling TP, Tchistiakova L, Olland S, Gill D, Jensen A, Barelle CJ (2012) Improving the pharmacokinetic properties of biologics by fusion to an anti-HSA shark VNAR domain. MAbs 4:673–685 (**Chapter 10**)

Müller N, Schneider B, Pfizenmaier K, Wajant H (2012) Superior serum half life of albumin tagged TNF ligands. Biochem Biophys Res Commun 396:793–799 (**Chapter 4**)

Mülsch A, Mordvintcev P, Vanin A, Busse R (1993) Formation and release of dinitrosyl iron complexes by endothelial cells. Biochem Biophys Res Commun 196:1303–1308 (**Chapter 8**)

Musante L, Bruschi M, Candiano G, Petretto A, Dimasi N, Del Boccio P, Urbani A, Rialdi G, Ghiggeri GM (2006) Characterization of oxidation end product of plasma albumin 'in vivo'. Biochem Biophys Res Commun 349:668–673 (**Chapter 3**)

Musante L, Candiano G, Petretto A, Bruschi M, Dimasi N, Caridi G, Pavone B, Del Boccio P, Galliano M, Urbani A, Scolari F, Vincenti F, Ghiggeri GM (2007) Active focal segmental glomerulosclerosis is associated with massive oxidation of plasma albumin. J Am Soc Nephrol 18:799–810 (**Chapter 3**)

Muzammil S, Kumar Y, Tayyab S (2000) Anion-induced stabilization of human serum albumin prevents the formation of intermediate during urea denaturation. Proteins 40:29–38 (**Chapter 2**)

Myers P, Minor RJ, Guerra RJ, Bates J, Harrison D (1990) Vasorelaxant properties of the endothelium-derived relaxing factor more closely resemble S-nitrosocysteine than nitric oxide. Nature 345:161–163 (**Chapter 8**)

Nagababu E, Ramasamy S, Rifkind JM, Jia Y, Alayash AI (2002) Site-specific cross-linking of human and bovine hemoglobins differentially alters oxygen binding and redox side reactions producing rhombic heme and heme degradation. Biochemistry 41:7407–7415 (**Chapter 9**)

Nagumo K, Tanaka M, Chuang VT, Setoyama H, Watanabe H, Yamada N, Kubota K, Matsushita K, Yoshida A, Jinnouchi H, Anraku M, Kadowaki D, Ishima Y, Sasaki Y, Otagiri M, Maruyama T (2014) Cys34-cysteinylated human serum albumin is a sensitive plasma marker in oxidative stress-related chronic diseases. PLoS ONE 9, e85216 (**Chapter 3**)

Nakajou K, Watanabe H, Kragh-Hansen U, Maruyama T, Otagiri M (2003) The effect of glycation on the structure, function and biological fate of human serum albumin as revealed by recombinant mutants. Biochim Biophys Acta 1623:88–97 (**Chapter 1**)

Nakamura H, Hoshino Y, Okuyama H, Matsuo Y, Yodoi J (2009) Thioredoxin 1 delivery as new therapeutics. Adv Drug Deliv Rev 61:303–309 (**Chapter 4**)

Nakaya H, Shimizu T, Isobe K, Tensho K, Okabe T, Nakamura Y, Nawata M, Yoshikawa H, Takaoka K, Wakitani S (2005) Microbubble-enhanced ultrasound exposure promotes uptake of methotrexate into synovial cells and enhanced antiinflammatory effects in the knees of rabbits with antigen-induced arthritis. Arthritis Rheum 52:2559–2566 (**Chapter 7**)

Naldi M, Baldassarre M, Nati M, Laggetta M, Giannone FA, Domenicali M, Bernardi M, Caraceni P, Bertucci C (2015) Mass spectrometric characterization of human serum albumin dimer: a new potential biomarker in chronic liver diseases. J Pharm Biomed Anal 112:169–175 (**Chapter 3**)

Narazaki R, Otagiri M (1997) Covalent binding of a bucillamine derivative with albumin in sera from healthy subjects and patients with various diseases. Pharm Res 14:351–353 (**Chapter 11**)

Narazaki R, Hamada M, Harada K, Otagiri M (1996) Covalent binding between bucillamine derivatives and human serum albumin. Pharm Res 13:1317–1321 (**Chapter 11**)

Narazaki R, Maruyama T, Otagiri M (1997) Probing the cysteine 34 residue in human serum albumin using fluorescence techniques. Biochim Biophys Acta 1338:275–281 (**Chapter 8**)

Natanson C, Kern SJ, Lurie P, Banks SM, Wolfe SM (2008) Cell-free hemoglobin-based blood substitutes and risk of myocardial infarction and death. J Am Med Assoc 299:2304–2312 (**Chapter 9**)

Nathan C (1992) Nitric oxide as a secretory product of mammalian cells. FASEB J 6:3051–3064 (**Chapter 8**)

Neumann E, Frei E, Funk D, Becker MD, Schrenk H-H, Müller-Ladner U, Fiehn C (2010) Native albumin for targeted drug delivery. Expert Opin Drug Deliv 7:915–925 (**Chapter 1**)

Newton EE, Wu Z, Simister NE (2005) Characterization of basolateral-targeting signals in the neonatal Fc receptor. J Cell Sci 118:2461–2469 (**Chapter 10**)

Ng YY, Hou CC, Wang W, Huang XR, Lan HY (2005) Blockade of NFkappaB activation and renal inflammation by ultrasound-mediated gene transfer of Smad7 in rat remnant kidney. Kidney Int Suppl 67:S83–S91 (**Chapter 7**)

Nguyen HH, Ko S (2010) Preparation of size-controlled BSA nanoparticles by intermittent addition of desolvating agent. IFMBE Proc 27:231–234 (**Chapter 5**)

Nguyen A, Reyes AE, Zhang M, McDonald P, Wong WLT, Damico LA, Dennis MS (2006) The pharmacokinetics of an albumin-binding Fab (AB.Fab) can be modulated as a function of affinity for albumin. Protein Eng Des Sel 19:291–297 (**Chapter 10**)

Nichols JW, Bae YH (2014) EPR: evidence and fallacy. J Control Release 190:451–464 (**Chapter 6**)

Nishida K, Watanabe H, Ogaki S, Kodama A, Tanaka R, Imafuku T, Ishima Y, Chuang VT, Toyoda M, Kondoh M, Wu Q, Fukagawa M, Otagiri M, Maruyama T (2015) Renoprotective effect of long acting thioredoxin by modulating oxidative stress and macrophage migration inhibitory factor against rhabdomyolysis-associated acute kidney injury. Sci Rep 5:14471 (**Chapters 3 and 4**)

Nolte MW, Nichols TC, Mueller-Cohrs J, Merricks EP, Pragst I, Zollner S, Dickneite G (2012) Improved kinetics of rIX-FP, a recombinant fusion protein linking factor IX with albumin, in cynomolgus monkeys and hemophilia B dogs. J Thromb Haemost 10:1591–1599 (**Chapters 4 and 10**)

Nord K, Gunneriusson E, Ringdahl J, Stahl S, Uhlen M, Nygren PA (1997) Binding proteins selected from combinatorial libraries of an alpha-helical bacterial receptor domain. Nat Biotechnol 15:772–777 (**Chapter 10**)

Nozaki T, Ogawa R, Watanabe A, Nishio R, Fuse H, Kondo T (2006) Ultrasound-mediated gene transfection: problems to be solved and future possibilities. J Med Ultrason 33:135–142 (**Chapter 7**)

Nygren PA, Uhlen M, Flodby P, Andersson R, Wigzell H (1991) In vivo stabilization of a human recombinant CD4 derivative by fusion to a serum-albumin-binding receptor. Vaccines 91:363–368 (**Chapter 10**)

Nyman DW, Campbell KJ, Hersh E, Long K, Richardson K, Trieu V, Desai N, Hawkins MJ, Von Hoff DD (2005) Phase I and pharmacokinetics trial of ABI-007, a novel nanoparticle formulation of paclitaxel in patients with advanced nonhematologic malignancies. J Clin Oncol 23:7785–7793 (**Chapter 6**)

Ober RJ, Radu CG, Ghetie V, Ward ES (2001) Differences in promiscuity for antibody-FcRn interactions across species: implications for therapeutic antibodies. Int Immunol 13:1551–1559 (**Chapter 10**)

Ober RJ, Martinez C, Lai X, Zhou J, Ward ES (2004a) Exocytosis of IgG as mediated by the receptor, FcRn: an analysis at the single-molecule level. Proc Natl Acad Sci U S A 101:11076–11081 (**Chapter 10**)

Ober RJ, Martinez C, Vaccaro C, Zhou J, Ward ES (2004b) Visualizing the site and dynamics of IgG salvage by the MHC class I-related receptor, FcRn. J Immunol 172:2021–2029 (**Chapter 10**)

O'Brien AJ, Fullerton JN, Massey KA, Auld G, Sewell G, James S, Newson J, Karra E, Winstanley A, Alazawi W, Garcia-Martinez R, Cordoba J, Nicolaou A, Gilroy DW (2014) Immunosuppression in acutely decompensated cirrhosis is mediated by prostaglandin E2. Nat Med 20:518–523 (**Chapter 3**)

O'Connor-Semmes RL, Lin J, Hodge RJ, Andrews S, Chism J, Choudhury A, Nunez DJ (2014) GSK2374697, a novel albumin-binding domain antibody (AlbudAb), extends systemic exposure of exendin-4: first study in humans – PK/PD and safety. Clin Pharmacol Ther 96:704–712 (**Chapter 10**)

Oettl K, Stadlbauer V, Petter F, Greilberger J, Putz-Bankuti C, Hallström S, Lackner C, Stauber RE (2008) Oxidative damage of albumin in advanced liver disease. Biochim Biophys Acta 1782:469–473 (**Chapter 3**)

Oettl K, Birner-Gruenberger R, Spindelboeck W, Stueger HP, Dorn L, Stadlbauer V, Putz-Bankuti C, Krisper P, Graziadei I, Vogel W, Lackner C, Stauber RE (2013) Oxidative albumin damage in chronic liver failure: relation to albumin binding capacity, liver dysfunction and survival. J Hepatol 59:978–983 (**Chapter 3**)

Oganesyan V, Damschroder MM, Cook KE, Li Q, Gao CS, Wu HR, Dall'Acqua WF (2014) Structural insights into neonatal Fc receptor-based recycling mechanisms. J Biol Chem 289:7812–7824 (**Chapters 1 and 10**)

O'Grady JG, Gimson AE, O'Brien CJ, Pucknell A, Hughes RD, Williams R (1988) Controlled trials of charcoal hemoperfusion and prognostic factors in fulminant hepatic failure. Gastroenterology 94:1186–1192 (**Chapter 11**)

Ohtani W, Masaki A, Ikeda Y, Hirose M, Chuganji M, Takeshima K, Kondo M, Sumi A, Ohmura T (1997) Structure of recombinant human serum albumin from Pichia pastoris. Yakugaku Zasshi 117:220–232 (**Chapter 11**)

Ohtani W, Nawa Y, Takeshima K, Kamuro H, Kobayashi K, Ohmura T (1998a) Physicochemical and immunochemical properties of recombinant human serum albumin from Pichia pastoris. Anal Biochem 256:56–62 (**Chapter 11**)

Ohtani W, Ohda T, Sumi A, Kobayashi K, Ohmura T (1998b) Analysis of Pichia pastoris components in recombinant human serum albumin by immunological assays and by HPLC with pulsed amperometric detection. Anal Chem 70:425–429 (**Chapter 11**)

Okabayashi K, Nakagawa Y, Hayasuke N, Ohi H, Miura M, Ishida Y, Shimizu M, Murakami K, Hirabayashi K, Minamino H et al (1991) Secretory expression of the human serum albumin gene in the yeast, Saccharomyces cerevisiae. J Biochem 110:103–110 (**Chapter 11**)

Olivieri JR, Craievich AF (1995) The subdomain structure of human serum albumin in solution under different pH conditions studied by small angle X-ray scattering. Eur Biophys J 24:77–84 (**Chapter 1**)

Onishi S, Itoh S, Isobe K, Ochi M, Kunikata T, Imai T (1989) Effect of the binding of bilirubin to either the first class or the second class of binding sites of the human serum albumin molecule on its photochemical reaction. Biochem J 257:711–714 (**Chapter 11**)

Osborn BL, Sekut L, Corcoran M, Poortman C, Sturm B, Chen G, Mather D, Lin HL, Parry TJ (2002) Albutropin: a growth hormone-albumin fusion with improved pharmacokinetics and pharmacodynamics in rats and monkeys. Eur J Pharmacol 456:149–158 (**Chapters 2, 4 and 10**)

Ozaki K, Makino H, Aoki M, Miyake T, Yasumasa N, Osako MK, Nakagami H, Rakugi H, Morishita R (2012) Therapeutic effect of ribbon-type nuclear factor-kappaB decoy oligonucleotides in a rat model of inflammatory bowel disease. Curr Gene Ther 12:484–492 (**Chapter 7**)

Paal K, Muller J, Hegedus L (2001) High affinity binding of paclitaxel to human serum albumin. Eur J Biochem 268:2187–2191 (**Chapter 6**)

Palmer R, Ferrige A, Moncada S (1987) Nitric oxide release accounts for the biological activity of endothelium-derived relaxing factor. Nature 327:524–526 (**Chapter 8**)

Papadjopoulos D, Allen TM, Gabizon A, Mayhew E, Matthay K, Huang SK, Lee KD, Woodle MC, Lasic DD, Redemon C (1991) Sterically stabilized liposomes: improvements in pharmacokinetics and antitumor therapeutic efficiency. Proc Natl Acad Sci U S A 88:11460–11464 (**Chapter 5**)

Park DS, Petersen CE, Ha C, Harohalli K, Feix JB, Bhagavan NV (1999) Expression of a human serum albumin fragment (consisting of subdomains IA, IB, and IIA) and a study of its properties. IUBMB Life 48:169–174 (**Chapter 11**)

Park S, Choi S, Lee MG, Lim C, Oh J (2012) Retinol binding protein-albumin domain III fusion protein deactivates hepatic stellate cells. Mol Cells 34:17–22 (**Chapter 4**)

Patil GV (2003) Biopolymer albumin for diagnosis and in drug delivery. Drug Dev Res 58:219–247 (**Chapter 5**)

Pearce LB, Gawryl MS, Rentko VT, Moon-Massat PF, Rausch CW (2006) HBOC-201 (hemoglobin glutamer-250) (bovine), Hemopure®): clinical studies. In: Winslow RM (ed) Blood substitutes. Elsevier, San Diego, pp 437–450 (**Chapter 9**)

Pergola PE, Krauth M, Huff JW, Ferguson DA, Ruiz S, Meyer CJ, Warnock DG (2011) Effect of bardoxolone methyl on kidney function in patients with T2D and Stage 3b-4 CKD. Am J Nephrol 33:469–476 (**Chapter 3**)

Peters T Jr (1996) All about albumin: biochemistry, genetics and medical application. Academic, San Diego (**Chapters 1, 2, 3, 5, 8 and 11**)

Petersen CE, Ha CE, Harohalli K, Feix JB, Bhagavan NV (2000) A dynamic model for bilirubin binding to human serum albumin. J Biol Chem 275:20985–20995 (**Chapter 11**)

Petersen SS, Klaning E, Ebbesen MF, Andersen B, Cameron J, Sorensen ES, Howard KA (2015) Neonatal Fc receptor binding tolerance toward the covalent conjugation of payloads to cysteine 34 of human albumin variants. Mol Pharm 13(2):677–682 (**Chapter 10**)

Petitpas I, Bhattacharya AA, Twine S, East M, Curry S (2001a) Crystal structure analysis of warfarin binding to human serum albumin: anatomy of drug site I. J Biol Chem 276:22804–22809 (**Chapter 11**)

Petitpas I, Grüne T, Bhattacharya A, Curry S (2001b) Crystal structures of human serum albumin complexed with monounsaturated and polyunsaturated fatty acids. J Mol Biol 314:955–960 (**Chapter 8**)

Petitpas I, Petersen CE, Ha CE, Bhattacharya AA, Zunszain PA, Ghuman J, Bhagavan NV, Curry S (2003) Structural basis of albumin-thyroxine interactions and familial dysalbuminemic hyperthyroxinemia. Proc Natl Acad Sci U S A 100:6440–6445 (**Chapter 11**)

Petkova SB, Akilesh S, Sproule TJ, Christianson GJ, Al Khabbaz H, Brown AC, Presta LG, Meng YG, Roopenian DC (2006) Enhanced half-life of genetically engineered human IgG1 antibodies in a humanized FcRn mouse model: potential application in humorally mediated autoimmune disease. Int Immunol 18:1759–1769 (**Chapter 10**)

Petrelli F, Borgonovo K, Barni S (2010) Targeted delivery for breast cancer therapy: the history of nanoparticle-albumin-bound paclitaxel. Expert Opin Pharmacother 11:1413–1432 (**Chapter 10**)

Piscaglia F, Nolsoe C Fau, Dietrich CF, Dietrich Cf Fau, Cosgrove DO, Cosgrove Do Fau, Gilja OH, Gilja Oh Fau, Bachmann Nielsen M, Bachmann Nielsen M Fau, Albrecht T, Albrecht T Fau, Barozzi L, Barozzi L Fau, Bertolotto M, Bertolotto M Fau, Catalano O, Catalano O Fau, Claudon M, Claudon M Fau, Clevert DA, Clevert Da Fau, Correas JM, Correas Jm Fau, D'Onofrio M, D'Onofrio M Fau, Drudi FM, Drudi Fm Fau, Eyding J, Eyding J Fau, Giovannini M, Giovannini M Fau, Hocke M, Hocke M Fau, Ignee A, Ignee A Fau, Jung EM, Jung Em Fau, Klauser AS, Klauser As Fau, Lassau N, Lassau N Fau, Leen E, Leen E Fau, Mathis G, Mathis G Fau, Saftoiu A, Saftoiu A Fau, Seidel G, Seidel G Fau, Sidhu PS, Sidhu Ps Fau, ter Haar G, ter Haar G Fau, Timmerman D, Timmerman D Fau, Weskott HP, Weskott HP (2011) The EFSUMB guidelines and recommendations on the clinical practice of contrast enhanced ultrasound (CEUS): update 2011 on non-hepatic applications (**Chapter 7**)

Pislaru SV, Pislaru C, Kinnick RR, Singh R, Gulati R, Greenleaf JF, Simari RD (2003) Optimization of ultrasound-mediated gene transfer: comparison of contrast agents and ultrasound modalities. Eur Heart J 24:1690–1698 (**Chapter 7**)

Pohl J, Ring A, Stremmel W (2002) Uptake of long-chain fatty acids in HepG2 cells involves caveolae: analysis of a novel pathway. J Lipid Res 43:1390–1399 (**Chapter 8**)

Poli G (2000) Pathogenesis of liver fibrosis: role of oxidative stress. Mol Asp Med 21:49–98 (**Chapter 3**)

Poole RM, Nowlan ML (2014) Albiglutide: first global approval. Drugs 74:929–938 (**Chapter 10**)

Porter T. Microvascular reperfusion utilizing sonothrombolysis in acute myocardial infarction (MRUSMI TRIAL). ClinicalTrials.gov. https://clinicaltrials.gov/ct2/show/NCT02170103 (**Chapter 7**)

Powner MB, McKenzie JA, Christianson GJ, Roopenian DC, Fruttiger M (2014) Expression of neonatal Fc receptor in the eye. Invest Ophthalmol Vis Sci 55:1607–1615 (**Chapter 10**)

Prabhat P, Gan Z, Chao J, Ram S, Vaccaro C, Gibbons S, Ober RJ, Ward ES (2007) Elucidation of intracellular recycling pathways leading to exocytosis of the Fc receptor, FcRn, by using multifocal plane microscopy. Proc Natl Acad Sci U S A 104:5889–5894 (**Chapter 10**)

Pratley RE, Nauck MA, Barnett AH, Feinglos MN, Ovalle F, Harman-Boehm I, Ye J, Scott R, Johnson S, Stewart M, Rosenstock J, HARMONY 7 Study Group (2014) Once-weekly albiglutide versus once-daily liraglutide in patients with type 2 diabetes inadequately controlled on oral drugs (HARMONY 7): a randomised, open-label, multicentre, non-inferiority phase 3 study. Lancet Diabetes Endocrinol 2:289–297 (**Chapter 4**)

Proetzel G, Roopenian DC (2014) Humanized FcRn mouse models for evaluating pharmacokinetics of human IgG antibodies. Methods 65:148–153 (**Chapter 10**)

Purcell M, Neault JF, Tajmir-Riahi HA (2000) Interaction of taxol with human serum albumin. Biochim Biophys Acta 1478:61–68 (**Chapter 6**)

Qin S, Caskey CF, Ferrara KW (2009) Ultrasound contrast microbubbles in imaging and therapy: physical principles and engineering. Phys Med Biol 54:R27–R57. doi:10.1088/0031-9155/1054/1086/R1001, Epub 2009 Feb 1019 (**Chapter 7**)

Qiu Y, Luo Y, Zhang Y, Cui W, Zhang D, Wu J, Zhang J, Tu J (2010) The correlation between acoustic cavitation and sonoporation involved in ultrasound-mediated DNA transfection with polyethylenimine (PEI) in vitro. J Control Release 145:40–48 (**Chapter 7**)

Qiu Y, Zhang C, Tu J, Zhang D (2012) Microbubble-induced sonoporation involved in ultrasound-mediated DNA transfection in vitro at low acoustic pressures. J Biomech 45:1339–1345 (**Chapter 7**)

Queiroz RG, Varca GHC, Kadlubowski S, Ulanski P, Lugão AB (2016) Radiation-synthesized protein-based drug carriers: size-controlled BSA nanoparticles. Int J Biol Macromol 85:82–91 (**Chapter 5**)

Quinlan GJ, Martin GS, Evans TW (2005) Albumin: biochemical properties and therapeutic potential. Hepatology 41:1211–1219 (**Chapter 1**)

Requejo R, Hurd TR, Costa NJ, Murphy MP (2010) Cysteine residues exposed on protein surfaces are the dominant intramitochondrial thiol and may protect against oxidative damage. FEBS J 277:1465–1480 (**Chapter 3**)

Richards DA, Braiteh FS, Garcia AA (2014) A phase 1 study of MM-111, a bispecific HER2/HER3 antibody fusion protein, combined with multiple treatment regimens in patients with advanced HER-positive solid tumors. J Clin Oncol 32:651 (**Chapter 4**)

Rifai K, Ernst T, Kretschmer U, Bahr MJ, Schneider A, Hafer C, Haller H, Manns MP, Fliser D (2003) Prometheus – a new extracorporeal system for the treatment of liver failure. J Hepatol 39:984–990 (**Chapter 11**)

Rink J, Ghigo E, Kalaidzidis Y, Zerial M (2005) Rab conversion as a mechanism of progression from early to late endosomes. Cell 122:735–749 (**Chapter 10**)

Roche M, Rondeau P, Singh NR, Tarnus E, Bourdon E (2008) The antioxidant properties of serum albumin. FEBS Lett 582:1783–1787 (**Chapter 1**)

Roda A, Cappelleri G, Aldini R, Roda E, Barbara L (1982) Quantitative aspects of the interaction of bile acids with human serum albumin. J Lipid Res 23:490–495 (**Chapter 11**)

Ródenas J, Mitjavila M, Carbonell T (1998) Nitric oxide inhibits superoxide production by inflammatory polymorphonuclear leukocytes. Am J Physiol 274:827–830 (**Chapter 8**)

Rohlfs RJ, Bruner E, Chiu A, Gonzales A, Gonzales ML, Magde D, Magde MD Jr, Vandegriff KD, Winslow RM (1998) Arterial blood pressure responses to cell-free hemoglobin solutions and the reaction with nitric oxide. J Biol Chem 273:12128–12134 (**Chapter 9**)

Rollett A, Reiter T, Nogueira P, Cardinale M, Loureiro A, Gomes A, Cavaco-Paulo A, Moreira A, Carmo AM, Guebitz GM (2012) Folic acid-functionalized human serum albumin nanocapsules for targeted drug delivery to chronically activated macrophages. Int J Pharm 427:460–466 (**Chapters 1 and 5**)

Roopenian DC, Christianson GJ, Sproule TJ, Brown AC, Akilesh S, Jung N, Petkova S, Avanessian L, Choi EY, Shaffer DJ, Eden PA, Anderson CL (2003) The MHC class I-like IgG receptor controls perinatal IgG transport, IgG homeostasis, and fate of IgG-Fc-coupled drugs. J Immunol 170:3528–3533 (**Chapter 10**)

Roopenian DC, Low BE, Christianson GJ, Proetzel G, Sproule TJ, Wiles MV (2015) Albumin-deficient mouse models for studying metabolism of human albumin and pharmacokinetics of albumin-based drugs. MAbs 7:344–351 (**Chapter 10**)

Roos ST, Juffermans LJM, Slikkerveer J, Unger EC, Porter TR, Kamp O (2014) Sonothrombolysis in acute stroke and myocardial infarction: a systematic review. IJC Heart Vessel 4:1–6 (**Chapter 7**)

Ross PD, Shrake A (1988) Decrease in stability of human albumin with increase in protein concentration. J Biol Chem 263:11196–11202 (**Chapter 2**)

Rustgi VK (2009) Albinterferon alfa-2b, a novel fusion protein of human albumin and human interferon alfa-2b, for chronic hepatitis C. Curr Med Res Opin 25:991–1002 (**Chapter 4**)

Ryan AJ, Chung C-W, Curry S (2011a) Crystallographic analysis reveals the structural basis of the high-affinity binding of iophenoxic acid to human serum albumin. BMC Struct Biol 11:18 (**Chapter 1**)

Ryan AJ, Ghuman J, Zunszain PA, Chung C-W, Curry S (2011b) Structural basis of binding of fluorescent, site-specific dansylated amino acids to human serum albumin. J Struct Biol 174:84–91 (**Chapter 1**)

Sadler PJ, Tucker A, Viles JH (1994) Involvement of a lysine residue in the N-terminal Ni2+ and Cu2+ binding site of serum albumins. Comparison with Co2+, Cd2+ and Al3+. Eur J Biochem 220:193–200 (**Chapter 3**)

Sakai H (2012) Present situation of the development of cellular-type hemoglobin-based oxygen carrier (hemoglobin-vesicles). Curr Drug Discov Technol 9:188–193 (**Chapter 9**)

Sakai T, Kawaguchi M, Kosuge Y (2009) siRNA-mediated gene silencing in the salivary gland using in vivo microbubble-enhanced sonoporation. Oral Dis 15:505–511 (**Chapter 7**)

Sakata M, Kawaguchi T, Taniguchi E, Nakayama A, Ishizaki S, Sonaka I, Maganuma M, Nakamura T, Itou M, Oriishi T, Abe M, Yanagimoto C, Koga H, Harada M, Sakamoto T, Oda S, Sata M (2010) Redox state of albumin is not associated with colloid osmotic pressure. Mol Med Rep 3:685–687 (**Chapter 3**)

Saleh MN, Khazaeli MB, Wheeler RH, Dropcho E, Liu T, Urist M, Miller DM, Lawson S, Dixon P, Russell CH et al (1992) Phase I trial of the murine monoclonal anti-GD2 antibody 14G2a in metastatic melanoma. Cancer Res 52:4342–4347 (**Chapter 10**)

San BH, Moh SH, Kim KK (2012) The effect of protein shells on the antioxidation activity of protein-encapsulated platinum nanoparticles. J Mater Chem 22:1774–1780 (**Chapter 9**)

Sand KM, Bern M, Nilsen J, Dalhus B, Gunnarsen KS, Cameron J, Grevys A, Bunting K, Sandlie I, Andersen JT (2014a) Interaction with both domain I and III of albumin is required for optimal pH-dependent binding to the neonatal Fc receptor (FcRn). J Biol Chem 289:34583–34594 (**Chapter 10**)

Sand KM, Dalhus B, Christianson GJ, Bern M, Foss S, Cameron J, Sleep D, Bjoras M, Roopenian DC, Sandlie I, Andersen JT (2014b) Dissection of the neonatal Fc receptor (FcRn)-albumin interface using mutagenesis and anti-FcRn albumin-blocking antibodies. J Biol Chem 289:17228–17239 (**Chapter 10**)

Sand KMK, Bern M, Nilsen J, Noordzij HT, Sandlie I, Andersen JT (2015) Unraveling the interaction between FcRn and albumin: opportunities for design of albumin-based therapeutics. Front Immunol 5:1–21 (**Chapters 1 and 10**)

Santagostino E, Negrier C, Klamroth R, Tiede A, Pabinger-Fasching I, Voigt C, Jacobs I, Morfini M (2012) Safety and pharmacokinetics of a novel recombinant fusion protein linking coagulation factor IX with albumin (rIX-FP) in hemophilia B patients. Blood 120:2405–2411 (**Chapters 4 and 10**)

Santagostino E, Martinowitz U, Lissitchkov T, Pan-Petesch B, Hanabusa H, Oldenburg J, Boggio L, Negrier C, Pabinger I, von Depka Prondzinski M, Altisent C, Castaman G, Yamamoto K, Alvarez-Roman MT, Voigt C, Blackman N, Jacobs I (2016) Long acting recombinant coagulation factor IX albumin fusion protein (rIX-FP) in hemophilia B: results of a phase 3 trial. Blood 127:1761–1769, pii: blood-2015-09-669234 (**Chapters 4 and 10**)

Santra MK, Banerjee A, Rahaman O, Panda D (2005) Unfolding pathways of human serum albumin: evidence for sequential unfolding and folding of its three domains. Int J Biol Macromol 37:200–204 (**Chapter 2**)

Sarav M, Wang Y, Hack BK, Chang A, Jensen M, Bao LH, Quigg RJ (2009) Renal FcRn reclaims albumin but facilitates elimination of IgG. J Am Soc Nephrol 20:1941–1952 (**Chapter 10**)

Sasaki Y, Nishina T, Yasui H, Goto M, Muro K, Tsuji A, Koizumi W, Toh Y, Hara T, Miyata Y (2014) Phase II trial of nanoparticle albumin-bound paclitaxel as second-line chemotherapy for unresectable or recurrent gastric cancer. Cancer Sci 105:812–817 (**Chapter 6**)

Scheffel U, Rhodes BA, Natarajan TK, Wagner HN Jr (1972) Albumin microspheres for study of the reticuloendothelial system. J Nucl Med 13:498–503 (**Chapter 5**)

Schiff PB, Fant J, Horwitz SB (1979) Promotion of microtubule assembly in vitro by taxol. Nature 277:665–667 (**Chapter 6**)

Schlachetzki F, Zhu CN, Pardridge WM (2002) Expression of the neonatal Fc receptor (FcRn) at the blood-brain barrier. J Neurochem 81:203–206 (**Chapter 10**)

Schlapschy M, Theobald I, Mack H, Schottelius M, Wester HJ, Skerra A (2007) Fusion of a recombinant antibody fragment with a homo-amino-acid polymer: effects on biophysical properties and prolonged plasma half-life. Protein Eng Des Sel 20:273–284 (**Chapter 10**)

Schmidt MM, Townson SA, Andreucci AJ, King BM, Schirmer EB, Murillo AJ, Dombrowski C, Tisdale AW, Lowden PA, Masci AL, Kovalchin JT, Erbe DV, Wittrup KD, Furfine ES, Barnes TM (2013) Crystal structure of an HSA/FcRn complex reveals recycling by competitive mimicry of HSA ligands at a pH-dependent hydrophobic interface. Structure 21:1966–1978 (**Chapter 10**)

Schnitzer JE (1992) gp60 is an albumin-binding glycoprotein expressed by continuous endothelium involved in albumin transcytosis. Am J Physiol 262:H246–H254 (**Chapter 6**)

Schulte S (2013) Innovative coagulation factors: albumin fusion technology and recombinant single-chain factor VIII. Thromb Res 131:S2–S6 (**Chapter 2**)

Schulte S (2014) Challenges for new haemophilia products from a manufacturer's perspective. Thromb Res 134(Suppl 1):S72–S76 (**Chapter 2**)

Schultze HE, Heremans JF (1966) Molecular biology of human proteins: with special reference to plasma proteins. Elsevier, New York, p 1 (**Chapter 10**)

Seige M, Kreymann B, Jeschke B, Schweigart U, Kopp KF, Classen M (1999) Long-term treatment of patients with acute exacerbation of chronic liver failure by albumin dialysis. Transplant Proc 31:1371–1375 (**Chapter 11**)

Sen S, Williams R, Jalan R (2002) The pathophysiological basis of acute-on-chronic liver failure. Liver 22(Suppl 2):5–13 (**Chapter 3**)

Shah C, Bell S, Locke I, Chowdrey H, Gordge M (2007) Interactions between cell surface protein disulphide isomerase and S-nitrosoglutathione during nitric oxide delivery. Nitric Oxide 16:135–142 (**Chapter 8**)

Shakil AO, Kramer D, Mazariegos GV, Fung JJ, Rakela J (2000) Acute liver failure: clinical features, outcome analysis, and applicability of prognostic criteria. Liver Transpl 6:163–169 (**Chapter 11**)

Shani M, Barash I, Nathan M, Ricca G, Searfoss GH, Dekel I, Faerman A, Givol D, Hurwitz DR (1992) Expression of human serum albumin in the milk of transgenic mice. Transgenic Res 1:195–208 (**Chapter 11**)

Sharma AK, Thanikachalam PV, Rajput SK (2016) Albiglutide: is a better hope against diabetes mellitus? Biomed Pharmacother 77:120–128 (**Chapter 2**)

Shearer WT, Bradshaw RA, Gurd FR, Peters T Jr (1967) The amino acid sequence and copper(II)-binding properties of peptide (1–24) of bovine serum albumin. J Biol Chem 242:5451–5459 (**Chapter 11**)

Sheffield WP, Marques JA, Bhakta V, Smith IJ (2000) Modulation of clearance of recombinant serum albumin by either glycosylation or truncation. Thromb Res 99:613–621 (**Chapter 11**)

Sheffield WP, Mamdani A, Hortelano G, Gataiance S, Eltringham-Smith L, Begbie ME, Leyva RA, Liaw PS, Ofosu FA (2004) Effects of genetic fusion of factor IX to albumin on in vivo clearance in mice and rabbits. Br J Haematol 126:565–573 (**Chapter 4**)

Sheffield WP, Gataiance S, Eltringham-Smith LJ (2007) Combined administration of barbourin-albumin and hirudin-albumin fusion proteins limits fibrin(ogen) deposition on the rabbit balloon-injured aorta. Thromb Res 119:195–207 (**Chapter 4**)

Sheikh S, Pallagatti S, Singh B, Puri N, Singh R, Kalucha A (2011) Sonoporation, a redefined ultrasound modality as therapeutic aid: a review. J Clin Exp Dent 3:e228–e234 (**Chapter 7**)

Sheikov N, McDannold N, Sharma S, Hynynen K (2008) Effect of focused ultrasound applied with an ultrasound contrast agent on the tight junctional integrity of the brain microvascular endothelium. Ultrasound Med Biol 34:1093–1104 (**Chapter 7**)

Shen ZP, Brayman AA, Chen L, Miao CH (2008) Ultrasound with microbubbles enhances gene expression of plasmid DNA in the liver via intraportal delivery. Gene Ther 15:1147–1155 (**Chapter 7**)

Shen Z, Li Y, Kohama K, Oneill B, Bi J (2011) Improved drug targeting of cancer cells by utilizing actively targetable folic acid-conjugated albumin nanospheres. Pharmacol Res 63:51–58 (**Chapter 5**)

Shi S, Liu J, Joshi SB, Krasnoperov V, Gill P, Middaugh CR, Volkin DB (2012) Biophysical characterization and stabilization of the recombinant albumin fusion protein sEphB4-HSA. J Pharm Sci 101:1969–1984 (**Chapter 2**)

Shimamura M, Sato N, Taniyama Y, Yamamoto S, Endoh M, Kurinami H, Aoki M, Ogihara T, Kaneda Y, Morishita R (2004) Development of efficient plasmid DNA transfer into adult rat central nervous system using microbubble-enhanced ultrasound. Gene Ther 11:1532–1539 (**Chapter 7**)

Shimamura M, Sato N, Taniyama Y, Kurinami H, Tanaka H, Takami T, Ogihara T, Tohyama M, Kaneda Y, Morishita R (2005) Gene transfer into adult rat spinal cord using naked plasmid DNA and ultrasound microbubbles. J Gene Med 7:1468–1474 (**Chapter 7**)

Shimoishi K, Anraku M, Kitamura K, Tasaki Y, Taguchi K, Hashimoto M, Fukunaga E, Maruyama T, Otagiri M (2007) An oral adsorbent, AST-120 protects against the progression of oxidative stress by reducing the accumulation of indoxyl sulfate in the systemic circulation in renal failure. Pharm Res 24:1283–1289 (**Chapter 3**)

Shrake A, Finlayson JS, Ross PD (1984) Thermal stability of human albumin measured by differential scanning calorimetry. I. Effects of caprylate and N-acetyltryptophanate. Vox Sang 47:7–18 (**Chapter 2**)

Shultz SC, Grady B, Cole F, Hamilton I, Burhop K, Malcolm DS (1993) A role for endothelin and nitric oxide in the pressor response to diaspirin cross-linked hemoglobin. J Lab Clin Med 122:301–308 (**Chapter 9**)

Simard JR, Zunszain PA, Hamilton JA, Curry S (2006) Location of high and low affinity fatty acid binding sites on human serum albumin revealed by NMR drug-competition analysis. J Mol Biol 361:336–351 (**Chapter 1**)

Simionescu M, Gafencu A, Antohe F (2002) Transcytosis of plasma macromolecules in endothelial cells: a cell biological survey. Microsc Res Tech 57:269–288 (**Chapter 6**)

Simister NE, Mostov KE (1989) An Fc receptor structurally related to MHC class I antigens. Nature 337:184–187 (**Chapter 10**)

Simister NE, Rees AR (1985) Isolation and characterization of an Fc receptor from neonatal rat small intestine. Eur J Immunol 15:733–738 (**Chapter 10**)

Simister NE, Story CM, Chen HL, Hunt JS (1996) An IgG-transporting Fc receptor expressed in the syncytiotrophoblast of human placenta. Eur J Immunol 26:1527–1531 (**Chapter 10**)

Simon D, Stamler J, Jaraki O, Keaney J, Osborne J, Francis S, Singel D, Loscalzo J (1993) Antiplatelet properties of protein S-nitrosothiols derived from nitric oxide and endothelium-derived relaxing factor. Arterioscler Thromb 13:791–799 (**Chapter 8**)

Simon M, Frey R, Zangemeister-Wittke U, Pluckthun A (2013) Orthogonal assembly of a designed ankyrin repeat protein-cytotoxin conjugate with a clickable serum albumin module for half-life extension. Bioconjug Chem 24:1955–1966 (**Chapter 10**)

Sinclair AM, Elliott S (2005) Glycoengineering: the effect of glycosylation on the properties of therapeutic proteins. J Pharm Sci 94:1626–1635 (**Chapter 4**)

Singh HD, Wang G, Uludağ H, Unsworth LD (2010) Poly-L-lysine-coated albumin nanoparticles: stability, mechanism for increasing. In vitro enzymatic resilience and siRNA release characteristics. Acta Biomater 6:4277–4284 (**Chapter 5**)

Sirsi S, Borden M (2009) Microbubble compositions, properties and biomedical applications. Bubble Sci Eng Technol 1:3–17 (**Chapter 7**)

Sleep D (2014) Albumin and its application in drug delivery. Expert Opin Drug Deliv 12:793–812 (**Chapter 1**)

Smith BJ, Popplewell A, Athwal D, Chapman AP, Heywood S, West SM, Carrington B, Nesbitt A, Lawson AD, Antoniw P, Eddelston A, Suitters A (2001) Prolonged in vivo residence times of antibody fragments associated with albumin. Bioconjug Chem 12:750–756 (**Chapter 10**)

Snyder SR, Welty EV, Walder RY, Williams LA, Walder JA (1987) HbXL99α: a hemoglobin derivative that is cross-linked between the α subunits is useful as a blood substitute. Proc Natl Acad Sci U S A 84:7280–7284 (**Chapter 9**)

Socinski MA, Vinnichenko I, Okamoto I, Hon JK, Hirsh V (eds) (2010) Results of a randomized, phase 3 trial of nab-Paclitaxel (nab-P) and Carboplatin (C) compared with cremophor-based Paclitaxel (P) and carboplatin as first-line therapy in advanced non-small cell lung cancer (NSCLC). Proceedings of the 46th American Society of Clinical Oncology Annual Meeting (ASCO); 2010 June 4–8. Chicago (**Chapter 6**)

Socinski MA, Bondarenko I, Karaseva NA, Makhson AM, Vynnychenko I, Okamoto I, Hon JK, Hirsh V, Bhar P, Zhang H, Iglesias JL, Renschler MF (2012) Weekly nab-paclitaxel in combination with carboplatin versus solvent-based paclitaxel plus carboplatin as first-line therapy in patients with advanced non-small-cell lung cancer: final results of a phase III trial. J Clin Oncol 30:2055–2062 (**Chapter 6**)

Sockolosky JT, Szoka FC (2015) The neonatal Fc receptor, FcRn, as a target for drug delivery and therapy. Adv Drug Deliv Rev 91:109–124 (**Chapter 10**)

Sogami M, Era S, Nagaoka S, Kuwata K, Kida K, Shigemi J, Miura K, Suzuki E, Muto Y, Tomita E (1985) High-performance liquid chromatographic studies on non-mercapt in equilibrium with mercapt conversion of human serum albumin. II. J Chromatogr 332:19–27 (**Chapter 3**)

Song J, Chappell JC, Qi M, VanGieson EJ, Kaul S, Price RJ (2002) Influence of injection site, microvascular pressure and ultrasound variables on microbubble-mediated delivery of microspheres to muscle. J Am Coll Cardiol 39:726–731 (**Chapter 7**)

Soppimath KS, Aminabhavi TM, Kulkarni AR, Rudzinski WE (2001) Biodegradable polymeric nanoparticles as drug delivery devices. J Control Release 7:1–20 (**Chapter 5**)

Sort P, Navasa M, Arroyo V, Aldeguer X, Planas R, Ruiz-del-Arbol L, Castells L, Vargas V, Soriano G, Guevara M, Ginès P, Rodés J (1999) Effect of intravenous albumin on renal impairment and mortality in patients with cirrhosis and spontaneous bacterial peritonitis. N Engl J Med 341:403–409 (**Chapter 3**)

Sparreboom A, Verweij J, van der Burg ME, Loos WJ, Brouwer E, Vigano L, Locatelli A, de Vos AI, Nooter K, Stoter G, Gianni L (1998) Disposition of Cremophor EL in humans limits the potential for modulation of the multidrug resistance phenotype in vivo. Clin Cancer Res 4:1937–1942 (**Chapter 6**)

Sparreboom A, Scripture CD, Trieu V, Williams PJ, De T, Yang A, Beals B, Figg WD, Hawkins M, Desai N (2005) Comparative preclinical and clinical pharmacokinetics of a cremophor-free, nanoparticle albumin-bound paclitaxel (ABI-007) and paclitaxel formulated in Cremophor (Taxol). Clin Cancer Res 11:4136–4143 (**Chapter 6**)

Spiekermann GM, Finn PW, Ward ES, Dumont J, Dickinson BL, Blumberg RS, Lencer WI (2002) Receptor-mediated immunoglobulin G transport across mucosal barriers in adult life: functional expression of FcRn in the mammalian lung. J Exp Med 196:303–310 (**Chapter 10**)

Spinella R, Sawhney R, Jalan R (2016) Albumin in chronic liver disease: structure, functions and therapeutic implications. Hepatol Int 10:124–132 (**Chapter 1**)
Squires JE (2002) Artificial blood. Science 295:1002–1005 (**Chapter 9**)
Stadler J, Bergonia H, Di Silvio M, Sweetland M, Billiar T, Simmons R, Lancaster JJ (1993) Nonheme iron-nitrosyl complex formation in rat hepatocytes: detection by electron paramagnetic resonance spectroscopy. Arch Biochem Biophys 302:4–11 (**Chapter 8**)
Stamler J, Singel D, Loscalzo J (1992a) Biochemistry of nitric oxide and its redox-activated forms. Science 258:1898–1902 (**Chapter 8**)
Stamler J, Jaraki O, Osborne J, Simon D, Keaney J, Vita J, Singel D, Valeri C, Loscalzo J (1992b) Nitric oxide circulates in mammalian plasma primarily as an S-nitroso adduct of serum albumin. Proc Natl Acad Sci U S A 89:7674–7677 (**Chapter 8**)
Stamler J, Osborne J, Jaraki O, Rabbani L, Mullins M, Singel D, Loscalzo J (1993) Adverse vascular effects of homocysteine are modulated by endothelium-derived relaxing factor and related oxides of nitrogen. J Clin Invest 91:308–318 (**Chapter 8**)
Stange J, Ramlow W, Mitzner S, Schmidt R, Klinkmann H (1993) Dialysis against a recycled albumin solution enables the removal of albumin-bound toxins. Artif Organs 17:809–813 (**Chapter 11**)
Stauber RE, Spindelboeck W, Haas J, Putz-Bankuti C, Stadlbauer V, Lackner C, Oettl K (2014) Human nonmercaptalbumin-2: a novel prognostic marker in chronic liver failure. Ther Apher Dial 18:74–78 (**Chapter 3**)
Stehle G, Sinn H, Wunder A, Schrenk HH, Stewart JC, Hartung G, Maier-Borst W, Heene DL (1997a) Plasma protein (albumin) catabolism by the tumor itself – implications for tumor metabolism and the genesis of cachexia. Crit Rev Oncol Hematol 26:77–100 (**Chapters 6 and 10**)
Stehle G, Wunder A, Sinn H, Schrenk HH, Schutt S, Frei E, Hartung G, Maier-Borst W, Heene DL (1997b) Pharmacokinetics of methotrexate-albumin conjugates in tumor-bearing rats. Anticancer Drugs 8:835–844 (**Chapter 10**)
Stewart MJ (2003) Contrast echocardiography. Heart 89:342–348 (**Chapter 7**)
Stockmann HB, IJzermans JN (2002) Prospects for the temporary treatment of acute liver failure. Eur J Gastroenterol Hepatol 14:195–203 (**Chapter 11**)
Stork R, Campigna E, Robert B, Müller D, Kontermann RE (2009) Biodistribution of a bispecific single-chain diabody and its half-life extended derivatives. J Biol Chem 284:25612–25619 (**Chapter 4**)
Storm T, Tranebjaerg L, Frykholm C, Birn H, Verroust PJ, Neveus T, Sundelin B, Hertz JM, Holmstrom G, Ericson K, Christensen EI, Nielsen R (2013) Renal phenotypic investigations of megalin-deficient patients: novel insights into tubular proteinuria and albumin filtration. Nephrol Dial Transplant 28:585–591 (**Chapter 10**)
Stride E, Saffari N (2003) Microbubble ultrasound contrast agents: a review. Proc Inst Mech Eng H J Eng Med 217:429–447 (**Chapter 7**)
Stubauer G, Giuffrè A, Sarti P (1999) Mechanism of S-nitrosothiol formation and degradation mediated by copper ions. J Biol Chem 274:28128–28133 (**Chapter 8**)
Subramanian GM, Fiscella M, Lamousé-Smith A, Zeuzem S, McHutchison JG (2007) Albintereron alpha-2b: a genetic fusion protein for the treatment of chronic hepatitis C. Nat Biotechnol 25:1411–1419 (**Chapters 2, 4 and 10**)
Sudlow G, Birkett DJ, Wade DN (1975) The characterization of two specific drug binding sites on human serum albumin. Mol Pharmacol 11:824–832 (**Chapter 1**)
Sugio S, Kashima A, Mochizuki S, Noda M, Kobayashi K (1999) Crystal structure of human serum albumin at 2.5 Å resolution. Prot Eng 12:827–835 (**Chapters 1, 2 and 10**)
Suk JS, Xu Q, Kim N, Hanes J, Ensign LM (2016) PEGylation as a strategy for improving nanoparticle-based drug and gene delivery. Adv Drug Deliv Rev 99:28–51, in Press (**Chapter 5**)
Sundar S, Kundu J, Kundu SC (2010) Biopolymeric nanoparticles. Sci Technol Adv Mater 11:1–13 (**Chapter 5**)
Sung C, Nardelli B, Lafleur DW, Blatter E, Corcoran M, Olsen HS, Birse CE, Pickeral OK, Zhang JL, Shah D, Moody G, Gentz S, Beebe L, Moore PA (2003) An IFN-beta-albumin fusion

protein that displays improved pharmacokinetic and pharmacodynamic properties in nonhuman primates. J Interf Cytokine Res 23:25–36 (**Chapter 10**)

Suzuki E, Yasuda K, Takeda N, Sakata S, Era S, Kuwata K, Sogami M, Miura K (1992) Increased oxidized form of human serum albumin in patients with diabetes mellitus. Diabetes Res Clin Pract 18:153–158 (**Chapter 3**)

Suzuki Y, Suda K, Matsuyama Y, Era S, Soejima A (2014) Close relationship between redox state of human serum albumin and serum cysteine levels in non-diabetic CKD patients with various degrees of renal function. Clin Nephrol 82:320–325 (**Chapter 3**)

Syed S, Schuyler PD, Kulczycky M, Sheffield WP (1997) Potent antithrombin activity and delayed clearance from the circulation characterize recombinant hirudin genetically fused to albumin. Blood 89:3243–3252 (**Chapter 10**)

Taguchi K, Urata Y, Anraku M, Watanabe H, Kawai K, Komatsu T, Tsuchida E, Maruyama T, Otagiri M (2010) Superior plasma retention of a cross-linked human serum albumin dimer in nephrotic rats as a new type of plasma expander. Drug Metab Dispos 38:2124–2129 (**Chapter 2**)

Taguchi K, Chuang VT, Yamasaki K, Urata Y, Tanaka R, Anraku M, Seo H, Kawai K, Maruyama T, Komatsu T, Otagiri M (2015a) Cross-linked human serum albumin dimer has the potential for use as a plasma-retaining agent for the fatty acid-conjugated antidiabetic drugs. J Pharm Pharmacol 67:255–263 (**Chapter 2**)

Taguchi K, Yamasaki K, Seo H, Otagiri M (2015b) Potential use of biological proteins for liver failure therapy. Pharmaceutics 7:255–274 (**Chapter 1**)

Takahashi M, Kido K, Aoi A, Furukawa H, Ono M, Kodama T (2007) Spinal gene transfer using ultrasound and microbubbles. J Control Release Off J Control Release Soc 117:267–272 (**Chapter 7**)

Takeuchi H, Ohmori K, Kondo I, Oshita A, Shinomiya K, Yu Y, Takagi Y, Mizushige K, Kangawa K, Kohno M (2003) Potentiation of C-type natriuretic peptide with ultrasound and microbubbles to prevent neointimal formation after vascular injury in rats. Cardiovasc Res 58:231–238 (**Chapter 7**)

Tam SH, McCarthy SG, Brosnan K, Goldberg KM, Scallon BJ (2013) Correlations between pharmacokinetics of IgG antibodies in primates vs. FcRn-transgenic mice reveal a rodent model with predictive capabilities. MAbs 5:397–405 (**Chapter 10**)

Tanaka R, Watanabe H, Kodama A, Chuang VT, Ishima Y, Hamasaki K, Tanaka K, Mizushima T, Otagiri M, Maruyama T (2013) Long-acting human serum albumin-thioredoxin fusion protein suppresses bleomycin-induced pulmonary fibrosis progression. J Pharmacol Exp Ther 345:271–283 (**Chapters 2 and 4**)

Tanaka R, Ishima Y, Enoki Y, Kimachi K, Shirai T, Watanabe H, Chuang VT, Maruyama T, Otagiri M (2014a) Therapeutic impact of human serum albumin-thioredoxin fusion protein on influenza virus-induced lung injury mice. Front Immunol 5:561 (**Chapter 4**)

Tanaka R, Ishima Y, Maeda H, Kodama A, Nagao S, Watanabe H, Chuang VT, Otagiri M, Maruyama T (2014b) Albumin fusion prolongs the antioxidant and anti-inflammatory activities of thioredoxin in mice with acetaminophen-induced hepatitis. Mol Pharm 11:1228–1238 (**Chapter 4**)

Taniyama Y, Tachibana K, Hiraoka K, Aoki M, Yamamoto S, Matsumoto K, Nakamura T, Ogihara T, Kaneda Y, Morishita R (2002a) Development of safe and efficient novel nonviral gene transfer using ultrasound: enhancement of transfection efficiency of naked plasmid DNA in skeletal muscle. Gene Ther 9:372–380 (**Chapter 7**)

Taniyama Y, Tachibana K, Hiraoka K, Namba T, Yamasaki K, Hashiya N, Aoki M, Ogihara T, Yasufumi K, Morishita R (2002b) Local delivery of plasmid DNA into rat carotid artery using ultrasound. Circulation 105:1233–1239 (**Chapter 7**)

Tavares BG, Aguiar MO, Tsutsui JM, Garcia D, De Oliveira AG, Ramires JAF, Filho RK, Porter T, Wilson M (eds) (2015) Safety and feasibility of diagnostic ultrasound high mechanical index impulses in restoring epicardial flow in acute ST segment elevation myocardial infarction in humans. Am Soc Echocardiogr, June 10–14, 2015; 2015 September 10–12, 2015. Seattle (**Chapter 7**)

Taxol P (2000) Taxol® (paclitaxel) for injection, BMS insert (**Chapter 6**)

ten Tije AJ, Verweij J, Loos WJ, Sparreboom A (2003) Pharmacological effects of formulation vehicles: implications for cancer chemotherapy. Clin Pharmacokinet 42:665–685 (**Chapter 6**)

Tenten V, Menzel S, Kunter U, Sicking EM, van Roeyen CRC, Sanden SK, Kaldenbach M, Boor P, Fuss A, Uhlig S, Lanzmich R, Willemsen B, Dijkman H, Grepl M, Wild K, Kriz W, Smeets B, Floege J, Moeller MJ (2013) Albumin is recycled from the primary urine by tubular transcytosis. J Am Soc Nephrol 24:1966–1980 (**Chapter 10**)

Terawaki H, Yoshimura K, Hasegawa T, Matsuyama Y, Negawa T, Yamada K, Matsushima M, Nakayama M, Hosoya T, Era S (2004) Oxidative stress is enhanced in correlation with renal dysfunction: examination with the redox state of albumin. Kidney Int 66:1988–1993 (**Chapter 3**)

Terawaki H, Matsuyama Y, Era S, Matsuo N, Ikeda M, Ogura M, Yokoyama K, Yamamoto H, Hosoya T, Nakayama M (2007) Elevated oxidative stress measured as albumin redox state in continuous ambulatory peritoneal dialysis patients correlates with small uraemic solutes. Nephrol Dial Transplant 22:968 (**Chapter 3**)

Terawaki H, Takada Y, Era S, Funakoshi Y, Nakayama K, Nakayama M, Ogura M, Ito S, Hosoya T (2010) The redox state of albumin and serious cardiovascular incidence in hemodialysis patients. Ther Apher Dial 14:465–471 (**Chapter 3**)

Terawaki H, Era S, Nakayama M, Hosoya T (2011) Decrease in reduced-form albumin among chronic kidney disease patients: new insights in cardiovascular complications. Ther Apher Dial 15:156–160 (**Chapter 3**)

Tesar DB, Tiangco NE, Bjorkman PJ (2006) Ligand valency affects transcytosis, recycling and intracellular trafficking mediated by the neonatal Fc receptor. Traffic 7:1127–1142 (**Chapter 10**)

The Albumin Website: http://albumin.org. Accessed Sept 2015 (**Chapter 1**)

Tiessen RG, Castaigne JP, Dreyfus JF, Nemansky M, Kruizinga HH, van Vliet AA (2008) Pharmacokinetics and tolerability of a novel long-acting glucagon-like peptide-1 analog, CJC-1131, in healthy and diabetic subjects. Int J Clin Pharmacol Ther 46:443–452 (**Chapter 2**)

Tijink BM, Laeremans T, Budde M, Stigter-van Walsum M, Dreier T, de Haard HJ, Leemans CR, van Dongen GA (2008) Improved tumor targeting of anti-epidermal growth factor receptor nanobodies through albumin binding: taking advantage of modular nanobody technology. Mol Cancer Ther 7:2288–2297 (**Chapter 10**)

Tiruppathi C, Song W, Bergenfeldt M, Sass P, Malik AB (1997) Gp60 activation mediates albumin transcytosis in endothelial cells by tyrosine kinase-dependent pathway. J Biol Chem 272:25968–25975 (**Chapter 6**)

Tiruppathi C, Naqvi T, Wu Y, Vogel SM, Minshall RD, Malik AB (2004) Albumin mediates the transcytosis of myeloperoxidase by means of caveolae in endothelial cells. Proc Natl Acad Sci U S A 101:7699–7704 (**Chapter 6**)

Tiwari B, Junghans RP (2005) Functional analysis of the mouse Fcgrt 5′ proximal promoter. Biochim Biophys Acta 1681:88–98 (**Chapter 10**)

Tolmachev V, Orlova A, Pehrson R, Galli J, Baastrup B, Andersson K, Sandstrom M, Rosik D, Carlsson J, Lundqvist H, Wennborg A, Nilsson FY (2007) Radionuclide therapy of HER2-positive microxenografts using a Lu-177-labeled HER2-specific affibody molecule. Cancer Res 67:2773–2782 (**Chapter 10**)

Tomita D, Kimura T, Hosaka H, Daijima Y, Haruki R, Böttcher C, Komatsu T (2013) Covalent core–shell architecture of hemoglobin and human serum albumin as an artificial O_2 carrier. Biomacromolecules 14:1816–1825 (**Chapter 9**)

Tomkin GH (2009) Albiglutide, an albumin-based fusion of glucagon-like peptide 1 for the potential treatment of type 2 diabetes. Curr Opin Mol Ther 11:579–588 (**Chapter 4**)

Topczewska-Bruns J, Pawlak D, Tankiewicz A, Chabielska E, Buczko W (2003) Kynurenine metabolism in central nervous system in experimental chronic renal failure. Adv Exp Med Biol 527:177–182 (**Chapter 2**)

Torres MJ, Turell L, Botti H, Antmann L, Carballal S, Ferrer-Sueta G, Radi R, Alvarez B (2012) Modulation of the reactivity of the thiol of human serum albumin and its sulfenic derivative by fatty acids. Arch Biochem Biophys 521:102–110 (**Chapter 3**)

Treat LH, McDannold N, Vykhodtseva N, Zhang Y, Tam K, Hynynen K (2007) Targeted delivery of doxorubicin to the rat brain at therapeutic levels using MRI-guided focused ultrasound. Int J Cancer 121:901–907 (**Chapter 7**)

Trey C, Burns DG, Saunders SJ (1966) Treatment of hepatic coma by exchange blood transfusion. N Engl J Med 274:473–481 (**Chapter 11**)

Trynda-Lemiesz L (2004) Paclitaxel-HSA interaction. Binding sites on HSA molecule. Bioorg Med Chem 12:3269–3275 (**Chapter 6**)

Tsipotis E, Shuja A, Jaber BL (2015) Albumin dialysis for liver failure: a systematic review. Adv Chronic Kidney Dis 22:382–390 (**Chapter 11**)

Tsutsumi Y, Maruyama T, Takadate A, Goto M, Matsunaga H, Otagiri M (1999) Interaction between two dicarboxylate endogenous substances, bilirubin and an uremic toxin, 3-carboxy-4-methyl-5-propyl-2-furanpropanoic acid, on human serum albumin. Pharm Res 16:916–923 (**Chapter 11**)

Turell L, Botti H, Carballal S, Ferrer-Sueta G, Souza JM, Durán R, Freeman BA, Radi R, Alvarez B (2008) Reactivity of sulfenic acid in human serum albumin. Biochemistry 47:358–367 (**Chapter 3**)

Turell L, Radi R, Alvarez B (2013) The thiol pool in human plasma: the central contribution of albumin to redox processes. Free Radic Biol Med 65:244–253 (**Chapter 3**)

Turell L, Botti H, Bonilla L, Torres MJ, Schopfer F, Freeman BA, Armas L, Ricciardi A, Alvarez B, Radi R (2014) HPLC separation of human serum albumin isoforms based on their isoelectric points. J Chromatogr B Anal Technol Biomed Life Sci 944:144–151 (**Chapter 3**)

Tzaban S, Massol RH, Yen E, Hamman W, Frank SR, Lapierre LA, Hansen SH, Goldenring JR, Blumberg RS, Lencer WI (2009) The recycling and transcytotic pathways for IgG transport by FcRn are distinct and display an inherent polarity. J Cell Biol 185:673–684 (**Chapter 10**)

Ulbrich K, Michaelis M, Rothweiler F, Knobloch T, Sithisarn P, Cinat J, Kreuter J (2011) Interaction of folate-conjugated human serum albumin (HSA) nano-particles with tumor cells. Int J Pharm 406:128–134 (**Chapter 5**)

Unger E, Porter T, Lindner J, Grayburn P (2014) Cardiovascular drug delivery with ultrasound and microbubbles. Adv Drug Deliv Rev 72:110–126 (**Chapter 7**)

Urban MW, Fatemi M, Greenleaf JF (2010) Modulation of ultrasound to produce multifrequency radiation force. J Acoust Soc Am 127:1228–1238 (**Chapter 7**)

van Tellingen O, Huizing MT, Panday VR, Schellens JH, Nooijen WJ, Beijnen JH (1999) Cremophor EL causes (pseudo-) non-linear pharmacokinetics of paclitaxel in patients. Br J Cancer 81:330–335 (**Chapter 6**)

van Zuylen L, Karlsson MO, Verweij J, Brouwer E, de Bruijn P, Nooter K, Stoter G, Sparreboom A (2001) Pharmacokinetic modeling of paclitaxel encapsulation in Cremophor EL micelles. Cancer Chemother Pharmacol 47:309–318 (**Chapter 6**)

Vandegriff KD, Malavalli A, Wooldbridge J, Lohman W, Winslow RM (2003) MP4, a new nonvasoactive PEG-Hb conjugate. Transfusion 43:509–516 (**Chapter 9**)

Varca GHC, Queiroz RG, Lugão AB (2016) Irradiation as an alternative route for protein crosslinking: cosolvent free BSA nanoparticles. Radiat Phys Chem 124:111–115 (**Chapter 5**)

Veronese FM, Pasut G (2005) PEGylation, successful approach to drug delivery. Drug Discov Today 10:1451–1458 (**Chapter 5**)

Vidarsson G, Stemerding AM, Stapleton NM, Spliethoff SE, Janssen H, Rebers FE, de Haas M, van de Winkel JG (2006) FcRn: an IgG receptor on phagocytes with a novel role in phagocytosis. Blood 108:3573–3579 (**Chapter 10**)

Viuff D, Antunes F, Evans L, Cameron J, Dyrnesli H, Thue Ravn B, Stougaard M, Thiam K, Andersen B, Kjaerulff S, Howard KA (2016) Generation of a double transgenic humanized neonatal Fc receptor (FcRn)/albumin mouse to study the pharmacokinetics of albumin-linked drugs. J Control Release Off J Control Release Soc 223:22–30 (**Chapter 10**)

Vogt W (1995) Oxidation of methionyl residues in proteins: tools, targets, and reversal. Free Radic Biol Med 18:93–105 (**Chapter 2**)

Volk LD, Flister MJ, Bivens CM, Stutzman A, Desai N, Trieu V, Ran S (2008) Nab-paclitaxel efficacy in the orthotopic model of human breast cancer is significantly enhanced by concurrent anti–vascular endothelial growth factor A therapy. Neoplasia 10:613–623 (**Chapter 6**)

Volk LD, Flister MJ, Chihade D, Desai N, Trieu V, Ran S (2011) Synergy of nab-paclitaxel and bevacizumab in eradicating large orthotopic breast tumors and preexisting metastases. Neoplasia 13:327–338 (**Chapter 6**)

Volovat C, Gladkov OA, Bondarenko IM, Barash S, Buchner A, Bias P, Adar L, Avisar N (2014) Efficacy and safety of balugrastim compared with pegfilgrastim in patients with breast cancer receiving chemotherapy. Clin Breast Cancer 14:101–108 (**Chapter 2**)

Von Hoff DD, Ramanathan RK, Borad MJ, Laheru DA, Smith LS, Wood TE, Korn RL, Desai N, Trieu V, Iglesias JL, Zhang H, Soon-Shiong P, Shi T, Rajeshkumar NV, Maitra A, Hidalgo M (2011) Gemcitabine plus nab-paclitaxel is an active regimen in patients with advanced pancreatic cancer: a phase I/II trial. J Clin Oncol 29:4548–4554 (**Chapters 6 and 10**)

Von Hoff DD, Ervin T, Arena FP, Chiorean EG, Infante J, Moore M, Seay T, Tjulandin SA, Ma WW, Saleh MN, Harris M, Reni M, Dowden S, Laheru D, Bahary N, Ramanathan RK, Tabernero J, Hidalgo M, Goldstein D, Van Cutsem E, Wei X, Iglesias J, Renschler MF (2013) Increased survival in pancreatic cancer with nab-paclitaxel plus gemcitabine. N Engl J Med 369:1691–1703 (**Chapter 6**)

Wagner LM, Yin H, Eaves D, Currier M, Cripe TP (2014) Preclinical evaluation of nanoparticle albumin-bound paclitaxel for treatment of pediatric bone sarcoma. Pediatr Blood Cancer 61:2096–2098 (**Chapter 6**)

Waldmann TA, Terry WD (1990) Familial hypercatabolic hypoproteinemia. A disorder of endogenous catabolism of albumin and immunoglobulin. J Clin Invest 86:2093–2098 (**Chapter 10**)

Walker A, Dunlevy G, Rycroft D, Topley P, Holt LJ, Herbert T, Davies M, Cook F, Holmes S, Jespers L, Herring C (2010) Anti-serum albumin domain antibodies in the development of highly potent, efficacious and long-acting interferon. Protein Eng Des Sel 23:271–278 (**Chapter 10**)

Wang XH (2011) Role of constituents of optison in optison-mediated gene transfection enhancement in skeletal muscle in vivo. J Ultrasound Med Off J Am Inst Ultrasound Med 30:325–332 (**Chapter 7**)

Wang W, Ou Y, Shi Y (2004) AlbuBNP, a recombinant B-type natriuretic peptide and human serum albumin fusion hormone, as a long-term therapy of congestive heart failure. Pharm Res 21:2105–2111 (**Chapter 4**)

Wang F, Wu M, Liu W, Shen Q, Sun H, Chen S (2013a) Expression, purification, and lipolytic activity of recombinant human serum albumin fusion proteins with one domain of human growth hormone in Pichia pastoris. Biotechnol Appl Biochem 60:405–411 (**Chapter 4**)

Wang Z-M, Ho JX, Ruble JR, Rose J, Rüker F, Ellenburg M, Murphy R, Click J, Soistman E, Wilkerson L, Carter DC (2013b) Structural studies of several clinically important oncology drugs in complex with human serum albumin. Biochim Biophys Acta 1830:5356–5374 (**Chapter 1**)

Wang M, Zhi D, Wang H, Ru Y, Ren H, Wang N, Liu Y, Li Y, Li H (2016a) TAT-HSA-α-MSH fusion protein with extended half-life inhibits tumor necrosis factor-α in brain inflammation of mice. Appl Microbiol Biotechnol 100:5353–5361 (**Chapter 4**)

Wang M, Zhi D, Xian J, Ru Y, Wei S, Wang N, Liu Y, Wang H, Pei Y, Song M, Li Y, Li H (2016b) Functional expression of human serum albumin-tandem thrombopoietin mimetic peptide fusion protein as a novel thrombopoietin analog in Pichia pastoris. Biotechnol Lett 38:779–785, in press (**Chapter 4**)

Wani MA, Haynes LD, Kim J, Bronson CL, Chaudhury C, Mohanty S, Waldmann TA, Robinson JM, Anderson CL (2006) Familial hypercatabolic hypoproteinemia caused by deficiency of the neonatal Fc receptor, FcRn, due to a mutant beta2-microglobulin gene. Proc Natl Acad Sci U S A 103:5084–5089 (**Chapter 10**)

Ward ES, Gussow D, Griffiths AD, Jones PT, Winter G (1989) Binding activities of a repertoire of single immunoglobulin variable domains secreted from Escherichia coli. Nature 341:544–546 (**Chapter 10**)

Ward M, Wu J, Chiu JF (2000) Experimental study of the effects of optison concentration on sonoporation in vitro. Ultrasound Med Biol 26:1169–1175 (**Chapter 7**)

Ward ES, Zhou J, Ghetie V, Ober RJ (2003) Evidence to support the cellular mechanism involved in serum IgG homeostasis in humans. Int Immunol 15:187–195 (**Chapter 10**)

Ward ES, Martinez C, Vaccaro C, Zhou J, Tang Q, Ober RJ (2005) From sorting endosomes to exocytosis: association of Rab4 and Rab11 GTPases with the Fc receptor, FcRn, during recycling. Mol Biol Cell 16:2028–2038 (**Chapter 10**)

Wartlick H, Michaelis K, Balthasar S, Strebhardt K, Kreuter J, Langer K (2004) Highly specific HER2-mediated cellular uptake of antibody-modified nanoparticles in tumor cells. J Drug Target 12:461–471 (**Chapter 5**)

Watanabe H, Tanase S, Nakajou K, Maruyama T, Kragh-Hansen U, Otagiri M (2000) Role of arg-410 and tyr-411 in human serum albumin for ligand binding and esterase-like activity. Biochem J 349(Pt 3):813–819 (**Chapter 2**)

Watanabe H, Yamasaki K, Kragh-Hansen U, Tanase S, Harada K, Suenaga A, Otagiri M (2001a) In vitro and in vivo properties of recombinant human serum albumin from Pichia pastoris purified by a method of short processing time. Pharm Res 18:1775–1781 (**Chapters 2 and 11**)

Watanabe H, Kragh-Hansen U, Tanase S, Nakajou K, Mitarai M, Iwao Y, Maruyama T, Otagiri M (2001b) Conformational stability and warfarin-binding properties of human serum albumin studied by recombinant mutants. Biochem J 357:269–274 (**Chapter 2**)

Watanabe H, Noguchi T, Miyamoto Y, Kadowaki D, Kotani S, Nakajima M, Miyamura S, Ishima Y, Otagiri M, Maruyama T (2012) Interaction between two sulfate-conjugated uremic toxins, p-cresyl sulfate and indoxyl sulfate, during binding with human serum albumin. Drug Metab Dispos 40:1423–1428 (**Chapter 11**)

Weber C, Coester C, Kreuter J, Langer K (2000) Desolvation process and surface characteristics of protein nanoparticles. Int J Pharm 194:91–102 (**Chapter 5**)

Wei K, Jayaweera AR, Firoozan S, Linka A, Skyba DM, Kaul S (1998) Basis for detection of stenosis using venous administration of microbubbles during myocardial contrast echocardiography: bolus or continuous infusion? J Am Coll Cardiol 32:252–260 (**Chapter 7**)

Weimer T, Wormsbacher W, Kronthaler U, Lang W, Liebing U, Schulte S (2008) Prolonged in-vivo half-life of factor VIIa by fusion to albumin. Thromb Haemost 99:659–667 (**Chapter 4**)

Weiss RB, Donehower RC, Wiernik PH, Ohnuma T, Gralla RJ, Trump DL, Baker JR Jr, Van Echo DA, Von Hoff DD, Leyland-Jones B (1990) Hypersensitivity reactions from taxol. J Clin Oncol 8:1263–1268 (**Chapter 6**)

Werle M, Bernkop-Schnurch A (2006) Strategies to improve plasma half life time of peptide and protein drugs. Amino Acids 30:351–367 (**Chapter 4**)

Wernick NL, Haucke V, Simister NE (2005) Recognition of the tryptophan-based endocytosis signal in the neonatal Fc receptor by the mu subunit of adaptor protein-2. J Biol Chem 280:7309–7316 (**Chapter 10**)

Weser U, Schubotz LM (1981) Imidazole-bridged copper complexes as Cu_2Zn_2-superoxide dismutase models. J Mol Catal 13:249–261 (**Chapter 9**)

West AP Jr, Bjorkman PJ (2000) Crystal structure and immunoglobulin G binding properties of the human major histocompatibility complex-related Fc receptor(,). Biochemistry 39:9698–9708 (**Chapter 10**)

Whitehead RP, Jacobson J, Brown TD, Taylor SA, Weiss GR, Macdonald JS (1997) Phase II trial of paclitaxel and granulocyte colony-stimulating factor in patients with pancreatic carcinoma: a Southwest Oncology Group study. J Clin Oncol 15:2414–2419 (**Chapter 6**)

Winslow RM (2003) Current status of blood substitute research: towards a new paradigm. J Intern Med 253:508–517

Wu Z, Simister NE (2001) Tryptophan- and dileucine-based endocytosis signals in the neonatal Fc receptor. J Biol Chem 276:5240–5247 (**Chapter 9**)

Wu J, Ross JP, Chiu JF (2002) Reparable sonoporation generated by microstreaming. J Acoust Soc Am 111:1460–1464 (**Chapter 7**)

Wu M, Shen Q, Yang Y, Zhang S, Qu W, Chen J, Sun H, Chen S (2013) Disruption of YPS1 and PEP4 genes reduces proteolytic degradation of secreted HSA/PTH in Pichia pastoris GS115. J Ind Microbiol Biotechnol 40:589–599 (**Chapter 4**)

Wu J, Xie F, Kumar T, Liu J, Lof J, Shi W, Everbach EC, Porter TR (2014) Improved sonothrombolysis from a modified diagnostic transducer delivering impulses containing a longer pulse duration. Ultrasound Med Biol 40:1545–1553 (**Chapter 7**)

Wunder A, Muller-Ladner U, Stelzer E, Neumann E, Sinn H, Gay S, Fiehn C (2003) Albumin-based drug delivery as novel therapeutic approach for rheumatoid arthritis. Arthritis Res Ther 5:S4–S4 (**Chapter 10**)

Xenariou S, Griesenbach U, Liang HD, Zhu J, Farley R, Somerton L, Singh C, Jeffery PK, Ferrari S, Scheule RK, Cheng SH, Geddes DM, Blomley M, Alton EW (2007) Use of ultrasound to enhance nonviral lung gene transfer in vivo. Gene Ther 14:768–774 (**Chapter 7**)

Xiong Y, Liu ZZ, Georgieva R, Smuda K, Steffen A, Sendeski M, Voigt A, Patzak A, Bäumler H (2013) Nonvasoconstrictive hemoglobin particles as oxygen carriers. ACS Nano 7:7454–7461 (**Chapter 9**)

Xu X, Yang J, Liu Y, Shan C, Wang Q, Chen Z, Cheng Y (2015) The induction of prolonged myelopoietic effects in monkeys by GW003, a recombinant human granulocyte colony-stimulating factor genetically fused to recombinant human albumin. J Pharm Sci 104:760–767 (**Chapter 4**)

Yamada K, Yokomaku K, Haruki R, Taguchi K, Nagao S, Maruyama T, Otagiri M, Komatsu T (2016) Influence of molecular structure on O_2-binding properties and blood circulation of hemoglobin-albumin clusters. PLoS ONE 11:e0149526, in press (**Chapter 9**)

Yamamoto S, Kazama JJ, Omori K, Matsuo K, Takahashi Y, Kawamura K, Matsuto T, Watanabe H, Maruyama T, Narita I (2015) Continuous reduction of protein-bound uraemic toxins with improved oxidative stress by using the oral charcoal adsorbent AST-120 in haemodialysis patients. Sci Rep 5:14381 (**Chapter 3**)

Yamasaki K, Maruyama T, Kragh-Hansen U, Otagiri M (1996) Characterization of site I on human serum albumin: concept about the structure of a drug binding site. Biochim Biophys Acta 1295:147–157 (**Chapter 1**)

Yamasaki K, Chuang VTG, Maruyama T, Otagiri M (2013) Albumin-drug interaction and its clinical implication. Biochim Biophys Acta 1830:5435–5443 (**Chapter 1**)

Yang GG, Xu XY, Ding Y, Cui QQ, Wang Z, Zhang QY, Shi SH, Lv ZY, Wang XY, Zhang JH, Zhang RG, Xu CS (2015) Linker length affects expression and bioactivity of the onconase fusion protein in Pichia pastoris. Genet Mol Res 14:19360–19370 (**Chapter 4**)

Yasuda H (2008) Solid tumor physiology and hypoxia-induced chemo/radio-resistance: novel strategy for cancer therapy: nitric oxide donor as a therapeutic enhancer. Nitric Oxide 19:205–216 (**Chapter 8**)

Yazaki PJ, Kassa T, Cheung CW, Crow DM, Sherman MA, Bading JR, Anderson ALJ, Colcher D, Raubitschek A (2008) Biodistribution and tumor imaging of an anti-CEA single-chain antibody-albumin fusion protein. Nucl Med Biol 35:151–158 (**Chapter 10**)

Yeh C, Altaf SA, Hoag SW (1997) Theory of force transducer design optimization for die wall stress measurement during tablet compaction: optimization and validation of split-web die using finite element analysis. Pharm Res 14:1161–1170 (**Chapter 10**)

Yogasundaram H, Bahniuk MS, Singh HD, Aliabadi HM, Uludağ H, Unsworth LD (2012) BSA nanoparticles for siRNA delivery: coating effects on nanoparticle properties, plasma protein adsorption, and in vitro siRNA delivery. Int J Biomater 2012:584060 (**Chapter 5**)

Yu S, Yao P, Jiang M, Zhang G (2006) Nanogels prepared by self-assembly of oppositely charged globular proteins. Biopolymers 83:148–158 (**Chapter 5**)

Yu X, Menard M, Prechl J, Bhakta V, Sheffield WP, Lazarus AH (2016) Monovalent Fc receptor blockade by an anti-Fcγ receptor/albumin fusion protein ameliorates murine ITP with abrogated toxicity. Blood 127:132–138 (**Chapter 4**)

Yuan F, Dellian M, Fukumura D, Leunig M, Berk DA, Torchilin VP, Jain RK (1995) Vascular permeability in a human tumor xenograft: molecular size dependence and cutoff size. Cancer Res 55:3752–3756 (**Chapter 6**)

Yuan Q, Li L, Pian Y, Hao H, Zheng Y, Zang Y, Jiang H, Jiang Y (2016) Preliminary investigation of human serum albumin-Vb inhibition on toxic shock syndrome induced by staphylococcus enterotoxin B in vitro and in vivo. Toxicon 113:55–59 (**Chapter 4**)

Zalevsky J, Chamberlain AK, Horton HM, Karki S, Leung IWL, Sproule TJ, Lazar GA, Roopenian DC, Desjarlais JR (2010) Enhanced antibody half-life improves in vivo activity. Nat Biotechnol 28:157–159 (**Chapter 10**)

Zauner W, Farrow NA, Haines AM (2001) In vitro uptake of polystyrene microspheres: effect of particles size, cell line and cell density. J Control Release 71:39–51 (**Chapter 5**)

Zeuzem S, Sulkowski MS, Lawitz EJ, Rustgi VK, Rodriguez-Torres M, Bacon BR, Grigorescu M, Tice AD, Lurie Y, Cianciara J, Muir AJ, Cronin PW, Pulkstenis E, Subramanian GM, McHutchison JG, ACHIEVE-1 Study Team (2010) Albinterferon alfa-2b was not inferior to pegylated interferon-α in a randomized trial of patients with chronic hepatitis C virus genotype 1. Gastroenterology 139:1257–1266 (**Chapter 4**)

Zhang N, Palmer AF (2010) Polymerization of human hemoglobin using the crosslinker 1,11-bis(maleimido)triethylene glycol for use as an oxygen carrier. Biotechnol Prog 26:1481–1485 (**Chapter 9**)

Zhang Y, Wilcox D (2002) Thermodynamic and spectroscopic study of Cu(II) and Ni(II) binding to bovine serum albumin. J Biol Inorg Chem 7:327–337 (**Chapter 8**)

Zhang S, Wang G, Lin X, Chatzinikolaidou M, Jennissen H, Laub M (2008a) Polyethylenimine-coated albumin nanoparticles for BMP-2 delivery. Biotechnol Prog 24:945–956 (**Chapter 5**)

Zhang Y, Bhatt VS, Sun G, Wang PG, Palmer AF (2008b) Site-selective glycosylation of hemoglobin on Cys 93. Bioconjug Chem 19:2221–2230 (**Chapter 9**)

Zhang Q, Lei J, Ding Y, Chen Y, Qu L, Chen S, Jin J (2009) Expression and purification of IFNbeta-HSA fusion protein in Pichia pastoris. Sheng Wu Gong Cheng Xue Bao 25:1746–1752 (**Chapter 4**)

Zhang S, Kucharski C, Doschak MR, Sebald W, Uludag H (2010) Polyethylenimine–PEG coated albumin nanoparticles for BMP-2 delivery. Biomaterials 31:952–963 (**Chapter 5**)

Zhang C, Awasthi N, Schwarz MA, Hinz S, Schwarz RE (2013a) Superior antitumor activity of nanoparticle albumin-bound paclitaxel in experimental gastric cancer. PLoS ONE 8, e58037 (**Chapter 6**)

Zhang L, Marrano P, Kumar S, Leadley M, Elias E, Thorner P, Baruchel S (2013b) Nab-Paclitaxel is an active drug in preclinical model of pediatric solid tumors. Clin Cancer Res 19:5972–5983 (**Chapter 6**)

Zhang L, Wang L, Meng Z, Gan H, Gu R, Wu Z, Gao L, Zhu X, Sun W, Li J, Zheng Y, Dou G (2014) A novel exendin-4 human serum albumin fusion protein, E2HSA, with an extended half-life and good glucoregulatory effect in healthy rhesus monkeys. Biochem Biophys Res Commun 445:511–516 (**Chapter 4**)

Zhao HL, Xue C, Wang Y, Sun B, Yao XQ, Liu ZM (2009) Elimination of the free sulfhydryl group in the human serum albumin (HSA) moiety of human interferon-alpha2b and HSA fusion protein increases its stability against mechanical and thermal stresses. Eur J Pharm Biopharm 72:405–411 (**Chapters 2 and 4**)

Zhao D, Zhao X, Zu Y, Li J, Zhang Y, Jiang R, Zhang Z (2010) Preparation, characterization, and in vitro targeted delivery of folate-decorated paclitaxel-loaded bovine serum albumin nanoparticles. Int J Nanomedicine 5:669–677 (**Chapter 5**)

Zhao HL, Xue C, Du JL, Ren M, Xia S, Cheng YG, Liu ZM (2012a) Sustained and cancer cell targeted cytosolic delivery of Onconase results in potent antitumor effects. J Control Release 159:346–352 (**Chapter 4**)

Zhao HL, Xue C, Du JL, Ren M, Xia S, Liu ZM (2012b) Balancing the pharmacokinetics and pharmacodynamics of interferon-α2b and human serum albumin fusion protein by proteolytic or reductive cleavage increases its in vivo therapeutic efficacy. Mol Pharm 9:664–670 (**Chapter 4**)

Zhao J, Si Y, Cheng M, Yang Y, Niu Y, Li X, Liu X, Yang W (2013a) Albumin fusion of interleukin-28B: production and characterization of its biological activities and protein stability. PLoS ONE 8, e64301 (**Chapter 4**)

Zhao S, Zhang Y, Tian H, Chen X, Cai D, Yao W, Gao X (2013b) Extending the serum half-life of G-CSF via fusion with the domain III of human serum albumin. Biomed Res Int 2013:107238 (**Chapter 4**)

Zhen Y, Stenmark H (2015) Cellular functions of Rab GTPases at a glance. J Cell Sci 128:3171–3176 (**Chapter 10**)

Zhu XP, Meng G, Dickinson BL, Li XT, Mizoguchi E, Miao LL, Wang YS, Robert C, Wu BY, Smith PD, Lencer WI, Blumberg RS (2001) MHC class I-related neonatal Fc receptor for IgG is functionally expressed in monocytes, intestinal macrophages, and dendritic cells. J Immunol 166:3266–3276 (**Chapter 10**)

Zhu RY, Xin X, Dai HY, Li Q, Lei JY, Chen Y, Jin J (2012) Expression and purification of recombinant human serum albumin fusion protein with VEGF165b in Pichia pastoris. Protein Expr Purif 85:32–37 (**Chapter 4**)

Zollner S, Schuermann D, Raquet E, Mueller-Cohrs J, Weimer T, Pragst I, Dickneite G, Schulte S (2014) Pharmacological characteristics of a novel, recombinant fusion protein linking coagulation factor VIIa with albumin (rVIIa-FP). J Thromb Haemost 12:220–228 (**Chapter 4**)

Zunszain PA, Ghuman J, Komatsu T, Tsuchida E, Curry S (2003) Crystal structural analysis of human serum albumin complexed with hemin and fatty acid. BMC Struct Biol 3:6 (**Chapter 11**)

Zunszain PA, Ghuman J, McDonagh AF, Curry S (2008) Crystallographic analysis of human serum albumin complexed with 4Z,15E-bilirubin-IXalpha. J Mol Biol 381:394–406 (**Chapters 1, 8, and 11**)

Index

M. Otagiri, V.T.G. Chuang (eds.), *Albumin in Medicine*,
DOI 10.1007/978-981-10-2116-9

O

P

MIX
Papier aus verantwortungsvollen Quellen
Paper from responsible sources
FSC® C105338

If you have any concerns about our products,
you can contact us on
ProductSafety@springernature.com

In case Publisher is established outside the EU,
the EU authorized representative is:
Springer Nature Customer Service Center GmbH
Europaplatz 3, 69115 Heidelberg, Germany

Printed by Libri Plureos GmbH
in Hamburg, Germany